Algebra C(

PRENTICE HALL SERIES IN MATHEMATICS FOR MIDDLE SCHOOL TEACHERS

JOHN BEEM *Geometry Connections*
ASMA HARCHARRAS and DORINA MITREA *Calculus Connections*
IRA J. PAPICK *Algebra Connections*
DEBRA A. PERKOWSKI and MICHAEL PERKOWSKI *Data Analysis and Probability Connections*

Algebra Connections
Mathematics for Middle School Teachers

Ira J. Papick
Mathematics Department
University of Missouri-Columbia

Upper Saddle River, New Jersey 07458

Library of Congress Cataloging-in-Publication Data

Papick, Ira J.
  Connecting middle school and college mathematics (CM)^2: algebra connections / Ira J. Papick.
    p. cm.—(Prentice Hall series in mathematics for middle school teachers)
  Includes bibliographical references and index.
  ISBN 0-13-144928-1
  1. Algebra—study and teaching (Middle school) 2. Geometry—Study and teaching (Middle school) I. Title: Algebra connections. II. Title. III. Series.

QA159.P37 2007
512′.0071′2—dc22

2005051528

Editor in Chief: *Sally Yagan*
Executive Acquisitions Editor: *Petra Recter*
Project Manager: *Michael Bell*
Production Management: *Progressive Publishing Alternatives*
Assistant Managing Editor: *Bayani Mendoza de Leon*
Senior Managing Editor: *Linda Mihatov Behrens*
Executive Managing Editor: *Kathleen Schiaparelli*
Manufacturing Manager: *Alexis Heydt-Long*
Manufacturing Buyer: *Maura Zaldivar*
Director of Marketing: *Patrice Jones*
Art Director: *Jayne Conte*
Cover Designer: *Bruce Kenselaar*
Art Studio/Formatter: *Laserwords*
Editorial Assistant/Supplement Editor: *Joanne Wendelken*
Cover Image: © *Stockbyte*

©2007 Pearson Education, Inc.
Pearson Prentice Hall
Pearson Education, Inc.
Upper Saddle River, New Jersey 07458

All rights reserved. No part of this book may be reproduced, in any form or by any other means, without permission in writing from the publisher.

Pearson Prentice Hall™ is a trademark of Pearson Education, Inc.
Development of these materials was supported by a grant from the National Science Foundation (ESI 0101822).

Printed in the United States of America

10 9 8 7 6 5 4 3 2 1

ISBN: 0-13-144928-1

Pearson Education LTD., *London*
Pearson Education Australia PTY, Limited, *Sydney*
Pearson Education Singapore, Pte. Ltd.
Pearson Education North Asia Ltd., *Hong Kong*
Pearson Education Canada, Ltd., *Toronto*
Pearson Educación de Mexico, S.A. de C.V.
Pearson Education—Japan, *Tokyo*
Pearson Education Malaysia, Pte. Ltd.

Contents

	Preface	vii
1	**Patterns**	**1**
	1.1 Classroom Connections: Representing Patterns	1
	1.2 Reflections on Classroom Connections: Representing Patterns	3
	1.3 Arithmetic Sequences	14
	1.4 Classroom Connections: A Quadratic Sequence	16
	1.5 Reflections on Classroom Connections: A Quadratic Sequence	16
	1.6 Finite Arithmetic Sequences	21
	1.7 Geometric Sequences	23
	1.8 Mathematical Induction	29
	1.9 Classroom Connection: Counting Tools	34
	1.10 The Binomial Theorem	47
	1.11 The Fibonacci Sequence	51
2	**Arithmetic and Algebra of the Integers**	**63**
	2.1 A Few Mathematical Questions Concerning the Periodical Cicadas	64
	2.2 Classroom Connections: Multiples and Divisors	64
	2.3 Reflections on Classroom Connections: Multiples and Divisors	65
	2.4 Multiples and Divisors	70
	2.5 Least Common Multiple and Greatest Common Divisor	73
	2.6 The Fundamental Theorem of Arithmetic	75
	2.7 Revisiting the LCM and GCD	82
	2.8 Relations and Results Concerning LCM and GCD	89
3	**The Division Algorithm and the Euclidean Algorithm**	**95**
	3.1 Measuring Integer Lengths and the Division Algorithm	95
	3.2 The Euclidean Algorithm	102
	3.3 Applications of the Representation GCD $(a,b) = ax + by$	106
	3.4 Place Value	110
	3.5 Prime Thoughts	125
4	**Arithmetic and Algebra of the Integers Modulo n**	**145**
	4.1 Classroom Connections: Divisibility Tests	146
	4.2 Reflections on Classroom Connections: Justifying the Divisibility Tests	146
	4.3 Clock Addition	148
	4.4 Modular Arithmetic	151
	4.5 Comparing Arithmetic Properties of $\mathbf{Z}$ and $\mathbf{Z}_n$	164
	4.6 Multiplicative Inverses in $\mathbf{Z}_n$	169

vi Contents

    4.7   Elementary Applications of Modular Arithmetic 174
    4.8   Fermat's Little Theorem and Wilson's Theorem 188
    4.9   Linear Equations Defined over $\mathbf{Z}_n$ 193
    4.10 Extended Studies: The Chinese Remainder Theorem 200
    4.11 Extended Studies: Quadratic Equations Defined over $\mathbf{Z}_n$ 203

5 Algebraic Modeling in Geometry: The Pythagorean Theorem and More   215
    5.1   The Significance of Daryl's Measurements and Related Geometry . 216
    5.2   Classroom Connections: The Pythagorean Theorem 217
    5.3   Reflections on Classroom Connections: The Pythagorean Theorem and Its Converse 217
    5.4   Computing Distance in Two-Dimensional and Three-Dimensional Euclidean Space: The Distance Formula 226
    5.5   An Extension of the Pythagorean Theorem: The Law of Cosines . . 227
    5.6   Integer Distances in the Plane 229
    5.7   Pythagorean Triples: Positive Integer Solutions to $x^2 + y^2 = z^2$ . . 230
    5.8   Extended Studies: Further Investigations into Integer Distance Point Sets—A Theorem of Erdös 237
    5.9   Extended Studies: Additional Questions Concerning Pythagorean Triples 240
    5.10 Fermat's Last Theorem 250

6 Arithmetic and Algebra of Matrices   253
    6.1   Classroom Connections: Systems of Linear Equations 254
    6.2   Reflections on Classroom Connections: Systems of Linear Equations 255
    6.3   Rational and Irrational Numbers 261
    6.4   Systems of Linear Equations 268
    6.5   Polynomial Curve Fitting: An Application of Systems of Linear Equations. 282
    6.6   Matrix Arithmetic and Matrix Algebra 286
    6.7   Multiplicative Inverses: Solving the Matrix Equation $AX = B$ 298
    6.8   Coding with Matrices 306

Glossary   **311**

References   **321**

Answers to (Most) Odd-Numbered Exercises   **323**

Photo Credits   **345**

Index   **347**

Preface

Improving the quality of mathematics education for middle school students is of critical importance, and increasing opportunities for students to learn important mathematics under the leadership of well-prepared and dedicated teachers is essential. New standards-based curriculum and instruction models, coupled with on-going professional development and teacher preparation, are foundational to this change.

These sentiments are eloquently articulated in the Glenn Commission Report: *Before It's Too Late: A Report to the Nation from the National Commission on Mathematics and Science Teaching for the 21st Century* (U.S. Department of Education, 2000). In fact, the principal message of the Glenn Commission Report is that America's students must improve their mathematics and science performance if they are to be successful in our rapidly changing technological world. To this end, the Report recommends that we greatly intensify our focus on improving the quality of mathematics and science teaching in grades K–12 by bettering the quality of teacher preparation, and it also stresses the necessity of developing creative plans to attract and retain substantial numbers of future mathematics and science teachers.

Some fifteen years ago, mathematics teachers, mathematics educators, and mathematicians collaborated to develop the architecture for standards-based reform, and their recommendations for the improvement of school mathematics, instruction, and assessment were articulated in three seminal documents published by the National Council of Teachers of Mathematics (*Curriculum and Evaluation Standards for School Mathematics* [1989], *Professional Standards for School Mathematics* [1991], and *Assessment Standards in School Mathematics* [1995]; more recently, these three documents were updated and combined into the single book, *NCTM Principles and Standards for School Mathematics, a.k.a. PSSM* [2000]).

The vision of school mathematics laid out in these three foundational documents was outstanding in spirit and content, yet abstract in practice. Concrete exemplary models reflecting the standards were needed and implementing the recommendations would be unrealizable without significant commitment of resources. Recognizing the opportunity for stimulating improvement in student learning, the National Science Foundation (NSF) made a strong commitment to bring life to the documents' messages and supported several K–12 mathematics curriculum development projects (standards-based curriculum), as well as other related dissemination and implementation projects.

Standards-based middle school curricula are designed to engage students in a variety of mathematical experiences, including thoughtfully planned explorations that provide and reinforce fundamental skills while illuminating the power and utility of mathematics in our world. These materials integrate central concepts in algebra, geometry, data analysis and probability, and mathematics of change, and they focus on important unifying ideas such as proportional reasoning.

The mathematical content of standards-based middle grade mathematics materials is challenging and relevant to our technological world. Its effective classroom implementation is dependent upon teachers having strong and appropriate mathematical preparation. *The Connecting Middle School and College Mathematics Project* $(CM)^2$ is a three-year (2001–2004) National Science Foundation funded project addressing the need for improved teacher qualifications and viable recruitment plans for middle grade mathematics teachers through the development of four foundational mathematics courses with accompanying support materials and the creation and implementation of effective teacher recruitment models.

The $(CM)^2$ materials are built upon a framework laid out in the *CBMS Mathematical Education of Teachers Report* (MET) (2001). This report outlines recommendations for the mathematical preparation of middle grade teachers that differ significantly from those for the preparation of elementary teachers and provides guidance to those developing new programs. Our books are designed to provide middle grade mathematics teachers with a strong mathematical foundation and connect the mathematics they are learning with the mathematics they will be teaching. Their focus is on algebraic and geometric structures, data analysis and probability, and mathematics of change, and they employ standards-based middle grade mathematics curricular materials as a springboard to explore and learn mathematics in more depth. They have been extensively piloted in Summer Institutes, in courses offered at school-based sites, through a variety of professional development programs, and in both undergraduate and graduate semester courses offered at a number of universities throughout the nation.

This book is written as an introduction to some basic concepts of number theory and modern algebra that underlie middle grade arithmetic and algebra, and thus the approach differs from some traditional texts in these subjects. The primary goal is to help teachers (both in-service and pre-service) gain a fundamental understanding of the key mathematical ideas that they will be teaching, so that in turn they can help their students learn important mathematics.

Throughout the book, the reader will find a number of **Classroom Connections, Classroom Discussions**, and **Classroom Problems**. These instructional components are designed to deepen the connections between the algebra and number theory students are studying now and the algebra they will teach. The **Classroom Connections** are middle grade investigations that serve as launch pads to the college level **Classroom Discussions, Classroom Problems,** and other related collegiate mathematics. The **Classroom Discussions** are intended to be detailed mathematical conversations between college teacher and pre-service middle grade teachers, and are used to introduce and explore a variety of important concepts during class periods. The **Classroom Problems** are a collection of problems with complete or partially complete solutions and are meant to illustrate and engage pre-service teachers in various problem solving techniques and strategies. The continual process of connecting what they are learning in the college classroom to what they will be teaching in their own classroom provides teachers with real motivation to strengthen their mathematical content knowledge.

Many of my recent students studied from preliminary versions of these materials, and their thoughtful comments significantly shaped the contents of this book. I am most grateful to these present and future teachers and take great pride in their mathematical growth. I am also thankful for the insightful suggestions of the many mathematicians and mathematics educators who piloted these materials in their college classrooms or in professional development venues. I am especially appreciative to Professors Jennifer Bay-Williams, Kansas State University; Al Dixon, Western Michigan University; and Steve Ziebarth, College of the Ozarks, for their careful reviews of a preliminary version of this text. Their astute and detailed remarks notably improved the materials. Writing this book has been a great joy. The mathematical adventure was especially exciting and having the opportunity to work with outstanding graduate students was an incredible bonus. I am deeply thankful to David Barker for crafting a comprehensive first draft of Chapter 1. He and I spent countless hours discussing the learning and teaching of mathematics, and we learned a great deal from each other. I would also like to extend my sincere gratitude to graduate students Dustin Foster and Chris Thornhill, to middle grade teacher Paul Rahmoeller, and to post-doctoral fellow, Jason Aubrey for reading (re-reading, re-re-reading,...) over the manuscript, solving selected exercises, and making many valuable suggestions. Finally, I am most appreciative to Petra Recter at Pearson/Prentice-Hall for her expert assistance in bringing this book to print.

Ira J. Papick

Algebra Connections

Patterns

CHAPTER 1

1.1 CLASSROOM CONNECTIONS: REPRESENTING PATTERNS
1.2 REFLECTIONS ON CLASSROOM CONNECTIONS: REPRESENTING PATTERNS
1.3 ARITHMETIC SEQUENCES
1.4 CLASSROOM CONNECTIONS: A QUADRATIC SEQUENCE
1.5 REFLECTIONS ON CLASSROOM CONNECTIONS: A QUADRATIC SEQUENCE
1.6 FINITE ARITHMETIC SEQUENCES
1.7 GEOMETRIC SEQUENCES
1.8 MATHEMATICAL INDUCTION
1.9 CLASSROOM CONNECTION: COUNTING TOOLS
1.10 THE BINOMIAL THEOREM
1.11 THE FIBONACCI SEQUENCE

In a broad sense, the study of patterns and relationships is the essence of mathematics and, accordingly, it occupies a central position in school mathematics. Mathematicians seek to understand fundamental structures by searching for patterns and relationships within classes of examples and collections of data. Their investigations involve insightful questions and conjectures in unison with creative thinking and problem-solving strategies, and it is especially crucial for all students of mathematics to comprehend and embrace these habits of discovery.

1.1 CLASSROOM CONNECTIONS: REPRESENTING PATTERNS

We begin this chapter by looking at the Tiling Pools problem from the eighth grade module *Say It with Symbols* of the *Connected Mathematics* curriculum. As you work through this problem (as well as through other middle-school problems throughout this textbook), pay special attention to the following questions:

1. What strategies did you use to solve the problem, and what strategies do you think students will use?

2. What types of rules did you discover, and what types of rules do you think students will produce?
3. How did you justify your rules, and what types of justifications do you expect your students to give?
4. What counts as an acceptable justification at the middle-school level?

This problem and others provide the basis of many of our discussions throughout this chapter and illuminate many important ideas concerning patterns.

Tiling Pools

Hot tubs and in-ground swimming pools are sometimes surrounded by borders of tiles. This drawing shows a square hot tub with sides of length 5 feet surrounded by square border tiles. The border tiles measure 1 foot on each side. A total of 24 tiles are needed for the border.

20 Say It with Symbols

Reproduced from page 20 of *Say It with Symbols* in *Connected Mathematics*.

FIGURE 1.1.1

Problem 2.1

In this problem, you will explore this question: If a square pool has sides of length s feet, how many tiles are needed to form the border?

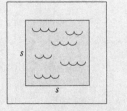

1 ft
1 ft
border tile

A. Make sketches on grid paper to help you figure out how many tiles are needed for the borders of square pools with sides of length 1, 2, 3, 4, 6, and 10 feet. Record your results in a table.

B. Write an equation for the number of tiles, N, needed to form a border for a square pool with sides of length s feet.

C. Try to write at least one more equation for the number of tiles needed for the border of the pool. How could you convince someone that your expressions for the number of tiles are equivalent?

Reproduced from page 21 of *Say It with Symbols* in *Connected Mathematics*.

FIGURE 1.1.2

Problem 2.1 Follow-Up

1. Make a table and a graph for each equation you wrote in part a of Problem 2.1. Do the table and the graph indicate that the equations are equivalent? Explain.

2. Is the relationship between the side length of the pool and the number of tiles linear, quadratic, exponential, or none of these? Explain your reasoning.

3. a. Write an equation for the area of the pool, A, in terms of the side length, s.
   b. Is the equation you wrote linear, quadratic, exponential, or none of these? Explain.

4. a. Write an equation for the combined area of the pool and its border, C, in terms of the side length, s.
   b. Is the equation you wrote linear, quadratic, exponential, or none of these? Explain.

Reproduced from page 21 of *Say It with Symbols* in *Connected Mathematics*.

FIGURE 1.1.3

1.2 REFLECTIONS ON CLASSROOM CONNECTIONS: REPRESENTING PATTERNS

It is common for students to think about and solve mathematics problems in a multitude of ways. For example, here are the thoughts of eighth graders Meaghan and Reese on the problem of determining the number of square tiles (1 foot by 1 foot) needed to form the boundary of a square pool of dimensions s feet by s feet (where s is a positive integer).

Meaghan. I drew out the first three pools and noticed that each time you add a foot to the side of the pool, the number of tiles goes up by 4.

Reese. I noticed that for a pool of any size you will always have a tile for each foot of the perimeter, or $4n$, and then you need 4 more tiles for the corners, so I added 4.

Meaghan and Reese have taken different approaches in solving this problem, and it is instructive to look at their responses in more detail.

Meaghan's Strategy. Meaghan initially drew pictures of square pools with side lengths 1, 2, and 3 feet. Once she drew these examples and calculated the number of tiles needed to surround the pools, she compared the results and conjectured a relationship between pools with consecutive integer side lengths.

4 Chapter 1 Patterns

8 tiles

12 tiles

16 tiles

By looking at the pattern 8 tiles, 12 tiles, and 16 tiles, she concluded that, "each time you add a foot to the side of the pool, the number of tiles goes up by 4." Hence, if you knew the number of tiles required to surround a square pool of length 9 feet (which can be determined from the previous cases), you could then find the number of tiles required to surround a square pool of length 10 feet by simply adding 4. Moreover, since you know the number of tiles for a square pool of length 1 foot, you can determine the number of tiles for all whole number length square pools. Why?

Classroom Problem. Using Meaghan's rule, determine how many tiles are needed for a square pool of length 9 feet. Represent your data in a table format (as here).

Side Length in Feet	Number of Tiles
1	8
2	12
3	16
4	
5	
6	
7	
8	
9	

The primary advantage of Meaghan's rule is that it is easy to calculate the number of tiles for a square pool of length n provided you know the number of tiles for a square pool of length $n - 1$, while the main disadvantage is that it is difficult to determine the number of tiles for larger length square pools (e.g., for a square pool of length 2,467 feet). ◆

Question. Meaghan arrived at her rule by inspecting some particular examples and did not show that her rule holds for all positive integer lengths. How would you justify the validity of Meaghan's rule for all positive integer lengths?

Section 1.2 Reflections on Classroom Connections: Representing Patterns

The kind of pattern that Meaghan observed—where after some explicit terms are specified, each subsequent term is defined in terms of a previous term or a combination of previous terms—is called a **recursive pattern**. The rule that describes the relationship between these consecutive terms is called a **recursive rule or formula**.

Representing Rules. The middle grades are an important time for students as they begin to develop the ideas of variable and function. The gradual transformation from describing rules using language to describing rules using symbols is a key transition during this time. Notation, which is intended to simplify thinking, can often be confusing to students during their initial exposure because they perceive the notion of *variable* in a variety of ways. Hence, an appropriate understanding of what these representations mean and how they are used is a must for students.

Although Meaghan described her rule in words, it is possible to express it in symbols. This kind of representation is especially useful for more complicated rules, since it compresses information into notation that is more workable. For example, if we let T_1 represent the number of tiles in a square pool of length 1 foot (the value of the pattern's first term), T_2 represent the number of tiles in a square pool of length 2 feet (the value of the pattern's second term), etc., then Meaghan's recursively defined rule for the pool problem could be stated as follows:

$$T_1 = 8$$
$$T_n = T_{n-1} + 4, (n > 1)$$

For this rule, T_n represents the number of tiles needed to surround a square pool with a side of length n (the value of the nth term of the pattern), and T_{n-1} is the number of tiles required for a square pool of length $n - 1$ (the value of the $(n - 1)^{th}$ term of the pattern).

Classroom Problem. Let's write a recursive rule for the pattern that occurs in the following problem.

Farmer Jim (or Jimbo as he is called by his closest friends) uses fence panels of the same length to create pens for his animals. He decides to arrange the pens in a single row with all the pens being connected as illustrated in the picture here.

The number of fence panels needed for these three pens is recorded in the following table.

Term (number of animal pens)	Value (number of panels required)
1	4
2	7
3	10

We see that it takes four panels to create the first animal pen, and this can be expressed in notation as $P_1 = 4$. Next, we need to find a relationship between consecutive terms of this pattern. If an additional pen is appended to the first pen, Farmer Jim will use one panel from the end of the first pen and add three more panels to get to the required four panels to complete the second pen. Similarly, three more panels are needed to create the third pen, and so the nth pen is built by adding three panels to the $(n-1)^{\text{th}}$ pen. This relationship can be expressed as

$$P_n = P_{n-1} + 3,$$

and so the complete recursive formula describing this situation is given by:

$$P_1 = 4$$
$$P_n = P_{n-1} + 3. \quad \blacklozenge$$

Reese's Strategy. Recall that Reese's approach to the Tiling Pools problem differed from Meaghan's strategy. He states, "For a pool of any size, you will always have a tile for each foot of the perimeter, or $4n$, and then you need four more tiles for the corners." Instead of comparing the number of tiles needed for a few different-length square pools (as Meaghan did), Reese developed a systematic way of counting the tiles needed for each square pool of length n (n a positive integer). The rule Reese developed establishes an explicit relationship between the length of a side of the pool and the number of tiles required to surround it.

n = positive integer length (in feet) of a square pool	$4n + 4$ = number of tiles in the boundary of the pool
1	$4 \cdot 1 + 4 = 8$
2	$4 \cdot 2 + 4 = 12$
3	$4 \cdot 3 + 4 = 16$
4	$4 \cdot 4 + 4 = 20$
5	$4 \cdot 5 + 4 = 24$

In mathematical terms, Reese's **explicit rule** defines a function T on the set of positive integers, given by $T(n) = 4n + 4$, where n is a positive integer length (in feet) of a square pool, and $T(n)$ is the total number of tiles needed for a square pool of length n.

In general, a function f whose domain is the positive integers (into any other set) is called an **infinite sequence** (or simply a **sequence**). The **range of a sequence**,

$$\text{Range of } f = \{f(n): n \text{ is a positive integer}\},$$

is usually written in the form

$$a_1, a_2, a_3, \ldots, a_n, \ldots,$$

Section 1.2 Reflections on Classroom Connections: Representing Patterns

where $f(n) = a_n$ for each positive integer n. The sequence that Reese discovered is

$$8, 12, 16, 20, 24, \ldots, 4n + 4, \ldots,$$

where n is a positive integer.

Convention. Since a sequence's domain is always the positive integers, it is common practice to identify a sequence f with its range: $a_1, a_2, a_3, \ldots, a_n, \ldots$

Using Reese's explicit rule, it is easy to calculate the number of tiles needed for a square pool of length n. This is a benefit of his rule over Meaghan's recursive rule. However, as we shall see when we study the Fibonacci sequence (Section 1.11), it is not always straightforward to determine an explicit rule for a given sequence.

An Explicit Rule for Farmer Jim. Let's return to the problem of Farmer Jim's livestock pens, but this time we try to describe the number of pens with an explicit rule rather than a recursive rule.

The following table consists of some conclusions we have drawn from looking at specific cases, which may be useful in formulating a specific rule.

Number of Pens	Panels Required
1	4
2	7
3	10
4	13
5	16

Looking at the table of values, it might be conjectured that the number of panels required for any (positive integer) number of pens n is $P(n) = 3n + 1$; however, we cannot draw this conclusion based solely on these few cases.

It is common for (middle grade) students to create rules based upon only a few cases, often a single case. They might look at the previous table and see that 3 pens require 10 panels and thus state that the general rule is $P(n) = n^2 + 1$, which as it turns out, only happens to work for this particular case. It is important for all students of mathematics to understand that they must justify general conclusions through valid arguments and not rely exclusively on the verification of a few cases. Since we cannot guarantee from the table's information that the explicit rule describing the pattern of this problem is $P(n) = 3n + 1$, let's turn to the context of the problem to assist us in justifying this rule.

Justification. One way to construct n pens is to put together n groups of three-sided pens and join them in the manner illustrated here (for $n = 4$). This grouping requires $3n$ panels and lacks one panel to close off the last pen. Hence, the expression $3n + 1$ gives the total number of panels required for n pens.

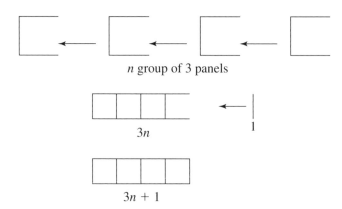

n group of 3 panels

$3n$

$3n + 1$

Recursive to Explicit. Another way to see the pens' construction is to start with a single pen. The recursive rule that we previously discovered stipulates that we need three panels to create each additional pen. Hence, if we want to produce three pens, we need to add two sets of three panels to an original pen:

So the expression $4 + 3(n - 1)$ gives the total number of panels required for *n* (positive integer) pens. Why? Note that this is equivalent to the formula we obtained earlier, since
$$4 + 3(n - 1) = 4 + 3n - 3 = 3n + 1.$$

Question. A student discovered the rule $P(n) = 2n + (n + 1)$ to calculate the total number of panels required for *n* (positive integer) pens. What reasoning do you think she used to develop this explicit rule?

Comparing Representations. We have seen that we can represent patterns in various ways. It is essential for teachers and students to understand the reasoning behind their development and whether or not they are equivalent.

Classroom Problem. Middle-grade students working the Tiling Pool problem gave the following (equivalent) explicit rules:

$T(n) = 4n + 4;$ $\quad T(n) = 4(n + 1);$ $\quad T(n) = 4(n + 2) - 4;$

$T(n) = 2(n + 2) + 2n;$ $\quad T(n) = 4\left[n + 2\left(\frac{1}{2}\right)\right];$ $\quad T(n) = (n + 2)^2 - n^2$

Discuss possible explanations for each of these equivalent (Check) explicit rules. ◆

For example, middle-grade students regularly offer the following (geometric) explanation for the last rule listed.

Looking at the figure of the pool, they notice that to obtain the number of tiles in the border, they can take the area of the large square, which is $(n + 2)^2$, and subtract the smaller inner square, which has an area of n^2. Hence, the number of tiles along the border is $(n + 2)^2 - n^2$.

Summary: Recursive and Explicit Rules. Middle-grade teachers and their students often make the assumption that the most valuable way to describe a sequence is by using an explicit rule; however, this is not always the case. In fact, the most advantageous representation is highly dependent upon the context in which the sequence appears and the kind of information that is to be gleaned from the description. Some sequences lend themselves naturally to explicit descriptions and others to recursive. Having both an explicit and recursive description of a given sequence provides the most versatility.

A question (raised in a middle-grade mathematics classroom) concerning the Tiling Pool problem illustrates the advantage of having both a recursive rule and an explicit rule. The teacher asked the class to find the number of tiles required to tile a square pool whose side length was 127. They all successfully substituted 127 into the explicit formula of $T(n) = 4n + 4$ to obtain the answer 512. She then asked them how many tiles were required for a pool of side length 128, and some responded by simply saying 4 more than 512 tiles. Unfortunately, not all students applied this reasoning, and those who did not proceeded to substitute 128 into the explicit rule. Although this certainly produced the correct answer, it demonstrated that some students in the class did not truly understand the recursive rule and the advantages of this type of reasoning.

EXERCISES 1.2

1. Work through the following Painting Faces problem from the *Patterns and Shapes in Numbers* unit from the middle-grade curriculum *MathScape* (answer all stated questions). Include, as part of your solutions, a recursive rule and an explicit rule describing the pattern in this problem.

2 Painting Faces

EXPLORING PATTERNS BY ORGANIZING DATA

Here is a problem about painting all sides of a three-dimensional shape. Sometimes using objects is helpful in solving problems like this. You can record the data you get when you solve for shorter lengths and organize it into a table to help you find a rule for any length.

Make a Table of the Data

How can you set up a table to record data about a pattern?

A company that makes colored rods uses a paint stamping machine to color the rods. The stamp paints exactly one square of area at a time. Every face of each rod has to be painted, so this length 2 rod would need 10 stamps of paint.

How many stamps would you need to paint rods from lengths 1 to 10? Record your answers in a table and look for a pattern.

Length 2 rod; each end equals 1 square.

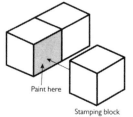
Paint here

Stamping block

How to Organize Your Data in a Table

1. Use the first column to show the data you begin with. Write the numbers in order from least to greatest. In this example, the lengths of the rods go in the first column.

2. Use the second column to show numbers that give information about the first sequence. Here, the numbers of stamps needed to paint each length of rod go in the second column.

Length of Rod	Stamps Needed
1	
2	
3	

PATTERNS IN NUMBERS AND SHAPES   LESSON 2

Reproduced from page 8 of *Patterns and Shapes in Numbers* in *MathScape*.

FIGURE 1.2.1

Section 1.2   Reflections on Classroom Connections: Representing Patterns

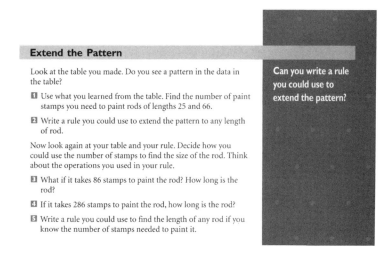

Reproduced from page 9 of *Patterns and Shapes in Numbers* in *MathScape*.

FIGURE 1.2.2

2. Given a circle with n designated points on it (n is a positive integer), determine if the explicit rule $R(n) = 2^{n-1}$ gives the maximum number of distinct nonoverlapping regions that are created by connecting the n points. If this formula is correct, give an argument that justifies it; if the formula is incorrect, provide a counterexample.

Cases $n = 1, 2,$ and 3.

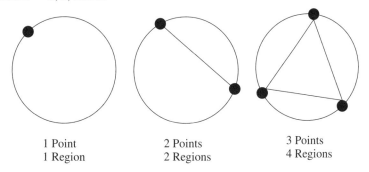

3. (Déjà vu problem—Have you recently seen a problem like this one?) A middle-grade cafeteria has square tables where students can eat lunch in groups of four. If six students want to eat at the same table, then they can push two square tables together to accommodate their group; even larger groups can be handled by joining together more tables in a straight line.

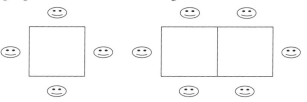

12 Chapter 1 Patterns

a. Fill in the missing data in the following table.

n = number of square tables pushed together in a line	$S(n)$ = number of students sitting at n tables
1	4
2	6
3	
4	
5	
6	

b. Write a recursive rule describing the table-joining pattern.
c. Write an explicit rule representing the table-joining pattern.
d. Determine the number of students who can sit together at a table formed by joining 17 square tables in a line.
e. In order to seat 51 students, what is the smallest number of square tables they could push together in a line? (Some chairs might not be used.)

4. **Clothespin problem:**[1] Heez A. Wasure is hanging clothes out on the clothesline to dry. Heez places one clothespin in the middle and two on the sides of each shirt. He links all of the shirts together so that he conserves clothespins. In the following picture he uses seven clothespins for three shirts.

a. How many clothespins would he need for 1 shirt? 2 shirts? 4 shirts? 5 shirts? 10 shirts? 11 shirts? 23 shirts? 76 shirts? 131 shirts? (He has a very long clothesline.)
b. Write a recursive rule describing the clothespin pattern.
c. Write an explicit rule representing the clothespin pattern.
d. Suppose that Heez used 77 clothespins for one line of shirts. How many shirts did he hang out to dry?

5. **Thumbtack problem:** Deck R. Ater hangs posters on the wall using one tack on each corner of the posters and one tack in the middle of each side. See the following example with four posters.

---
[1]Thanks to Professor John Lanin of the University of Missouri for problems 4 and 5 of this exercise set.

Section 1.2   Reflections on Classroom Connections: Representing Patterns   13

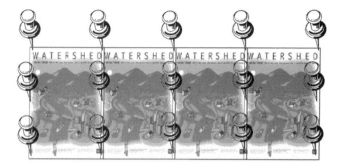

a. How many tacks will Deck need to hang 2 posters? 5 posters? 8 posters? 16 posters? 27 posters? 35 posters?
b. Write a recursive rule describing the thumbtack pattern.
c. Write an explicit rule representing the thumbtack pattern.
d. Suppose that Deck used 237 tacks. How many posters did he hang?
e. How many tacks will Deck need to hang 439 posters?

6. Work through the Crossing the River problem from the unit *Patterns and Shapes in Numbers* unit from the middle-grade curriculum *MathScape* (answer all stated questions). Include, as part of your solutions, a recursive rule and an explicit rule describing the pattern in this problem.

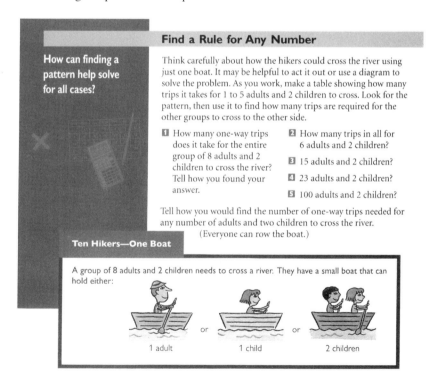

Reproduced from page 10 of *Patterns and Shapes in Numbers* in *MathScape*.

FIGURE 1.2.3

> **Use Your Method in Another Way**
>
> Use the pattern to find the number of adults who need to cross the river for each case.
>
> 1. It takes 13 trips to get all of the adults and the 2 children across the river.
> 2. It takes 41 trips to get all of the adults and the 2 children across the river.
> 3. It takes 57 trips to get all of the adults and the 2 children across the river.
>
> How can you work backward from what you know?

Reproduced from page 11 of *Patterns and Shapes in Numbers* in MathScape.

FIGURE 1.2.4

7. **a.** A sequence $a_1, a_2, a_3, \ldots$ is defined by the recursive formula

$$a_1 = 6$$
$$a_n = a_{n-1} + 7.$$

   Find the explicit formula that describes this sequence.
   b. A sequence $a_1, a_2, a_3, \ldots$ is defined by the explicit rule $a_n = 3n + 2$, where $n \geq 1$. Find a recursive formula that describes this sequence.
   c. A sequence $a_1, a_2, a_3, \ldots$ is defined by the explicit rule $a_n = 5 \cdot 3^{n-1}$. Find a recursive formula that describes this sequence.
8. Write the first 15 terms of the sequence defined by the following recursive formula:

$$a_1 = a_2 = 1$$
$$a_n = a_{n-1} + a_{n-2}.$$

This sequence is called the **Fibonacci sequence**. It has many fascinating connections to nature and a variety of interesting mathematical properties. It is not easy to find an explicit rule for the Fibonacci sequence (give it your best shot). In Section 1.11 we study this famous sequence in more detail and discuss the surprising explicit rule describing it.

1.3 ARITHMETIC SEQUENCES

The sequences that emerged in the Tiling Pool problem and in the Farmer Jim's Pen problem were, respectively,

$$8, 12, 16, \ldots, 4n + 4, \ldots \text{ and } 4, 7, 10, \ldots, 3n + 1, \ldots$$

Both of these sequences have the property that each term after the first term is determined by adding a fixed number to the previous term. In particular, 4 is the added fixed number for the first sequence, and 3 is the added fixed number for the second sequence.

Arithmetic Sequences. A sequence $a_1, a_2, a_3, \ldots$ possessing the property that each term a_i is obtained by adding a fixed real number d (called the **common difference**) to the previous term a_{n-1} (where $n > 1$) is called an **arithmetic sequence** (or **arithmetic**

progression). The previous sequences are examples of arithmetic sequences. Note that our definition of arithmetic sequence is a recursive rule, since the sequence is completely defined by the initial term a_1 and by the rule $a_n = a_{n-1} + d$ for all $n > 1$.

The previous arithmetic sequences are linear functions whose domains are the positive integers, i.e., functions defined for each positive integer n by the rule $f(n) = dn + c$ for some real numbers d and c (see Section 6.3 for a review of the real numbers). We next show that all such functions are arithmetic sequences.

Observation. A linear function f whose domain is the positive integers is an arithmetic sequence.

Analysis. Since f is a linear function defined on the positive integers, we know that $f(n) = dn + c$ for some real numbers d and c. Note that for each positive integer n,

$$f(n) - f(n-1) = (dn + c) - [d(n-1) + c] = dn + c - dn + d - c = d$$

Thus, a linear function f defined on the positive integers is an arithmetic sequence (by the definition of an arithmetic sequence).

Question. Is the converse of the previous observation valid? In particular, is it true that if f is an arithmetic sequence with common difference d (d some fixed real number), then f is a linear function whose domain is the positive integers?

Analysis. Since f is an arithmetic sequence with common difference d, we may write

$$f(1) = a_1, f(2) = a_1 + d, f(3) = a_1 + 2d, \ldots, f(n) = a_1 + (n-1)d, \ldots,$$

and hence f is a linear function defined on the positive integers by the rule

$$a_n = f(n) = a_1 + (n-1)d = dn + (a_1 - d). \qquad 1.3.1$$

Our work has led to the following complete characterization of arithmetic sequences.

1.3.2 Characterizing Arithmetic Sequences

A sequence f is an arithmetic sequence if and only if f is a linear function whose domain is the positive integers. Moreover, the common difference d is the slope of the line defined by f.

EXERCISES 1.3

1. Let f be an arithmetic sequence where $f(3) = 9$ and $f(8) = 24$. Describe the function f explicitly. Direction: Find the common difference d and the term a_1 by using the given information and the fact that $f(n) = a_1 + (n-1)d$ (cf. 1.3.1), where $a_1 = f(1)$.

2. Given the arithmetic sequence $21, 28, 35, \ldots, 707, \ldots$, find the number of terms in the sequence between 21 to 707 (including 21 and 707).

1.4 CLASSROOM CONNECTIONS: A QUADRATIC SEQUENCE

The Handshake problem from the *Frogs, Fleas, and Painted Cubes* unit from the *Connected Mathematics* curriculum leads to an interesting sequence that is not an arithmetic sequence. Let's work through this problem to get a feel for the mathematics involved. See Figures 1.4.1 below and 1.4.2 on the facing page.

3.1 Counting Handshakes

After a sporting event, the opposing teams often line up and shake hands. To celebrate their victory, members of the winning team may congratulate each other with a round of high fives. In this problem, you will explore the total number of handshakes or high fives that take place in several situations. You will consider three cases.

Case 1
Two teams with the same number of players shake hands.

Case 2
Two teams with different numbers of players shake hands. For example, although 5 players from each basketball team participate at one time, one team may have a total of 8 players and the other may have a total of 10 players.

Case 3
Members of the same team exchange high fives.

Reproduced from page 41 of *Frogs, Fleas, and Painted Cubes* in *Connected Mathematics*.

FIGURE 1.4.1

1.5 REFLECTIONS ON CLASSROOM CONNECTIONS: A QUADRATIC SEQUENCE

Counting Convention. In this problem, when two people shake hands, it is counted as one handshake.

The following table collects some data representing the number of handshakes $H(n)$ for a group of n people. Verify that this data is accurate.

Section 1.5 Reflections on Classroom Connections: A Quadratic Sequence

> **C.** Consider case 3, in which each member of a team gives a high five to each teammate.
>
> **1.** How many high fives will take place among an academic quiz team with 4 members?
>
> **2.** How many high fives will take place among a golf team with 12 members?
>
> **3.** Write an equation for the number of high fives, h, that will take place among a team with n members.

Reproduced from page 42 of *Frogs, Fleas, and Painted Cubes* in *Connected Mathematics*.

FIGURE 1.4.2

n people	$H(n)$ handshakes
1	0
2	1
3	3
4	6
5	10
6	15
7	21

Recursive Rule. This handshaking process defines a function H on the positive integers (i.e., a sequence). We can say that H is not an arithmetic sequence, since there is no common difference between consecutive terms. (Explain.) Rather than simply listing some terms from the handshake sequence H, let's develop a recursive rule describing this sequence.

Analysis. The problem was to determine how many handshakes there would be if every person shook hands with everyone else. When a new person is added to the group, the total number of handshakes is comprised of all the original handshakes plus the number of this new person's handshakes. Hence, if the nth person joins the group, there are $n - 1$ people currently in the room who have to shake hands with him. Thus, the recursive formula for this sequence is given by:

$$H_1 = 0$$
$$H_n = H_{n-1} + (n - 1),$$

where H_n represents the number of handshakes for a group of n people.

Explicit Rule. Finding an explicit formula for the handshaking sequence H is generally more difficult for middle-grade students to accomplish. They often stray away from the context of the problem and deal solely with numbers (as collected in

18   Chapter 1   Patterns

the previous table). This kind of strategy, based on finite collections of data, often leads to incorrect conclusions and does not effectively utilize the given context of the problem. In the following analysis, we employ the problem's framework to reach an explicit formula.

Analysis. If there are n people in the room, we know that each person has to shake the hand of everyone else. Thus, since each person must make $n - 1$ handshakes, there are a total of $n(n - 1)$ handshakes. Why? However, each handshake has been counted twice, once for each person involved in the handshake, and thus we must divide the expression by 2 to adjust for this overcount. Therefore, the following quadratic polynomial function describes the handshaking sequence H:

$$H(n) = \frac{n(n - 1)}{2} = \frac{n^2}{2} - \frac{n}{2}.$$

Recursive to Explicit. Another way to find the explicit formula for the handshaking problem is to use the recursive relationship $H_n = H_{n-1} + (n - 1)$ to determine the number of handshakes for a certain number of people. In particular,

$$H_2 = H_1 + 1 = 0 + 1$$
$$H_3 = H_2 + 2 = 1 + 2$$
$$H_4 = H_3 + 3 = 1 + 2 + 3$$
$$\vdots$$
$$H_n = H_{n-1} + (n - 1) = 1 + 2 + 3 + \ldots + (n - 1).$$

Therefore, the number of handshakes for n (positive integer) people is given by the rule
$$H_n = 1 + 2 + 3 + \ldots + (n - 1).$$

Our work is complete by showing that

$$1 + 2 + 3 + \ldots + n - 1 = \frac{n(n - 1)}{2} = H(n).$$

To verify the previous equality, we generalize an argument that the brilliant mathematician Carl Friedrich Gauss (see Chapter 2) gave as a child when asked to add all the whole numbers from 1 to 100. To the surprise of his teacher, Gauss accomplished the task quickly and reported the sum to be 5,050. His strategy was based upon a pattern he observed that involved pairing up the numbers to be added in the following manner:

$$1 + 2 + 3 + \ldots + 98 + 99 + 100.$$

$$101$$
$$101$$
$$101$$

Section 1.5   Reflections on Classroom Connections: A Quadratic Sequence   19

Since each of the bracketed pairs adds up to 101, and there are exactly 50 of these pairs, the total sum is $50 \cdot 101 = 5{,}050$.

Gauss's strategy can easily be extended to determine a general formula for the sum $1 + 2 + 3 + \ldots + n$. After we obtain this formula, we then apply it to our specific problem to show that $1 + 2 + 3 + \ldots + (n - 1) = \frac{n(n-1)}{2}$.

As in Gauss's strategy, we start by breaking the general sum up into pairs,

$$1 + 2 + 3 + \ldots + (n - 2) + (n - 1) + n,$$

with bracketed pair sums of $1+n$, $1+n$, $1+n$.

each having a sum of $n + 1$. If n is even, then, as Gauss observed, there are $\frac{n}{2}$ pairs, and so the total sum is given by $\frac{n}{2}(n + 1)$. If n is odd, then we must slightly modify our reasoning to obtain the same formula. A specific example helps us develop a general argument.

Classroom Problem.   Use a modified version of Gauss's strategy to calculate the sum $1 + 2 + 3 + 4 + 5 + 6 + 7$, and then apply your method to calculate the sum $1 + 2 + 3 + \ldots + n$, where n is odd. (In Theorem 1.6.1, we demonstrate how to derive the formula $1 + 2 + 3 + \ldots + n = \frac{n(n+1)}{2}$ without having to consider separate even and odd cases.)

Analysis.   Partitioning the specific sum into pairs gives

$$1 + 2 + 3 + \underbrace{4}_{\text{middle term}} + 5 + 6 + 7,$$

with bracketed pair sums of 8, 8, 8.

and thus,

$$1 + 2 + 3 + 4 + 5 + 6 + 7 = (\text{number of pairs}) \cdot (\text{sum of each pair})$$
$$+ \text{ middle term}$$
$$= 3 \cdot 8 + 4 = 28.$$

To compute the general sum $1 + 2 + 3 + \ldots + n$, where n is odd, we can proceed in an analogous fashion. Separating the sum into pairs plus a middle term gives

$$1 + 2 + 3 + \ldots + \underbrace{\frac{n+1}{2}}_{\text{middle term}} + \ldots + (n - 2) + (n - 1) + n,$$

with bracketed pair sums of $n+1$, $n+1$, $n+1$.

and so

$$1 + 2 + 3 + \ldots + n = \text{(number of pairs)} \text{(sum of each pair)} + \text{middle term}$$

$$= \frac{(n-1)}{2}(n+1) + \frac{(n+1)}{2} = \frac{(n+1)}{2}[(n-1) + 1]$$

$$= \frac{(n+1)n}{2}.$$

Therefore, for any positive integer n,

$$1 + 2 + 3 + \ldots + n = \frac{n(n+1)}{2}. \quad \textbf{1.5.1}$$

It is now a simple matter to complete our derivation of the explicit rule from the recursive in the handshaking problem. In particular, by substituting $(n-1)$ for n in the previous formula, it follows that

$$1 + 2 + 3 + \ldots + n - 1 = \frac{(n-1)[(n-1)+1]}{2} = \frac{(n-1)n}{2}$$

$$= \frac{n(n-1)}{2}. \quad \blacklozenge$$

Classroom Problem. Use Gauss's adding method to calculate the following sums (check your answers by pencil and paper arithmetic or a calculator):

1. $3 + 8 + 13 + 18 + 23 + 28 + 33 + 38 =$
2. $2 + 6 + 10 + 14 + 18 + 22 + 26 + 30 + 34 =$
3. $11 + 8 + 5 + 2 + (-1) + (-4) + (-7) + (-10) + (-13) + (-16) =$
4. $4 + \frac{14}{3} + \frac{16}{3} + 6 + \frac{20}{3} + \frac{22}{3} + 8 =$
5. $\frac{1}{2} + \frac{5}{4} + 2 + \frac{11}{4} + \frac{14}{4} + \frac{17}{4} = \quad \blacklozenge$

EXERCISES 1.5

1. The ancient Greek mathematicians studied number patterns that were associated with certain geometric figures. For example, they were intrigued by a sequence of numbers defined by the number of uniformly spaced dots used to form equilateral triangles, and they accordingly named these numbers **triangular numbers**. The first four triangular numbers, corresponding to the following dot diagrams, are 1, 3, 6, and 10.

Note that the nth triangular number T_n is given by the sum $T_n = 1 + 2 + 3 + \ldots + n$, and thus $T_n = \frac{n(n+1)}{2}$. Why?

a. Show that $9T_n + 1$ is the triangular number T_{3n+1}.
b. Show that $8T_n + 1$ is a perfect square (see Section 2.6).
c. Show that the sum of two consecutive triangular numbers $T_n + T_{n+1}$ is a perfect square.
d. Show that $T_{n-1} + 6T_n + T_{n+1} = (2n + 1)^2$.
e. Find three triangular numbers that are perfect squares.
f. Show that if T_n is a perfect square, then $T_{4n(n+1)}$ is also a perfect square.
Direction: $T_{4n(n+1)} = \frac{(4n^2+4n)(4n^2+4n+1)}{2} = (2n^2 + 2n)(2n + 1)^2 = 4\left(\frac{n(n+1)}{2}\right)$
• $(2n + 1)^2 = 4T_n(2n + 1)^2$. Verify each step and explain how the conclusion follows from the hypothesis.

1.6 FINITE ARITHMETIC SEQUENCES

In the previous problems, we saw that we can use Gauss's adding strategy to find the sum of "finite" arithmetic sequences. Note that a function f whose domain is the set

$$N_n = \{1, 2, 3, \ldots, n\},$$

where n is a positive integer, is called a **finite sequence**, and the range of a finite sequence, range of $f = \{f(t): t \in N_n\}$, is written as

$$\{a_1, a_2, a_3, \ldots, a_n\}.$$

Furthermore, a finite sequence of the form $a, a + d, a + 2d, \ldots, a + (n - 1)d$ is called a **finite arithmetic sequence.**

Classroom Problems.

a. Find the number of multiples of 7 between 18 and 711.
b. Find the number of multiples of 8 between 35 and 1,635.

Analysis of Problem. **a.** (We'll leave the analysis of b for your enjoyment.) Think of this problem in terms of arithmetic sequences. First note that 21 is the smallest multiple of 7 that is greater than 18, and 707 is the largest multiple of 7 that is less than 711. Thus, determining the number of multiples of 7 between 18 and 711 is equivalent to finding the number of terms in the following finite arithmetic sequence with common difference 7:

$$21, 28, 35, \ldots, 707.$$

The explicit rule describing the nth term of this sequence is $a_n = 21 + 7 \cdot (n - 1)$ (cf. 1.3.1), so solving the equation $707 = 21 + 7(n - 1)$ for n gives the desired answer. In particular, there are $n = 99$ multiples of 7 between 18 and 711. Verify (cf. Exercises 1.3, no. 2). ◆

We complete this section by deriving a formula (analogous to 1.5.1) for the sum of an arbitrary finite arithmetic sequence. Rather than constructing arguments based on even and odd numbers of terms, as we did for the formula $1 + 2 + 3 + \ldots + n = \frac{n(n+1)}{2}$, we use a variation of Gauss's idea to give a proof that handles both cases together.

1.6.1 Theorem

If $a_1, a_2, a_3, \ldots, a_n$ is a finite arithmetic sequence defined by

$$a_n = a_1 + (n-1)d, \text{ then } a_1 + a_2 + a_3 + \ldots + a_n = \frac{n(a_1 + a_n)}{2}.$$

Proof. To show that the previous formula is true, we must prove that

$$2(a_1 + a_2 + a_3 + \ldots + a_n) = n(a_1 + a_n).$$

Consider the following n pairings and resulting sums:

$$2(a_1 + a_2 + a_3 + \ldots + a_n) = (a_1 + a_2 + a_3 + \ldots + a_n) + (a_1 + a_2 + a_3 + \ldots + a_n) =$$

$$(a_1 + a_2 + a_3 + \ldots + a_n + a_1 + \ldots + a_{n-2} + a_{n-1} + a_n).$$

$$\underbrace{}_{a_1 + a_n}$$
$$\underbrace{}_{a_3 + a_{n-2}}$$
$$\underbrace{}_{a_2 + a_{n-1}}$$
$$\underbrace{}_{a_1 + a_n}$$

Observe that

$$a_2 + a_{n-1} = [a_1 + (2-1)d] + (a_1 + [(n-1) - 1]d) =$$
$$(a_1 + d) + [a_1 + (n-2)d] = a_1 + (a_1 + [d + (n-2)]d)$$
$$= a_1 + (a_1 + [1 + (n-2)]d) =$$
$$a_1 + [a_1 + (n-1)d] = a_1 + a_n,$$

and

$$a_3 + a_{n-2} = [a_1 + (3-1)d] + (a_1 + [(n-2) - 1]d) =$$
$$(a_1 + 2d) + [a_1 + (n-3)d] = a_1 + (a_1 + [2d + (n-3)]d)$$
$$= a_1 + (a_1 + [2 + (n-3)]d) =$$
$$a_1 + [a_1 + (n-1)d] = a_1 + a_n.$$

It follows in a similar manner that each of the n pairings has a sum of $a_1 + a_n$, and thus

$$2(a_1 + a_2 + a_3 + \ldots + a_n) = n(a_1 + a_n).$$

Therefore, $a_1 + a_2 + a_3 + \ldots + a_n = \frac{n(a_1 + a_n)}{2}$.

EXERCISES 1.6

1. Use Theorem 1.6.1 to find the sum of the following finite arithmetic sequences:
   a. $1, 3, 5, 7, 9, 11, 13, 15, 17$.
   b. $1, 3, 5, \ldots, (2n - 1)$.
2. Consider a right triangle with legs of length a and b and hypotenuse of length c.

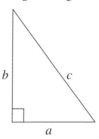

Suppose that a, b, c forms a finite arithmetic sequence (i.e., there is a real number d, such that $b = a + d$ and $c = a + 2d$).
Show that $a = 3d, b = 4d$, and $c = 5d$. Direction: By the Pythagorean Theorem (cf., Theorem 5.3.1), we know that $a^2 + (a + d)^2 = (a + 2d)^2$. Expand both sides of the equation and solve for a in terms of d.

1.7 GEOMETRIC SEQUENCES

We know that the nth term of an arithmetic sequence $a, a + d, a + 2d, \ldots, a + (n - 1)d, \ldots$ having common difference d is explicitly described by the rule

$$a_n = a + (n - 1)d.$$

In this section, we study another interesting class of sequences defined by the property that each term a_n is obtained by multiplying the previous term by a fixed real number r.

Classroom Discussion: Geometric Sequences. The Groups of People problem asks us to develop a rule that specifies the total number of groups of people that could be formed from a set of people of any given size. For example, there are eight possible groups of people that we can create from the three individuals Amelia (A), Beatrice (B), and Camilla (C), as follows:

$$\{\emptyset, \{A\}, \{B\}, \{C\}, \{A, B\}, \{A, C\}, \{B, C\}, \{A, B, C\}\}.$$

The empty set $\emptyset$ denotes the group of people consisting of no members.
If an additional person named Donald (D) joins Amelia (A), Beatrice (B), and Camilla (C), then the total number of groups of people would now be 16,

$$\{\emptyset, \{A\}, \{B\}, \{C\}, \{A,B\}, \{A,C\}, \{B,C\}, \{A,B,C\}, \{D\}, \{A,D\}, \{B,D\},$$
$$\{C,D\}, \{A,B,D\}, \{A,C,D\}, \{B,C,D\}, \{A,B,C,D\}\},$$

since the new collection of people would consist of the original eight groups plus the extra groups created by adding Donald.

Using the same reasoning as above, it follows in general that each time a person is added to a previous group of people, the total number of groups doubles. Hence, if we let P_n = the total number of groups of people that could be formed from a set of n people, then we have the following recursive rule defining the sequence $P_1, P_2, P_3, \ldots, P_n, \ldots$:

$$P_1 = 2$$
$$P_n = 2P_{n-1}.$$

Such a sequence, where each successive term is obtained by multiplying the previous term by a fixed (nonzero) number, is called a **geometric sequence**, and the fixed (nonzero) number multiplier is called the **common ratio**.

Using the recursive formula for the Groups of People problem, we can generate an explicit formula for this sequence by using an iterative process. In particular,

$$P_n = 2 \cdot P_{n-1}$$
$$P_n = 2 \cdot 2 \cdot P_{n-2}$$
$$P_n = 2 \cdot 2 \cdot 2 \cdot P_{n-3}$$
$$\vdots$$
$$P_n = 2 \cdot 2 \cdot \cdots \cdot P_1$$
$$P_n = \underbrace{2 \cdot 2 \cdot \cdots \cdot 2}_{n \text{ times}} = 2^n.$$

Hence, the total number of groups of people P_n that could be formed from a set of n people is given by the explicit rule $P_n = 2^n$. ◆

n^{th} **Term of a Geometric Sequence.** Note that, in general, the nth term of a geometric sequence $a, ra, r^2a, \ldots, r^{n-1}a, \ldots$ with a common ratio $r \neq 0$ is given by

$$a_n = r^{n-1}a.$$

Classroom Discussion: Pennies Add Up. Suppose you put 2 cents in a jar today, and each day thereafter you triple the amount that you put in the previous day. In recursive notation, this can be expressed as

$$M_1 = 2$$
$$M_n = 3M_{n-1},$$

where M_n equals the amount put in on day n.

a. How much would you put in the jar on the 17th day?
b. How much would you have saved after 17 days?

Analysis. **a.** Your daily savings form a geometric sequence having initial term $M_1 = 2$ and a common ratio $r = 3$. We know that the nth term of this sequence can be

expressed explicitly by the rule $M_n = 2 \cdot 3^{n-1}$. Therefore, on the 17th day, you would have put $M_{17} = 2 \cdot 3^{16} = 86,093,442$ cents in the jar (you must have won the lottery).
b. In this question, we are asked to determine the sum

$$M_1 + M_2 + M_3 + \ldots + M_{17} = 2 + 2 \cdot 3 + 2 \cdot 3^2 + \ldots + 2 \cdot 3^{16}.$$

One (laborious) way to find this sum is to calculate each term and then add all of the terms. Instead of a direct calculation, we employ a general method that arrives at this particular sum without all the messy calculations and gives the sum of any finite geometric sequence.

Sum of a Finite Geometric Sequence. The sum of a finite geometric sequence $a, ra, r^2 a, \ldots, r^{n-1} a$ is called a **finite geometric series** and is denoted by

$$S_n = a + ar + ar^2 + \ldots + ar^{n-1}.$$

Our goal is to find a practical formula for S_n just as we did for the sum of a finite arithmetic sequence.

Analysis. We'll start by multiplying both sides of the equality $S_n = a + ar + \ldots + ar^{n-1}$ by the common ratio r to obtain the equality

$$rS_n = ar + ar^2 + \ldots + ar^n.$$

Subtracting S_n from rS_n results in the following string of equalities:

$$rS_n - S_n = (ar + ar^2 + \ldots + ar^n) - (a + ar + \ldots + ar^{n-1})$$
$$= ar^n - a = a(r^n - 1).$$

Therefore,
$$S_n(r - 1) = a(r^n - 1),$$

and so

$$S_n = \frac{a(r^n - 1)}{(r - 1)}, \text{ if } r \neq 1. \qquad \textbf{1.7.1}$$

If $r = 1$, then the geometric sequence $a, 1a, 1^2 a, \ldots, 1^{n-1} a$ (which is also an arithmetic sequence with common difference 0) has sum $S_n = na$.

You can now use the concise formula given in 1.7.1 to determine how much you saved after 17 days. Specifically, you accumulated

$$S_{17} = 2 + 2 \cdot 3 + 2 \cdot 3^2 + \ldots + 2 \cdot 3^{16} =$$

$$\frac{2(3^{17} - 1)}{(3 - 1)} = 3^{17} - 1 = 129,140,162 \text{ cents (what a big jar you must have!)}. \quad \blacklozenge$$

The formula for the sum of a finite geometric series is useful in many areas of mathematics. In the following classroom problem, we see how it can be used to find an explicit rule from a recursive rule.

Classroom Problem. Find the explicit formula for the nth term of the sequence $a_1, a_2, a_3, \ldots$ defined by the recursive rule

$$a_1 = 1$$
$$a_n = 2a_{n-1} + 1.$$

Analysis. To find the explicit rule, we employ an iterative process as we did in the Groups of People problem. Justify each of the steps.

$$a_n = 2a_{n-1} + 1 = 2(2a_{n-2} + 1) + 1$$
$$= 2^2 a_{n-2} + 2 + 1 = 2^2(2a_{n-3} + 1) + 2 + 1$$
$$= 2^3 a_{n-3} + 2^2 + 2 + 1$$
$$\vdots$$
$$= 2^{n-1} + 2^{n-2} + 2^{n-3} + \ldots + 2^2 + 2 + 1$$
$$= \frac{1(2^n - 1)}{(2-1)} = 2^n - 1. \blacklozenge$$

EXERCISES 1.7

1. Work through the Requesting a Reward problem from the *Growing, Growing, Growing* unit from the middle-grade curriculum *Connected Mathematics* (answer all stated questions).

Reproduced from page 7 of *Growing, Growing, Growing* in *Connected Mathematics*.

FIGURE 1.7.1

Problem 1.2

A. Make a table showing the number of rubas the king will place on squares 1 through 16 of the chessboard.

B. How does the number of rubas change from one square to the next?

C. How many rubas will be on square 20? On square 30? On square 64?

D. What is the first square on which the king will place at least 1 million rubas?

E. If a Montarek ruba had the value of a U.S. penny, what would be the dollar values of the rubas on squares 10, 20, 30, 40, 50, and 60?

Problem 1.2 Follow-Up

1. Graph the (number of the square, number of rubas) data for squares 1 to 10. As the number of the square increases, how does the number of rubas change? What does this pattern of change tell you about the peasant's reward?

2. Write an equation for the relationship between the number of the square, n, and the number of rubas, r.

3. If a chessboard had 100 squares, how many rubas would be on square 100?

4. The pattern of change in the number of ballots in Problem 1.1 and the pattern of change in the number of rubas in this problem show **exponential growth**.
   a. How are the patterns of change in these two situations similar?
   b. Write an equation for the relationship between the number of ballots, b, and the number of cuts, n, in Problem 1.1.

Making a New Offer

When the king told the queen about the reward he had promised the peasant, the queen said, "You have promised her more money than we have in the entire royal treasury! You must convince her to accept a different reward."

After much contemplation, the king thought of a plan. He would create a new board with only 16 squares. He would place 1 ruba on the first square, 3 on the next, 9 on the next, and so on. Each square would have three times as many rubas as the previous square.

Reproduced from page 8 of *Growing, Growing, Growing* in *Connected Mathematics*.

FIGURE 1.7.2

Problem 1.3

A. In the table below, plan 1 is the reward requested by the peasant, and plan 2 is the king's new plan. Copy and complete the table to show the number of rubas on squares 1 to 16 for each plan.

Square	Number of rubas	
	Plan 1	Plan 2
1	1	1
2	2	3
3	4	9
4		

B. How is the pattern of change in the number of rubas under plan 2 similar to and different from the pattern of change in the number of rubas under plan 1?

C. Write an equation for the relationship between the number of the square, n, and the number of rubas, r, for plan 2.

D. Is the total reward under the king's plan greater than or less than the total reward under the peasant's plan? How did you decide?

Reproduced from page 9 of *Growing, Growing, Growing* in *Connected Mathematics*.

FIGURE 1.7.3

2. Suppose you invest \$1,000 at 8% simple interest (compounded at the end of each year). Let A_n denote the amount of money that you have at the end of n years. Since you are starting with \$1,000, it is reasonable to let $A_0 = 1,000$.
   a. How much money will you have at the end of 5 years (i.e., calculate A_5)?
   b. Describe the sequence $A_0, A_1, A_2, \ldots, A_n, \ldots$ with a recursive formula.
   c. Describe the sequence $A_0, A_1, A_2, \ldots, A_n, \ldots$ with an explicit formula.
   d. Use the explicit formula that you found in part c (and use your calculator) to compute A_{12}.
3. Suppose the lengths of the sides of a triangle form a finite geometric sequence a, ar, ar^2. Using the fact that $ar^2 < a + ar$ (see the triangle Inequality, Theorem 5.8.1), find the lower and upper bounds for r.

1.8 MATHEMATICAL INDUCTION

In the previous sections, we investigated a variety of sequences and found explicit rules describing those sequences. The contexts in which we were working made each of these rules feasible, however, we did not formally show that they were valid for all positive integers. Although the correctness of such formulae cannot be directly verified for each positive integer (there are infinitely many positive integers), there is an efficient methodology, called the **Principle of Mathematical Induction**, for addressing this kind of problem.

Recall that we used Gauss's childhood observation to assert that the formula for the sum of the first n consecutive positive integers is given by the rule:

$$S(n) = 1 + 2 + 3 + 4 + \ldots + n = \frac{n(n+1)}{2}.$$

We now want to prove that this formula holds for all positive integers n. First of all, we know that $S(1)$ is true, since $1 = \frac{1(1+1)}{2}$.

Question. If $S(k)$ is true for some positive integer k, does it then follow that $S(k+1)$ is true?

If this question has an affirmative answer, we can then conclude (without directly checking) that $S(2)$ is true. Why? Moreover, since $S(2)$ is true, it follows that $S(3)$ is true. Using this reasoning, we can conclude that $S(n)$ is true for all positive integers n. Hence, to complete this process we must verify that the posed question has an affirmative answer.

Analysis of Question. If $S(k)$ is true for some positive integer k, then

$$1 + 2 + 3 + 4 + \ldots + k = \frac{k(k+1)}{2} \text{ (hypothesized equality).}$$

We wish to show that $S(k+1)$ is true, and so we must prove that

$$1 + 2 + 3 + 4 + \ldots + k + (k+1)$$
$$= \frac{(k+1)[(k+1)+1]}{2} \text{ (implied equality).}$$

Note that if we add the quantity $(k+1)$ to both sides of the hypothesized equality, we get an equality whose left-hand side is identical to the left-hand side of the implied equality, i.e.,

$$1 + 2 + 3 + 4 + \ldots + k + (k+1) = \frac{k(k+1)}{2} + (k+1).$$

Moreover, the right-hand side of this equality also equals the right-hand side of the implied equality, since

$$\frac{k(k+1)}{2} + (k+1) = \frac{k(k+1)}{2} + \frac{2(k+1)}{2} = \frac{(k+1)(k+2)}{2}.$$

Hence, we have shown that $S(k)$ true $\Rightarrow S(k + 1)$ true, and therefore $S(n)$ is true for each positive integer n. In particular, the formula

$$1 + 2 + 3 + 4 + \ldots + n = \frac{n(n + 1)}{2}$$

holds for each positive integer n.

The Principle of Mathematical Induction is a powerful justification tool, and we now state it in complete generality.

1.8.1 The Principle of Mathematical Induction (Abbreviated, Induction)

For each positive integer n, let $S(n)$ be a statement that is either true or false.

The statement $S(n)$ is true for each positive integer n provided the following two conditions are satisfied:

a. **Basis step:** $S(1)$ is true.
b. **Inductive step:** If $S(k)$ is true for any positive integer k, then $S(k + 1)$ is true.

The hypothesis "If $S(k)$ is true for any positive integer k" is called the **inductive hypothesis**.

The Principle of Mathematical Induction is commonly illustrated using manipulatives such as dominoes, regularly shaped rectangular blocks, or even a "large" class of eager middle-grade students. For example, suppose we had an infinite collection of middle-grade students aligned in a regularly spaced, single-file, marching column so that if any one of the students falls backwards, it causes the next student to also fall backward (this is the inductive step). If the first student happens to fall backward (this is the basis step), then we can conclude (by induction) that all of the students fall backward. (Warning: Do not try this activity without parental permission, lots of padding, and an extremely large room.)

Classroom Discussion. In the Groups of People problem (see Section 1.7), we concluded that 2^n groups of people could be formed from a given set of n people. Equivalently, in the language of set theory, our conclusion is simply: There are exactly 2^n subsets for any given set with n elements.

You can use the Principle of Mathematical Induction to show that this statement is always true.

Proof. Let $S(n)$ be the statement that a set with n elements has exactly 2^n subsets.

Basis Step. Note that $S(1)$ is true, since an arbitrary set $A = \{a\}$ with 1 element has exactly two subsets $\emptyset$ and A.

Inductive Step. Assume that any set with k elements has exactly 2^k subsets (i.e., $S(k)$ is true). We wish to show that any set $A_{k+1} = \{a_1, a_2, a_3, \ldots, a_k, a_{k+1}\}$ with $k + 1$ elements has 2^{k+1} subsets (i.e., $S(k + 1)$ is true).

It follows from the inductive hypothesis that the set $A_k = \{a_1, a_2, a_3, \ldots, a_k\}$ has exactly 2^k subsets. The collection of all subsets of A_{k+1} consists of all the subsets of A_k plus each subset of A_k union with the set $\{a_{k+1}\}$. Hence, A_{k+1} has twice as many subsets as A_k, and so the total number of subsets of A_{k+1} is $2 \cdot 2^k = 2^{k+1}$.

Therefore, by induction, $S(n)$ is true for each positive integer n (i.e., a set with n elements has exactly 2^n subsets). ◆

Notation and Terminology. For a set X, the set of all subsets of X is called the **power set** of X and is denoted by $\mathcal{P}(X)$. We have just shown that if a set X has n elements, then $\mathcal{P}(X)$ has 2^n elements (hence the name *power set*). Note that

$$A \in \mathcal{P}(X) \Leftrightarrow A \subseteq X.$$

Classroom Problems.

1. True or false. If true, provide justification. If false, give a counterexample.

   a. If $X = \{1,2,3\}$, then $\{1,2\} \in \mathcal{P}(X)$.
   b. If $X = \{1,2,3\}$, then $\{1,2\} \subseteq \mathcal{P}(X)$.
   c. If $X = \{1,2,3\}$, then $\{\{1,2\}\} \subseteq \mathcal{P}(X)$.
   d. For any set $X, \emptyset \in \mathcal{P}(X)$.
   e. For any set $X, \emptyset \subseteq \mathcal{P}(X)$.
   f. For any set $X, \{\emptyset\} \subseteq \mathcal{P}(X)$.
   g. For any set $X, \{\emptyset\} \in \mathcal{P}(X)$.

2. If a set X has n elements, how many elements are in the set $\mathcal{P}(\mathcal{P}(X))$?

We complete this section with some additional examples involving induction proofs. You will see that for some statements $S(n)$, it may be necessary to have the basis step start at a positive integer greater than 1. ◆

EXAMPLES

1. Recall that for any positive integer n, the definition of n **factorial** is given by:

$$n! = n(n-1)(n-2)(n-3) \cdots 2 \cdot 1,$$

or the following recursive rule:

$$1! = 1$$
$$n! = n(n-1)!.$$

A Factorial Inequality. Notice that

$$1! < 2^1, 2! < 2^2, \text{ and } 3! < 2^3.$$

A novice student of mathematics might conclude from this data that $n! < 2^n$ for each positive integer n, but we need a proof to draw such a general conclusion. As a matter of fact, $n!$ is not less than 2^n for each positive integer n, since $4! > 2^4$, $5! > 2^5$, and $6! > 2^6$ (check these inequalities); so maybe it is the case that:

$$n! > 2^n \text{ for all positive integers } n \geq 4.$$

Let's establish the truth of this inequality by using induction.

Analysis. Let $S(n)$ denote the statement $n! > 2^n$.

We know that $S(4)$ is true, since $4! > 2^4$, and thus the basis step is complete. For the inductive step, if $k! > 2^k$ for some positive integer $k \geq 4$, then we must show that $(k+1)! > 2^{k+1}$. Multiplying both sides of the inequality $k! > 2^k$ by $(k+1)$ yields the inequality:

$$k!(k+1) = (k+1)! > 2^k(k+1).$$

Note that $(k+1) > 2$ for any positive integer $k > 1$, so $2^k(k+1) > 2^k \cdot 2 = 2^{k+1}$ for any positive integer $k > 1$. Hence,

$$(k+1)! > 2^k(k+1) > 2^{k+1},$$

and thus the inductive step has been established. Therefore, $n! > 2^n$ is true for each positive integer $n \geq 4$.

2. Our next example employs induction to establish a well-known geometric statement involving the interior angles of an n-sided convex polygon.

Let $S(n)$ be the statement: the sum of the interior angles of an n-sided convex polygon is $(n-2) \cdot 180$ degrees.

Proof. The basis step does not make sense for $n = 1$ or $n = 2$. (Why?) We'll start at $n = 3$.

Basis Step. It follows that $S(3)$ is true, since the sum of a triangle's interior angles (measured in degrees) is

$$180 = (3-2) \cdot 180.$$

Inductive Step. Assume that $S(k)$ is true and show that this implies that $S(k+1)$ is true. Hence (by the inductive hypothesis), we know that for some positive integer k, the sum of the interior angles of a k-sided convex polygon is $(k-2) \cdot 180$ degrees.

Suppose P is a $(k+1)$-sided convex polygon with vertices $v_1, v_2, v_3, \ldots, v_k, v_{k+1}$. Our goal is to show that the sum of the interior angles of P is $[(k+1) - 2] \cdot 180$ degrees.

Consider the k-sided convex polygon P' with vertices $v_1, v_2, v_3, \ldots, v_k$ (P' is convex, since if it was not convex, then this would imply that P was not convex). The six-sided convex polygon here illustrates such a situation when $k = 5$.

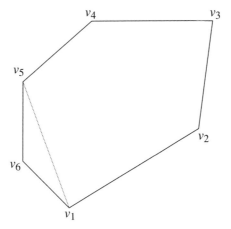

We know by the inductive assumption that the sum of the interior angles of the k-sided convex polygon P' is $(k - 2) \cdot 180$ degrees, so the sum of the interior angles of P equals the sum of the interior angles of P plus the sum of the interior angles of the triangle having vertices v_1, v_5, and v_6. Therefore,

$$\text{the sum of the interior angles of } P = [(k - 2) \cdot 180] + 180$$
$$= [(k - 2) + 1] \cdot 180$$
$$= (k - 1) \cdot 180 \text{ degrees.}$$

Thus, for each positive integer $n \geq 3$, the sum of the interior angles of an n-sided convex polygon is $(n - 2) \cdot 180$ degrees.

3. Our final example illustrates the importance of verifying both the basis step and the inductive step in a proof by mathematical induction.

Let $S(n)$ be the statement: $n = n + 1$.
Suppose we skipped the basis step and jumped right to the inductive step.

Inductive Step. Assume $S(k)$ is true for some positive integer k (i.e., $k = k + 1$). By adding 1 to each side of this equality, we obtain the equality $k + 1 = k + 2$, and so $S(k + 1)$ is true.

From this inductive step, some might be tempted to conclude that $S(n)$ is true for each positive integer n, but, in fact, this statement is not true for any positive integer n. Hence, to avoid reaching such absurd conclusions through mathematical induction, it is imperative to verify both steps in the induction process.

EXERCISES 1.8

1. Prove the following statements using the Principle of Mathematical Induction:
    a. For each positive integer n, $1 + 3 + 5 + \ldots + (2n - 1) = n^2$.
    b. For each positive integer n, $1^2 + 2^2 + 3^2 + \ldots + n^2 = \frac{n(n+1)(2n+1)}{6}$.
    c. For each positive integer n, $3 + 3^2 + 3^3 + \ldots + 3^n = \frac{3^{n+1}-3}{2}$.

34 Chapter 1 Patterns

d. If $r \neq 1$ is a real number, then for each positive integer n,

$$a + ar + ar^2 + \ldots + ar^n = a\left(\frac{r^{n+1} - 1}{r - 1}\right).$$

e. For each positive integer $n > 4$, $2^n > n^2$.
f. For each positive integer $n \geq 7$, $3^n < n!$.
g. For each positive integer n, 4 is a factor of $5^n - 1$.
h. For each positive integer n, the sum of the first n triangular numbers is given by the formula:

$$T_1 + T_2 + \ldots + T_n = \frac{n(n + 1)(n + 2)}{6}.$$

1.9 CLASSROOM CONNECTION: COUNTING TOOLS

We start this section by examining the Delivery Routes problem from the *Patterns and Symbols* module of the middle-grade curriculum *Mathematics in Context*. (See Figures 1.9.1 to 1.9.3 on pages 35–37). The content of this classroom investigation helps set the stage for a very interesting pattern that we explore throughout this section.

Classroom Discussion: Shortest Routes Problem. The following classroom problem considers the routes that can be taken from point A to point B and explores how to shorten these routes. Once you discover a shortest route between two points, the next step is to determine the total number of shortest routes between those points.

Let's consider the problem of determining the total number of distinct shortest routes from point A to point B in the following diagram.

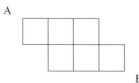

Analysis. There is only one choice for the shortest path from point A to points V_1 or V_4. However, to get to point V_5 on the shortest path, you can either go through point V_1 or through point V_4. Thus, there are two shortest paths to get from A to V_5.

```
         V₁   V₂   V₃
    A ┌────┬────┬────┐
  V₄  │ V₅ │ V₆ │ V₇ │ V₈
      └────┼────┼────┼────┐
           │ V₉ │V₁₀ │V₁₁ │
           └────┴────┴────┘
                              B
```

To get from point A to any other point V_i in the diagram by the shortest route, you must go through either the point directly above V_i or through the point left of V_i. This observation gives us a recursive rule to calculate the total number of shortest routes from point A to another point V_i in the diagram.

Namely, the total number of shortest paths from A to a given point V_i is equal to the sum of the number of shortest paths from A to the point directly above V_i

D. DELIVERY ROUTES

The ITALIAN RESTAURANT

The Italian restaurant Gino's (G on the map below) will bring pasta and pizza to your door. A new delivery girl, Lashanda, is sent to Barbara's house (B on the map). The instruction for the route could be: first go north; after passing a church, turn left; stop at the third house on the right.

Lashanda gets this note on a little piece of paper:

NNWW

1. If she follows the instruction will she deliver the pizza to the right address?

Use **Student Activity Sheet 5** for drawing routes in the town. Below is an example of a route from G to B.

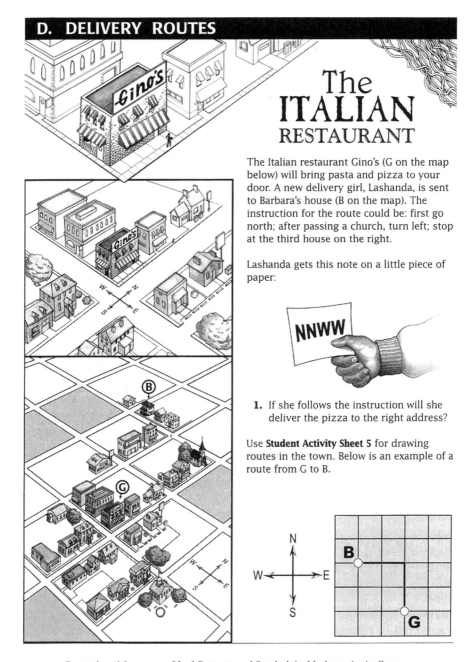

Reproduced from page 23 of *Patterns and Symbols* in *Mathematics in Context*.

FIGURE 1.9.1

36   Chapter 1   Patterns

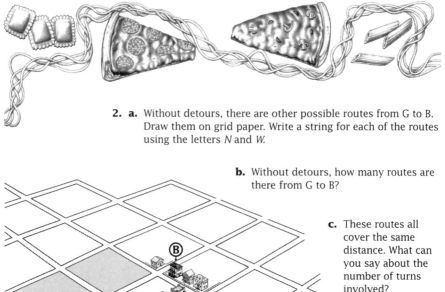

2. a. Without detours, there are other possible routes from G to B. Draw them on grid paper. Write a string for each of the routes using the letters *N* and *W*.

b. Without detours, how many routes are there from G to B?

c. These routes all cover the same distance. What can you say about the number of turns involved?

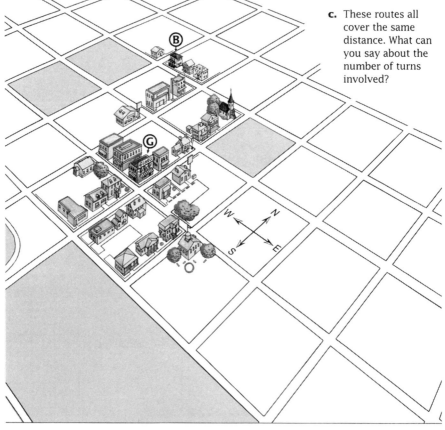

Reproduced from page 24 of *Patterns and Symbols* in *Mathematics in Context*.

FIGURE 1.9.2

Section 1.9 Classroom Connection: Counting Tools 37

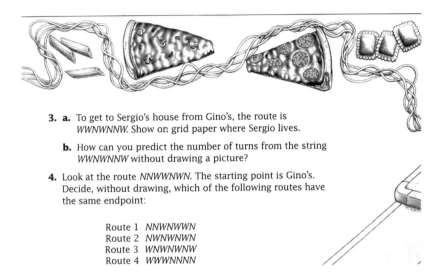

3. **a.** To get to Sergio's house from Gino's, the route is *WWNWNNW*. Show on grid paper where Sergio lives.

   b. How can you predict the number of turns from the string *WWNWNNW* without drawing a picture?

4. Look at the route *NNWWNWN*. The starting point is Gino's. Decide, without drawing, which of the following routes have the same endpoint:

   Route 1 *NNWNWWN*
   Route 2 *NWNWNWN*
   Route 3 *WNWNWNW*
   Route 4 *WWWNNNN*

Reproduced from page 25 of *Patterns and Symbols* in *Mathematics in Context*.

FIGURE 1.9.3

and to the left of V_i. For example, the total number of shortest routes from A to V_6 = (number of shortest routes from A to V_2) + (number of shortest routes from A to V_5) = 1 + 2 = 3.

Using this rule, we can calculate the total number of shortest paths from A to all of the points in the figure. (We say the number of shortest paths from A to A is 1.)

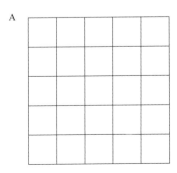

Using the recursive method previously described, let's determine the number of shortest paths from A to every other corner point on the following grid map.

Each number in the following array represents the number of shortest routes from point A to the corresponding corner point in the previous grid.

```
1   1   1    1    1    1
1   2   3    4    5    6
1   3   6   10   15   21
1   4  10   20   35   56
1   5  15   35   70  126
1   6  21   56  126  252    ◆
```

The numbers in this array, and those that would arise from larger grid maps, possess many interesting patterns. A 45 degree rotation of the array helps make them more apparent.

```
Row 0                             1
Row 1                           1   1
Row 2                         1   2   1
Row 3                       1   3   3   1
Row 4                     1   4   6   4   1
Row 5                   1   5  10  10   5   1
Row 6                 1   6  15  20  15   6   1
Row 7               1   7  21  35  35  21   7   1
Row 8             1   8  28  56  70  56  28   8   1
Row 9         1   9  36  84 126 126  84  36   9   1
```

If we define all numbers that are not in the triangle as 0, then the first 1 (from the left) in Row 1 is obtained by adding the two numbers 0 and 1 that are above it to the immediate left and right (i.e., 1 = 0 + 1). The second 1 in Row 1 is obtained by adding the two numbers 1 and 0, which are above 1 to the immediate left and right (i.e., 1 = 1 + 0). The numbers in each successive row are created similarly. For example, in Row 5, the first 5 (from the left) is the sum of 1 and 4, and the first 10 is the sum of 4 and 6.

The triangular array of numbers that we have just described was systematically investigated by the French mathematician Blaise Pascal (1623–1662), and for that reason it is called **Pascal's triangle**.

Postage stamp issued by Monaco in 1973 to celebrate Pascal's 350th birthday.

Classroom Problem.   Calculate Rows 10, 11, 12, and 13 of Pascal's triangle.   ◆

Counting Tools. It is helpful to review some elementary counting tools from college algebra. These tools assist us in our explorations of Pascal's triangle and in the Shortest Routes problem.

EXAMPLES

1. Sam's Sandwich Shop prides itself in the variety it offers its customers. Sam offers three different choices for bread (wheat, Italian, and rye) and four different choices of meat (turkey, ham, chicken, and roast beef).

Question. How many different sandwiches can be made with these breads and meats (one type of bread and meat per sandwich)?

Analysis. A tree diagram is a useful tool for analyzing (small number) counting problems. We have constructed one for this problem.

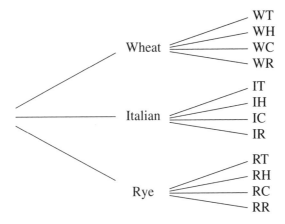

Since there are three bread choices and four meat choices, we see from the tree diagram that there are total of 3 • 4 = 12 sandwich possibilities.

1.9.1 Counting the Number of Ways Two Tasks Can Be Completed

Based on the analysis of the preceding Example 1, it follows that if there are m_1 ways to complete task M_1 (there are three ways to complete the task of choosing a bread) and m_2 ways to complete task M_2 (there are four ways to complete the task of choosing a meat), then there are $m_1 m_2$ different ways to complete tasks M_1 and M_2 together (there are 12 different ways to complete the tasks of choosing a bread and choosing a meat).

2. A middle school offers the following lunch choices in a typical school week.

Meal	Drink	Dessert
Pizza	Milk	Cookie
Tuna	Juice	Fruit
Macaroni and cheese		Jello
Ham sandwich		

Question. How many different lunch possibilities would there be if a student must choose one item from each category?

As in Example 1, you can construct a tree diagram to show that there are $4 \cdot 2 \cdot 3 = 24$ lunch possibilities. Verify.

The general counting process underlying the last two examples is called the **Fundamental Counting Principle** and is defined in the following general terms.

1.9.2 Fundamental Counting Principle

If there are m_1 ways to complete task M_1, m_2 to complete task $M_2, \ldots$, and m_n ways to complete task M_n, then there are $m_1 m_2 \cdots m_n$ different ways for the tasks $M_1, M_2, M_3, \ldots,$ and M_n to be collectively completed.

Let's use induction to prove the fundamental counting principle.

Proof. Let $S(n)$ be the statement: If there are m_1 ways to complete task M_1, m_2 ways to complete task $M_2, \ldots,$ and m_n ways to complete task M_n, then there are $m_1 m_2 \cdots m_n$ different ways for the tasks $M_1, M_2, M_3, \ldots,$ and M_n to be collectively completed.

Basis Step. $S(1)$ is trivially true, since there are m_1 ways to complete task M_1 (this is given).

Inductive Step. Suppose $S(k)$ is true and show that this implies that $S(k+1)$ is also true.

Assume that there are m_1 ways to complete task M_1, m_2 ways to complete task $M_2, \ldots, m_k$ ways to complete task M_k, and m_{k+1} ways to complete task M_{k+1}. Hence, by the inductive hypothesis, there are $m_1 m_2 \cdots m_k$ different ways for the tasks $M_1, M_2, M_3, \ldots,$ and M_k to be collectively completed. Thus, by the two-task case (cf., result 1.9.1), there are $(m_1 m_2 \ldots m_k) m_{k+1}$ different ways for the tasks $M_1, M_2, M_3, \ldots, M_k,$ and M_{k+1} to be completed together.

Therefore, by induction, $S(n)$ is true for each positive integer n.

EXAMPLES

1. A mathematics department consisting of 40 members wishes to elect a chair, vice chair, and a curriculum coordinator.

Question. How many different ways are there for this mathematics department to elect individuals to fill these three positions?

There are 40 possibilities for the chair, leaving 39 possibilities for the vice chair, and 38 possibilities for the curriculum coordinator. Hence, by the fundamental counting principle, there are $40 \cdot 39 \cdot 38 = 59{,}280$ different ways to elect a chair, vice chair, and a curriculum coordinator.

Two of these 59,280 distinct possibilities might be:

a. Carmen elected as chair, Stephen elected as vice chair, and Russ as curriculum coordinator; or

b. Stephen elected as chair, Russ elected as vice chair, and Carmen as curriculum coordinator.

Although both possibilities involve the same individuals, each one consists of a distinct **ordered** arrangement called a **permutation**. (Review: a permutation of r objects from a set of n objects is an ordered arrangement of the r objects.) The total number of permutations of 40 faculty members chosen three at a time is denoted by

$$_{40}P_3 = 59{,}280.$$

In general, the number of permutations for r distinct positions selected from a group of n objects $(n \geq r)$ is given by:

$$_nP_r = n(n-1) \cdots ((n-r)+1),$$

which can also be expressed as:

$$_nP_r = \frac{n!}{(n-r)!}.$$

2. After the department of mathematics elected a chair, vice chair, and a curriculum coordinator, they then wanted to elect a three member textbook selection committee. In this election, the order of the committee members is unimportant, so the total number of three person unordered committees that could be chosen from 40 faculty members is given by:

$$_nC_r = \frac{_{40}P_3}{\text{number of ways each three person committee can be arranged}} = \frac{_{40}P_3}{3!}.$$

Each unordered committee is called a **combination**. Review: A combination of r objects from a set A of n objects is a subset of A consisting of r objects.

In general, the number of combinations for r positions (without regard to order) selected from a group of n objects $(n \geq r)$ is denoted by $_nC_r$ or $\binom{n}{r}$, read as "n **choose** r". This is defined by:

$$_nC_r = \frac{_nP_r}{r!} = \frac{\frac{n!}{(n-r)!}}{r!}$$

$$= \frac{n!}{r!(n-r)!}.$$

Convention. It is helpful to define $0! = 1$, so for each integer $n \geq 0$, $\binom{n}{0} = 1$ and $\binom{n}{n} = 1$. Justify each equality.

Revisiting the Shortest Routes Problem. Let's find the number of shortest paths from point A to point B by counting combinations.

42 Chapter 1 Patterns

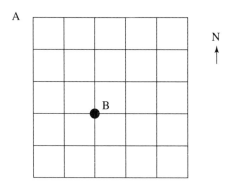

Analysis. The shortest route from point A to point B involves traveling east 2 blocks (not necessarily consecutively) and south 3 blocks (not necessarily consecutively) for a total of 5 blocks. Hence, the total number of shortest routes from A to B can be determined by counting the number of ways in which 2 eastern moves could be chosen from a total of 5 moves or by counting the number of ways 3 southern moves could be chosen from a total of 5 moves.

Thus, the total number of shortest routes from A to B equals

$$\binom{5}{2} = \frac{5!}{2!(5-2)!} = \frac{5!}{2!3!} = \frac{5 \cdot 4}{2 \cdot 1} = 10 \text{ (counting eastern moves)},$$

or, equivalently,

$$\binom{5}{3} = \frac{5!}{3!(5-3)!} = \frac{5!}{3!2!} = \frac{5 \cdot 4}{2 \cdot 1} = 10 \text{ (counting southern moves)}$$

Observation. By using the context of the Shortest Routes problem (and carefully inspecting the previous calculations), it is reasonable to assert that

$$\binom{n}{r} = \binom{n}{n-r}$$

is true for all positive integers $n \geq r$. Justify this equality.

The total number of shortest paths from A to every other corner point on a 5×5 grid map (determined by counting eastern moves) is given here in terms of the numbers $\binom{n}{r}$.

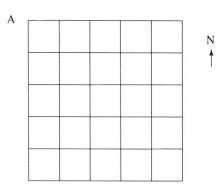

Each $\binom{n}{r}$ in the array represents the number of shortest routes from point A to the corner point corresponding to where $\binom{n}{r}$ is positioned in the array (we say there is $\binom{0}{0} = 1$ is the total number of shortest routes from A to A).

Number of shortest routes from A (eastern moves)

$\binom{0}{0}$ $\binom{1}{1}$ $\binom{2}{2}$ $\binom{3}{3}$ $\binom{4}{4}$ $\binom{5}{5}$

$\binom{1}{0}$ $\binom{2}{1}$ $\binom{3}{2}$ $\binom{4}{3}$ $\binom{5}{4}$ $\binom{6}{5}$

$\binom{2}{0}$ $\binom{3}{1}$ $\binom{4}{2}$ $\binom{5}{3}$ $\binom{6}{4}$ $\binom{7}{5}$

$\binom{3}{0}$ $\binom{4}{1}$ $\binom{5}{2}$ $\binom{6}{3}$ $\binom{7}{4}$ $\binom{8}{5}$

$\binom{4}{0}$ $\binom{5}{1}$ $\binom{6}{2}$ $\binom{7}{3}$ $\binom{8}{4}$ $\binom{9}{5}$

$\binom{5}{0}$ $\binom{6}{1}$ $\binom{7}{2}$ $\binom{8}{3}$ $\binom{9}{4}$ $\binom{10}{5}$

A 45 degree rotation of the previous array results in (a section of) Pascal's triangle (following).

Pascal's Triangle in terms of $\binom{n}{r}$

$\binom{0}{0}$

$\binom{1}{0}$ $\binom{1}{1}$

$\binom{2}{0}$ $\binom{2}{1}$ $\binom{2}{2}$

$\binom{3}{0}$ $\binom{3}{1}$ $\binom{3}{2}$ $\binom{3}{3}$

$\binom{4}{0}$ $\binom{4}{1}$ $\binom{4}{2}$ $\binom{4}{3}$ $\binom{4}{4}$

$\binom{5}{0}$ $\binom{5}{1}$ $\binom{5}{2}$ $\binom{5}{3}$ $\binom{5}{4}$ $\binom{5}{5}$

Note that $\binom{n}{r}$ is the number in the $(r + 1)^{th}$ position (from the left) in the nth row, where $(0 \leq r \leq n)$. (Recall that Pascal's triangle starts with Row 0.)

In terms of the Shortest Routes problem, notice that the rows of Pascal's triangle represent all the shortest routes from point A of a certain fixed length. For instance, the fourth row in the previous Pascal's triangle represents all of the paths from point A of length 4 blocks.

44 Chapter 1 Patterns

Summary: The Shortest Routes Problem, Pascal's Triangle, and Combinations. Our initial work on the Shortest Routes problem led us to Pascal's triangle, and our most recent work focused on a computational relationship between the total number of shortest routes and the notion of combinations. Using these connections, we saw that elements of Pascal's triangle can be expressed with the symbols $\binom{n}{r}$.

Classroom Problem. From the definition of Pascal's triangle, and from the fact that each number in the triangle is represented by $\binom{n}{r}$, we know that $\binom{2}{1}$ in Row 2 equals the sum of $\binom{1}{0}$ and $\binom{1}{1}$ from Row 1 (verify), and $\binom{4}{3}$ in Row 4 is the sum of $\binom{3}{2}$ and $\binom{3}{3}$ from Row 3 (verify). ◆

In general, $\binom{n+1}{r}$ in Row $(n + 1)$ is the sum of $\binom{n}{r-1}$ and $\binom{n}{r}$ from Row n. This defining property of Pascal's triangle is called **Pascal's formula**.

1.9.3 Pascal's Formula

$$\binom{n+1}{r} = \binom{n}{r-1} + \binom{n}{r}.$$

Let's show that Pascal's formula holds for all positive integers $1 \le r \le n$.

Justification of Pascal's Formula. We show that the equality's right-hand side equals the left-hand side.

$$\binom{n}{r-1} + \binom{n}{r} = \frac{n!}{(r-1)![n-(r-1)]!} + \frac{n!}{r!(n-r)!}.$$

We need a common denominator to add these fractions. Some simple observations (that we arrive at after some initial calculations) help us choose a denominator that makes the computations less complicated.

Looking at the denominator of the first fraction, we notice that multiplying it by r results in the expression (right side of the equality):

$$r(r-1)![(n-r)+1]! = r![(n-r)+1]!.$$

Focusing on the denominator of the second fraction, we see that multiplying it by $[(n - r) + 1]$ gives the same expression (right side of the equality):

$$[(n-r)+1][r!(n-r)!] = r![(n-r)+1]!.$$

Section 1.9   Classroom Connection: Counting Tools   45

We now have a (relatively simple) common denominator for the given fractions, and so we continue from where we left off.

$$\binom{n}{r-1} + \binom{n}{r} = \frac{n!}{(r-1)![n-(r-1)]!} + \frac{n!}{r!(n-r)!}$$

$$= \frac{r}{r}\left[\frac{n!}{(r-1)![(n-r)+1]!}\right] + \frac{[(n-r)+1]}{[(n-r)+1]}\left[\frac{n!}{r!(n-r)!}\right]$$

$$= \frac{rn!}{r![(n-r)+1]!} + \frac{[(n-r)+1]n!}{r![(n-r)+1]!}$$

$$= \frac{rn! + nn! - rn! + n!}{r![(n-r)+1]!}$$

$$= \frac{(n+1)n!}{r![(n-r)+1]!} = \frac{(n+1)!}{r![(n+1)-r]!} = \binom{n+1}{r}.$$

EXERCISES 1.9

In problems 1–5, answer each question and explain your reasoning.

1. **a.** How many different ways can you arrange the eight letters in the word **computer** so that each arrangement uses all eight letters?
   b. How many eight letter configurations can you make from the 26 letters of the alphabet without repeating letters?

2. How many four-digit numbers $n = a_3 \cdot 10^3 + a_2 \cdot 10^2 + a_1 \cdot 10^1 + a_0 \cdot 10^0$ are there, where $2 \leq a_3 \leq 5, 0 \leq a_2 \leq 9, 4 \leq a_1 \leq 5$, and $1 \leq a_0 \leq 8$?

3. How many five-digit odd numbers are there?

4. Determine the number of lines that pass through nine points in a plane, where no three of the points are collinear.

5. On a final exam, students had the choice of doing 10 out of 12 problems. How many different ways can they accomplish this?

6. The Cartesian product of sets A and B is the set of all ordered pairs (a,b), where a is in A and b is in B, i.e., $A \times B = \{(a,b) : a \in A \text{ and } b \in B\}$.
   For example, if $A = \{a,b,c\}$ and $B = \{1,2\}$, then $A \times B = \{(a,1),(a,2),(b,1),(b,2),(c,1),(c,2)\}$. The (x,y)-plane is the set $\mathbf{R} \times \mathbf{R}$, where $\mathbf{R}$ = real numbers.
   Determine the number of ordered pairs in $A \times B$, if A has m elements and B has n elements. Justify your answer.

7. **a.** Let $A = \{a,b,c\}$ and $B = \{1,2\}$. Determine the number of functions f defined from A into B. Justify your answer. For example, one of the functions from A into B is: $f(a) = 1, f(b) = 2, f(c) = 1$.
   b. Determine the number of functions defined from a set A with n elements into a set B with m elements. Explain your reasoning.

8. **a.** Let $A = \{a,b\}$. Consider an operation $*$ defined on the elements of A in the given "multiplication" table.

46 Chapter 1 Patterns

$*$	a	b
a	b	b
b	a	a

Note that $a * a = b, a * b = b, b * a = a, b * b = a$.

How many different "multiplication" tables can you construct with the elements of the set A?

$*$	a	b
a		
b		

b. How many different "multiplication" tables can you form from a set $A = \{a_1, a_2, \ldots, a_n\}$ with n elements?

c. An operation $*$ defined on a set A is called **commutative** if $x * y = y * x$ for each $x, y \in A$. In part a, observe that $*$ is not commutative since $a * b \neq b * a$.

Fill in the following "multiplication" table so that $*$ is commutative:

$*$	a	b
a		
b		

d. How many different commutative "multiplication" tables can you form with the elements of the set $A = \{a, b\}$?

e. How many different commutative "multiplication" tables can you make with the elements of a set $A = \{a_1, a_2, \ldots, a_n\}$ with n elements?

f. An operation $*$ defined on a set A is **associative** if $x * (y * z) = (x * y) * z$ for each $x, y, z \in A$.

Let $A = \{a, b\}$. Consider an operation $*$ defined on the elements of A in the given "multiplication" table.

$*$	a	b
a	a	b
b	b	a

Show that $*$ is associative of A.

g. Fill in the following "multiplication" table so that $*$ is not associative.

$*$	a	b
a		
b		

h. How many different expressions $x * (y * z)$ have to be inspected to determine if an operation $*$ on the set $A = \{a, b\}$ is associative?

i. How many different expressions $x * (y * z)$ have to be inspected to determine if an operation $*$ on the set $A = \{a_1, a_2, \ldots, a_n\}$ is associative?

9. Establish the following equalities:

a. $\binom{n+1}{n} = n + 1$ for each integer $n \geq 0$.

b. $\binom{n}{n-2} = \frac{n(n-1)}{2}$ for each positive integer $n > 1$.

10. a. Recall from Exercises 1.3 that the nth triangular number T_n is given by the sum

$$T_n = 1 + 2 + 3 + \ldots + n, \text{ and thus } T_n = \frac{n(n+1)}{2}.$$

Show that $\binom{n+1}{2} = T_n$.

b. Prove by induction that $\left(\frac{n(n+1)}{2}\right)^2 = 1^3 + 2^3 + 3^3 + \ldots + n^3$, and so $(T_n)^2 = \sum_{r=1}^{n} r^3$.

11. a. Explain why there are $\binom{12 + 4 - 1}{12}$ nonnegative integer solutions x_1, x_2, x_3, and x_4 to the equation $x_1 + x_2 + x_3 + x_4 = 12$. (Note: In this problem, $x_1 = 3, x_2 = 4, x_3 = 5, x_4 = 0$, and $x_1 = 5, x_2 = 4, x_3 = 0, x_4 = 3$ are viewed as different solutions.)

b. How many nonnegative integer solutions x_1, x_2, x_3, x_4, and x_5 are there to the equation $x_1 + x_2 + x_3 + x_4 + x_5 = 37$?

c. How many nonnegative integer solutions are there to the equation $x_1 + x_2 + x_3 + \ldots + x_n = r$, where r is a positive integer?

1.10 THE BINOMIAL THEOREM

In this brief section we apply our work on Pascal's triangle and on combinations to help us understand a basic result concerning binomial expansions.

In your previous algebra classes you have studied binomial expressions and are familiar with the following expansions that you painstakingly derived through multiplication.

$(x + y)^1 = (x + y)$

$(x + y)^2 = x^2 + 2xy + y^2$

$(x + y)^3 = (x + y)^2(x + y)$

$\qquad = (x^2 + 2xy + y^2)(x + y)$

$\qquad = (x^2 + 2xy + y^2)x + (x^2 + 2xy + y^2)y$

$\qquad = x^3 + 2x^2y + y^2x + x^2y + 2xy^2 + y^3$

$\qquad = x^3 + 3x^2y + 3xy^2 + y^3.$

$(x + y)^4 = (x + y)^3(x + y)$

$\qquad = (x^3 + 3x^2y + 3xy^2 + y^3)(x + y)$

$\qquad = (x^3 + 3x^2y + 3xy^2 + y^3)x + (x^3 + 3x^2y + 3xy^2 + y^3)y$

$\qquad =$ several steps to be completed by you

$\qquad = x^4 + 4x^3y + 6x^2y^2 + 4xy^3 + y^4.$

As you have undoubtedly noticed, higher-power binomial expressions involve a fair amount of algebra in order to arrive at the final expanded form. You may not have noticed that the coefficients of these successive powers are the numbers in Rows 1 through 4 of Pascal's triangle:

Binomial expansion		**Pascal's triangle**
$(x + y)^1 = x + y$	Row 1	1 1
$(x + y)^2 = x^2 + 2xy + y^2$	Row 2	1 2 1
$(x + y)^3 = x^3 + 3x^2y + 3xy^2 + y^3$	Row 3	1 3 3 1
$(x + y)^4 = x^4 + 4x^3y + 6x^2y^2 + 4xy^3 + y^4$	Row 4	1 4 6 4 1

Thus, these coefficients can also be expressed in terms of the symbols $\binom{n}{r}$.

$$\binom{1}{0} \quad \binom{1}{1}$$

$$\binom{2}{0} \quad \binom{2}{1} \quad \binom{2}{2}$$

$$\binom{3}{0} \quad \binom{3}{1} \quad \binom{3}{2} \quad \binom{3}{3}$$

$$\binom{4}{0} \quad \binom{4}{1} \quad \binom{4}{2} \quad \binom{4}{3} \quad \binom{4}{4}$$

Section 1.10 The Binomial Theorem

From this pattern, we might conjecture that

$$(x+y)^5 = \binom{5}{0}x^5 + \binom{5}{1}x^4y + \binom{5}{2}x^3y^2 + \binom{5}{3}x^2y^3$$
$$+ \binom{5}{4}xy^4 + \binom{5}{5}y^4 \text{ (verify this equality)},$$

and more generally that

$$(x+y)^n = \binom{n}{0}x^n + \binom{n}{1}x^{n-1}y + \binom{2}{2}x^{n-2}y^2 + \cdots + \binom{n}{n-2}x^2y^{n-2}$$
$$+ \binom{n}{n-1}xy^{n-1} + \binom{n}{n}y^n$$

$$= \sum_{r=0}^{n} \binom{n}{r}x^{n-r}y^r \text{ for each positive integer } n. \qquad 1.10.1$$

This general binomial expansion (1.10.1) is called the **Binomial Theorem**, and the coefficients $\binom{n}{r}$ are called **binomial coefficients**. We know that the binomial coefficients $\binom{n}{r}$ for $0 \leq r \leq n$ are the entries of the nth row of Pascal's triangle.

We now use mathematical induction to show that the Binomial Theorem is true.

Proof. Let $S(n)$ be the statement:

$$(x+y)^n = \sum_{r=0}^{n} \binom{n}{r}x^{n-r}y^r.$$

Basis Step. $S(1)$ is true since

$$\sum_{r=0}^{1} \binom{1}{r}x^{1-r}y^r = \binom{1}{0}x^1y^0 + \binom{1}{1}x^0y^1 = x+y.$$

Inductive Step. Assume that $S(k)$ is true and show that this implies that $S(k+1)$ is also true. More specifically, we want to prove that the following equality holds:

$$(x+y)^{k+1} = \sum_{r=0}^{k+1} \binom{k+1}{r}x^{(k+1)-r}y^r.$$

Starting with the left-hand side of the desired equality, we get:

$$(x+y)^{k+1} = (x+y)^k(x+y) = \left[\sum_{r=0}^{k}\binom{k}{r}x^{k-r}y^r\right]x$$

$$+ \left[\sum_{r=0}^{k}\binom{k}{r}x^{k-r}y^r\right]y \text{ (by inductive hypothesis)}$$

$$= \sum_{r=0}^{k}\binom{k}{r}x^{(k+1)-r}y^r + \sum_{r=0}^{k}\binom{k}{r}x^{k-r}y^{r+1}$$

$$= x^{k+1} + \left[\sum_{r=1}^{k}\binom{k}{r}x^{(k+1)-r}y^r\right] + \sum_{r=0}^{k}\binom{k}{r}x^{k-r}y^{r+1}.$$

In order to combine the two summations into one summation, it will be necessary to rewrite the second summation so that the powers of x and y match up with the powers of x and y in the first summation. In particular,

$$\sum_{r=0}^{k}\binom{k}{r}x^{k-r}y^{r+1} = \left[\sum_{r=1}^{k}\binom{k}{r-1}x^{(k+1)-r}y^r\right] + y^{k+1} \text{ (check)}.$$

Substituting the adjusted second summation into the original calculation results in the next string of equalities.

$$(x+y)^{k+1} = x^{k+1} + \left[\sum_{r=1}^{k}\binom{k}{r}x^{(k+1)-r}y^r\right] + \sum_{r=0}^{k}\binom{k}{r}x^{k-r}y^{r+1}$$

$$= x^{k+1} + \left[\sum_{r=1}^{k}\binom{k}{r}x^{(k+1)-r}y^r\right]$$

$$+ \left[\sum_{r=1}^{k}\binom{k}{r-1}x^{(k+1)-r}y^r\right] + y^{k+1}$$

$$= x^{k+1} + \left[\sum_{r=1}^{k}\left[\binom{k}{r} + \binom{k}{r-1}\right]x^{(k+1)-r}y^r\right] + y^{k+1}$$

$$= x^{k+1} + \left[\sum_{r=1}^{k}\binom{k+1}{r}x^{(k+1)-r}y^r\right] + y^{k+1}$$

(see 1.9.3 Pascal's Formula)

$$= \sum_{r=0}^{k}\binom{k+1}{r}x^{(k+1)-r}y^r \text{ (check)}.$$

Therefore, by induction, we have shown that

$$(x + y)^n = \sum_{r=0}^{n} \binom{n}{r} x^{n-r} y^r \text{ for each positive integer } n.$$

EXERCISES 1.10

1. Expand the following expressions using the Binomial Theorem.
   a. $(x + 5)^8$
   b. $(x - 3)^7$
   c. $(-2x + 5y)^5$
2. Show that the sum of the nth row of Pascal's triangle equals 2^n, i.e., demonstrate that
   $\sum_{r=0}^{n} \binom{n}{r} = 2^n$. Direction: Expand the expression $(1 + 1)^n$.
3. Show that $\sum_{r=0}^{n} (-1)^n \binom{n}{r} = 0$. Direction: Use a similar idea to the direction given in problem 2.

1.11 THE FIBONACCI SEQUENCE

Our work in this chapter has focused on several interesting patterns and relationships. A most intriguing pattern that arises naturally both in the "real world" and in the mathematical world is one that emerged from a population-growth problem that was posed by the thirteenth-century Italian mathematician Leonardo Pisano (nickname, Leonardo Fibonacci) in his treatise *Liber Abaci* (1202). In the following Classroom Connection, we investigate his problem and the sequence that emerges from it (called the Fibonacci sequence), and then we explore other facets of this fascinating sequence of integers.

1.11.1 Classroom Connection: Fibonacci's Rabbit Population Problem

Let's work through Fibonacci's Rabbit Population problem as it appears in Book 3 of the middle-grade curriculum, *MathThematics*. (see Figures 1.11.1 and 1.11.2 on pages 52 and 53). This is an idealized problem and is based upon the assumptions that the rabbits do not die during the year the population started and that newborn pairs of rabbits always consist of one male and one female.

1.11.2 Reflections on Classroom Connections: Fibonacci's Rabbit Population Problem

In order to clearly understand the problem, it is helpful to write out the month-by-month birth outcomes and then organize them in tabular form (as is done in the Classroom Connection). Other means of representation, such as family tree diagrams, are also quite useful for counting future generations.

Monthly rabbit pair population census:

Start of first month: There is **1 pair** of newborn rabbits.

52   Chapter 1   Patterns

Rotations and Rational Numbers

IN THIS SECTION
EXPLORATION 1
• Rotational Symmetry
EXPLORATION 2
• Rational and Irrational Numbers

WARM-UP EXERCISES

Rewrite each fraction with a denominator that is a power of 10.

1. $\dfrac{4}{5}$ $\dfrac{8}{10}$

2. $\dfrac{3}{25}$ $\dfrac{12}{100}$

3. $\dfrac{3}{4}$ $\dfrac{75}{100}$

Find each square root.

4. $\sqrt{9}$ 3
5. $\sqrt{64}$ 8

Nature's Sequences

••• Setting the Stage

SET UP Work with a partner.

His neighbors called him *Bigollone*, which means "the blockhead." His real name was Leonardo Fibonacci, and he was not a blockhead at all. He was a mathematician. He was born in Italy around 1170, but grew up in the city of Bougie on the Barbary Coast of North Africa.

▲
Roman and Arabic numerals

Fibonacci learned the Arabic number system in Africa. He realized that Arabic numerals were much easier to use than the Roman numerals then used in Europe. In 1202 Fibonacci published a book that introduced Arabic numerals to the European world. In his book he presented a mathematical problem that has fascinated people for centuries.

> Suppose two newborn rabbits, male and female, are put in a cage. How many pairs of rabbits will there be at the end of one year if this pair of rabbits produces another pair every month, and every new pair of rabbits produces another new pair every month? All rabbits must be two months old before they can produce more rabbits.

 Module 4   Patterns and Discoveries

Reproduced from page 256 of *Book 3* in *MathThematics*.

FIGURE 1.11.1

Section 1.11   The Fibonacci Sequence   53

Here is what happens in the first three months after the first pair of newborn rabbits are put in the cage:

Start	Start with **1st pair** of newborn rabbits.
Month 1	**1st pair** are growing.
Month 2	**1st pair** are adults. They produce **2nd pair** of rabbits.
Month 3	**1st pair** produce a **3rd pair** of rabbits.
	2nd pair are growing.

Think About It

1 Solve Fibonacci's rabbit problem. Start by creating a model or diagram that shows the total number of rabbit pairs over the first six months of the year. Find a way to show new-born, growing, and adult rabbits. Use your model to help complete a table like the one shown. See Additional Answers.

Month	Number of rabbit pairs			Total number of rabbit pairs
	Newborn	Growing	Adult	
1	1	0	0	1
2	0	1	?	?
3	?	?	?	?

2 The number pattern in the last column of the table you made in Question 1 is known as the *Fibonacci sequence.* The numbers in the sequence are sometimes called Fibonacci numbers.

Fibonacci sequence				
1	1	2	3	5...
↓	↓	↓		
1st term	2nd term	3rd term		

a. How are any two consecutive terms of the Fibonacci sequence used to find the next term in the sequence? Each term after the second is the sum of the two previous consecutive terms.
b. What is the answer to Fibonacci's rabbit problem? Explain how you got your answer.

2. b. 13 rabbits; The total number of pairs are terms of the Fibonacci sequence. The seventh term, which corresponds to the sixth month, is 13.

▶ Fibonacci's rabbit problem is not a realistic model of how rabbits reproduce. Nevertheless, in this section you'll see that the Fibonacci sequence has a strange way of showing up in nature.

Section 2   Rotations and Rational Numbers

Reproduced from page 257 of *Book 3* in *MathThematics.*

FIGURE 1.11.2

End of first month: The rabbits have mated but have not yet produced offspring, so there is still **1 pair**.

End of second month: The original pair produces a new pair, and there are now **2 pairs** of rabbits.

End of third month: The original pair produces another pair, and this brings the total to **3 pairs** of rabbits.

End of fourth month: The original pair produces another pair, the first offspring of the original pair also produces a pair, and thus there are now **5 pairs** of rabbits in the population.

End of fifth month: The original pair produces another pair, the first offspring of the original pair produces a pair, the second offspring of the original pair produces a pair, and so there are now **8 pairs** of rabbits.

It is becoming difficult to verbalize the monthly outcomes, so a table makes it easier for us to count and organize the population growth for the 12-month duration of this problem.

Monthly rabbit pair population census

End of nth month	Newborn rabbit pairs	Adult rabbit pairs	F_n = Total number of rabbit pairs at end of nth month
1	1	0	1
2	1	1	2
3	1	2	3
4	2	3	5
5	3	5	8
6	5	8	13
7	8	13	21
8	13	21	34
9	21	34	55
10	34	55	89
11	55	89	144
12	89	144	233

Number of rabbit pairs at end of 1 year: We now know that there are 233 pairs of rabbits at the end of 1 year.

When completing the previous table, you may have noticed that the number of rabbit couples at the end of the nth month ($n \geq 2$) equals the number of rabbit couples at the end of the $(n-1)^{th}$ month plus the number of rabbit couples at the end of the $(n-2)^{th}$ month. This relationship can be expressed symbolically as follows:

$$F_n = F_{n-1} + F_{n-2}, \text{ where } n \geq 2.$$

Note that $F_0 = 1$ since this rabbit population explosion started with one original pair of "vigorous" rabbits, and $F_1 = 1$ since the original pair does not produce offspring until the second month.

Summary: Fibonacci's Rabbit Population Problem. We noted that F_n (the total number of rabbit pairs in the nth month) equals F_{n-1} (the number of rabbit pairs in the $[n-1]^{th}$ month) plus the number of new pairs of rabbits born. Since new pairs of rabbits are only born to parents that are at least 1 month old, there will be F_{n-2} new pairs. So

$$F_n = F_{n-1} + F_{n-2}, \text{ where } F_0 = 1, F_1 = 1, \text{ and } n \geq 2. \qquad \textbf{1.11.3}$$

The Fibonacci Sequence. The recursive rule that emerged in the Rabbit Population problem defines a sequence that is called the **Fibonacci sequence**. In solving the rabbit problem, we found the first 13 terms of the sequence $(F_0, F_1, F_2, \ldots, F_{12})$. Here is a list of the first 22 terms (note that it is convenient to start this sequence at $n = 0$):

n	$F_n = F_{n-1} + F_{n-2}, n \geq 2$	n	$F_n = F_{n-1} + F_{n-2}, n \geq 2$
0	1	11	144
1	1	12	233
2	2	13	377
3	3	14	610
4	5	15	987
5	8	16	1,597
6	13	17	2,584
7	21	18	4,181
8	34	19	6,765
9	55	20	10,946
10	89	21	17,711

Rabbit Nightmare. Several years ago, I was in Pisa, Italy, attending a mathematics conference. My colleagues and I went to the Camposanto Monumentale in the Piazza dei Miracoli (near the leaning tower of Pisa) to see the famous statute of Leonardo Fibonacci. While gazing up at the impressive statue, one of my Italian friends made some interesting remarks about the Fibonacci sequence. We were all familiar with the standard properties of this remarkable sequence, but none of us were aware of the fascinating results that he brought to our attention. Throughout the day, I continued to think about the Fibonacci sequence and about my friend's comments. That evening, after a wonderful meal, I fell asleep pondering several questions related to the Fibonacci sequence.

It was about 3:00 AM when I abruptly awoke from a claustrophobic dream in which I had been trapped in a small rabbit warren with what seemed like millions of rabbits. They were crawling all over me, and I was having trouble breathing. Finally, I managed to escape the herd, and that's when I woke up.

Rabbit Nightmare

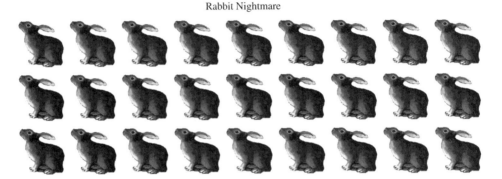

Picture courtesy of free clipart from freeclipartnow.com

I'm not sure how a dream analyst would interpret my dream, but I'm convinced it was connected to my daylong thoughts of the Fibonacci sequence. In any case, my strange rabbit nightmare does raise a question related to Fibonacci's Rabbit Population problem.

1.11.4 Question. Under the same assumptions as Fibonacci's Rabbit Population problem, approximately how many months would be needed to produce 10 million rabbits?

You could answer this question by using the recursive definition of the Fibonacci sequence (1.11.3), but this would be a laborious task. It might be simpler to answer the question if we had an explicit rule for F_n in terms of n. Lucky for us, Leonhard Euler (pronounced *Oil-er*), in 1765, discovered an explicit rule for the nth term of the Fibonacci sequence. As is often the case in mathematics, another mathematician rediscovered this rule. In fact, in 1843, Jacques Binet derived the following formula, which is called Binet's formula:

$$F_n = \frac{1}{\sqrt{5}}\left[\left(\frac{1+\sqrt{5}}{2}\right)^{n+1} - \left(\frac{1-\sqrt{5}}{2}\right)^{n+1}\right], \qquad \textbf{1.11.5}$$

for each integer $n \geq 0$.

At first glance, it doesn't seem that this formula could give the nth term of the Fibonacci sequence, since it is not even clear that it produces integer values for integers $n \geq 0$. Before we show that Binet's formula does, in fact, describe the

Fibonacci sequence, let's calculate the first three terms F_0, F_1, and F_2 using Binet's formula:

$$F_0 = \frac{1}{\sqrt{5}}\left[\left(\frac{1+\sqrt{5}}{2}\right)^{0+1} - \left(\frac{1-\sqrt{5}}{2}\right)^{0+1}\right]$$

$$= \frac{1}{\sqrt{5}}\left[\left(\frac{1+\sqrt{5}}{2}\right) - \left(\frac{1-\sqrt{5}}{2}\right)\right] = \frac{1}{\sqrt{5}}\left[\frac{2\sqrt{5}}{2}\right] = 1.$$

$$F_1 = \frac{1}{\sqrt{5}}\left[\left(\frac{1+\sqrt{5}}{2}\right)^{1+1} - \left(\frac{1-\sqrt{5}}{2}\right)^{1+1}\right]$$

$$= \frac{1}{\sqrt{5}}\left[\left(\frac{1+\sqrt{5}}{2}\right)^{2} - \left(\frac{1-\sqrt{5}}{2}\right)^{2}\right]$$

$$= \frac{1}{\sqrt{5}}\left[\left(\frac{1+2\sqrt{5}+5}{4}\right) - \left(\frac{1-2\sqrt{5}+5}{4}\right)\right]$$

$$= \frac{1}{\sqrt{5}}\left[\left(\frac{6+2\sqrt{5}}{4}\right) - \left(\frac{6-2\sqrt{5}}{4}\right)\right]$$

$$= \frac{1}{\sqrt{5}}\left[\frac{4\sqrt{5}}{4}\right] = 1.$$

$$F_2 = \frac{1}{\sqrt{5}}\left[\left(\frac{1+\sqrt{5}}{2}\right)^{2+1} - \left(\frac{1-\sqrt{5}}{2}\right)^{2+1}\right]$$

$$= \frac{1}{\sqrt{5}}\left[\left(\frac{1+\sqrt{5}}{2}\right)^{3} - \left(\frac{1-\sqrt{5}}{2}\right)^{3}\right]$$

$$= \frac{1}{\sqrt{5}}\left[\left(\frac{1+3\sqrt{5}+3\cdot 5+5\sqrt{5}}{8}\right) - \left(\frac{1-3\sqrt{5}+3\cdot 5-5\sqrt{5}}{8}\right)\right]$$

$$= \frac{1}{\sqrt{5}}\left[\left(\frac{16\sqrt{5}}{8}\right)\right] = 2.$$

A few preliminaries help simplify our proof of Binet's formula. First of all, let $u = \frac{1+\sqrt{5}}{2}$ and $v = \frac{1-\sqrt{5}}{2}$, so $F_n = \frac{1}{\sqrt{5}}[u^{n+1} - v^{n+1}]$. Next, note that u and

v are the two real roots of the quadratic equation $x^2 - x - 1 = 0$ (check), so $u^2 = u + 1$ and $v^2 = v + 1$.

Our proof employs the following equivalent form of mathematical induction called **strong induction** (sometimes called the **Second Principle of Mathematical Induction**).

1.11.6 Strong Induction

For each positive integer n, let $S(n)$ be a statement that is either true or false.

The statement $S(n)$ is true for each positive integer n provided the following two conditions are satisfied:

a. Basis step: $S(1)$ is true.

b. Inductive step: If $S(t)$ is true for all positive integers $t \leq k$, then $S(k+1)$ is true.

The hypothesis "If $S(t)$ is true for all positive integers $t \leq k$" is called the **inductive hypothesis**.

We are now ready to prove Binet's formula by strong induction.

Proof. Let $S(n)$ be the statement:

$$F_n = \frac{1}{\sqrt{5}}[u^{n+1} - v^{n+1}],$$

where F_n denotes the nth term of the Fibonacci sequence.

Basis Step. We just showed that $S(0)$ is true (for this result, we're starting induction at $n = 0$).

Inductive Step. Assume $S(t)$ is true for all positive integers $t \leq k$, and show that this implies that $S(k+1)$ is true, i.e., show that $F_{k+1} = \frac{u^{k+2} - v^{k+2}}{\sqrt{5}}$.

By the inductive hypothesis, we know that $F_{k-1} = \frac{(u^k - v^k)}{\sqrt{5}}$ and $F_k = \frac{(u^{k+1} - v^{k+1})}{\sqrt{5}}$. Hence, by the recursive definition of the Fibonacci sequence and by the identities $u^2 = u + 1$ and $v^2 = v + 1$, we see that

$$F_{k+1} = F_k + F_{k-1} = \frac{(u^{k+1} - v^{k+1})}{\sqrt{5}} + \frac{(u^k - v^k)}{\sqrt{5}}$$

$$= \frac{(u^{k+1} + u^k) - (v^{k+1} + v^k)}{\sqrt{5}}$$

$$= \frac{u^k(u+1) - v^k(v+1)}{\sqrt{5}} = \frac{u^k u^2 - v^k v^2}{\sqrt{5}} = \frac{u^{k+2} - v^{k+2}}{\sqrt{5}},$$

which is exactly what we wanted to prove.

Therefore, by strong induction, Binet's formula is true for all integers $n \geq 0$.

Answer to Question 1.11.4. Earlier we wondered, "How many months would be needed to produce 10 million rabbits?" Using a calculator and Binet's formula, it follows that

$$F_{34} = 9{,}227{,}465 \text{ and } F_{35} = 14{,}930{,}352 \text{ (verify both calculations)}.$$

Thus, it takes over 34 months to produce 10 million rabbits.

Classroom Problem: Landscape Block Problem. We want to line a driveway with 6 inch by 12 inch rectangular landscape blocks so that the height of the border is 12 inches tall. We would also like to make the block border by repeating a fixed block configuration of width $6n$ inches. The figures here illustrate that there is only one 6-inch-width configuration, two 12-inch-width configurations, three 18-inch-width configurations, and five 24-inch-width configurations.

B_n = number of block configurations of width $6n$ inches and height 12 inches.

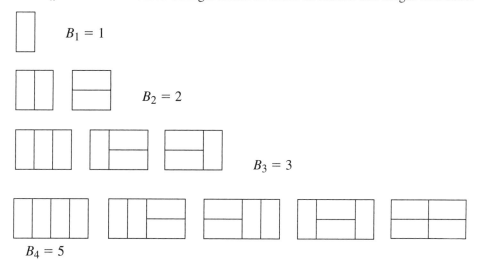

Determine the number of different 30-inch, 36-inch, and 42-inch width block configurations (i.e., determine B_5, B_6, and B_7). Write a recursive rule describing the sequence $B_1, B_2, B_3, \ldots$ ◆

As mentioned earlier, the Fibonacci sequence appears in a myriad of mathematical and physical contexts, and the mathematics literature (textbooks, research journals, puzzle books, etc.) abounds with interesting results and applications involving this intriguing sequence. We end this section with a few such examples.

Fibonacci Series. Let's determine a formula for the sum $S_n = F_0 + F_1 + \ldots + F_{n-1}$ of the first n terms of the Fibonacci sequence. By looking at some calculated sums, we can make a conjecture on a potential formula and then attempt to verify our conjecture by induction.

$S_1 = F_0 = 1$

$S_2 = F_0 + F_1 = 1 + 1 = 2$

$S_3 = F_0 + F_1 + F_2 = 1 + 1 + 2 = 4$

$S_4 = F_0 + F_1 + F_2 + F_3 = 1 + 1 + 2 + 3 = 7$

$S_5 = F_0 + F_1 + F_2 + F_3 + F_4 = 1 + 1 + 2 + 3 + 5 = 12$

$S_6 = F_0 + F_1 + F_2 + F_3 + F_4 + F_5 = 1 + 1 + 2 + 3 + 5 + 8 = 20$

$S_7 = F_0 + F_1 + F_2 + F_3 + F_4 + F_5 + F_6 = 1 + 1 + 2 + 3 + 5 + 8 + 13 = 33$

$S_8 = F_0 + F_1 + F_2 + F_3 + F_4 + F_5 + F_6 + F_7 = 1 + 1 + 2 + 3 + 5 + 8 + 13 + 21 = 54$

The above eight cases all satisfy the formula $S_n = F_{n+1} - 1$ (check), but a proof is needed to show the formula holds in general. Mathematical induction to the rescue!

Proof. Let $S(n)$ be the statement:

$$S_n = F_0 + F_1 + \ldots + F_{n-1} = F_{n+1} - 1.$$

Basis Step. Note that $S(1)$ is true, since $S_1 = F_0 = 1 = 2 - 1 = F_{1+1} - 1$.

Inductive Step. Assume $S(k)$ is true and show that this implies that $S(k + 1)$ is true, i.e., show that $S_{k+1} = F_0 + F_1 + \ldots + F_{(k+1)-1} = F_{(k+1)+1} - 1$. To reach this end, consider the following string of equalities:

$$S_{k+1} = (F_0 + F_1 + \ldots + F_{k-1}) + F_k$$

$$= S_k + F_k$$

$$= (F_{k+1} - 1) + F_k \text{ (by the inductive hypothesis)}$$

$$= (F_k + F_{k+1}) - 1$$

$$= F_{k+2} - 1 \text{ (by the recursive definition of the Fibonacci sequence)}.$$

Hence, $S(k + 1)$ is true, and so by induction, $S(n)$ is true for each positive integer n.

The Golden Ratio. Recall that a geometric sequence is a sequence $a_1, a_2, a_3, \ldots, a_n, a_{n+1}, \ldots$ having the property that for each positive integer n, $\frac{a_{n+1}}{a_n} = r \neq 0$, where r is a nonzero real number. We know that the Fibonacci sequence is not a geometric sequence, however, the ratios of successive terms do possess an interesting property. Let's calculate several successive ratios and see what pattern emerges from these computations.

Section 1.11 The Fibonacci Sequence 61

n	$\frac{F_{n+1}}{F_n}$	n	$\frac{F_{n+1}}{F_n}$	n	$\frac{F_{n+1}}{F_n}$	n	$\frac{F_{n+1}}{F_n}$
0	$\frac{1}{1} = 1$	3	$\frac{5}{3} \approx 1.667$	6	$\frac{21}{13} \approx 1.615$	9	$\frac{89}{55} \approx 1.618$
1	$\frac{2}{1} = 2$	4	$\frac{8}{5} = 1.6$	7	$\frac{34}{21} \approx 1.619$	10	$\frac{144}{89} \approx 1.618$
2	$\frac{3}{2} = 1.5$	5	$\frac{13}{8} = 1.625$	8	$\frac{55}{34} \approx 1.618$	11	$\frac{233}{144} \approx 1.618$

Our calculations seem to indicate that $\frac{F_{n+1}}{F_n} \approx 1.618$ as n gets larger and larger. Actually, we can make a more precise statement with the aid of calculus. In particular, it can be shown that as n gets arbitrarily large, the ratios $\frac{F_{n+1}}{F_n}$ get arbitrarily close to the number $\phi = \frac{1+\sqrt{5}}{2} \approx 1.61803398875$ (ϕ is the Greek letter phi). In calculus terminology, this conclusion would be stated as

$$\lim_{n \to \infty} \frac{F_{n+1}}{F_n} = \frac{1+\sqrt{5}}{2} = \phi,$$

and it is said that the sequence $\frac{F_1}{F_0}, \frac{F_2}{F_1}, \frac{F_3}{F_2}, \ldots, \frac{F_n}{F_{n-1}}, \frac{F_{n+1}}{F_n}, \ldots$ converges to $\phi = \frac{1+\sqrt{5}}{2}$. The number $\phi = \frac{1+\sqrt{5}}{2} \approx 1.61803398875$ is called the **golden ratio** (or **golden mean**, or **golden section**) and is the positive root of the equation $x^2 - x - 1 = 0$. Recall that the roots of this equation figured prominently in Binet's formula (1.11.5).

A rectangle whose length-to-width ratio is approximately equal to the golden ratio is called a **golden rectangle**. These rectangles are thought to be aesthetically pleasing and appear in everyday objects (e.g., grocery items such as cereal boxes), and in many famous works of art (e.g., various proportions of the Parthenon in Athens, Greece; some of Leonardo da Vinci's paintings; Michelangelo's magnificent sculpture of David, etc.).

Classroom Problem. You need a tape measure for this activity. Check to see if your body and your friends' bodies are in golden-ratio proportions. Calculate the following ratios:

1. $\frac{\text{height (length from top of head to floor)}}{\text{length from belly button to floor}} \approx \phi$?

2. $\frac{\text{length from belly button to floor}}{\text{length from top of head to belly button}} \approx \phi$?

3. $\frac{\text{length from top of head to bottom of chin}}{\text{length of head from ear to ear}} \approx \phi$?

4. $\dfrac{\text{length of one arm}}{\text{length from shoulder to shoulder}} \approx \phi?$ ◆

EXERCISES 1.11

1. Prove the following statement using induction (regular or strong induction):

$$F_n < 2^n \text{ for each integer } n \geq 1.$$

2. In each part, find a formula for the given series of Fibonacci numbers, and then use induction (regular or strong induction) to prove that your formula is true.
   a. The alternating series of the first n Fibonacci numbers:

   $$A_n = F_0 - F_1 + F_2 - F_3 + \ldots + (-1)^{n-1} F_{n-1}, \text{ where } n \geq 1.$$

   b. The series of the first n Fibonacci numbers with even subscripts:

   $$E_n = F_0 + F_2 + F_4 + \ldots + F_{2n-2}, \text{ where } n \geq 1.$$

   c. The series of the first n Fibonacci numbers with odd subscripts:

   $$O_n = F_1 + F_3 + F_5 + \ldots + F_{2n-1}, \text{ where } n \geq 1.$$

Arithmetic and Algebra of the Integers

CHAPTER 2

2.1 A FEW MATHEMATICAL QUESTIONS CONCERNING THE PERIODICAL CICADAS
2.2 CLASSROOM CONNECTIONS: MULTIPLES AND DIVISORS
2.3 REFLECTIONS ON CLASSROOM CONNECTIONS: MULTIPLES AND DIVISORS
2.4 MULTIPLES AND DIVISORS
2.5 LEAST COMMON MULTIPLE AND GREATEST COMMON DIVISOR
2.6 THE FUNDAMENTAL THEOREM OF ARITHMETIC
2.7 REVISITING THE LCM AND GCD
2.8 RELATIONS AND RESULTS CONCERNING LCM AND GCD

"Mathematics is the queen of the sciences and number theory is the queen of mathematics." These famous words of Carl Friedrich Gauss (1777–1855) fittingly set the stage for our investigations into the arithmetic and algebra of the integers (i.e., the theory of numbers). Most of our work centers on basic applications of the Fundamental Theorem of Arithmetic and how these ideas underlie many important mathematical concepts.

FIGURE 2.1 German postage stamp issued in 1955 to memorialize the centennial of Gauss's death.

Gene the Entomologist. Missouri is known as the Show Me State, but it also has been called the Cave State. There are over 5,000 known caves in Missouri, and many of them are open for public spelunking. Our home is situated on an old tobacco farm near the ridge of a cave system known as Boone Cave. We love the peace and quiet of country living, although my wife could do without the snakes, spiders, and other n-legged creatures (where $0 \leq n \leq 8$) sharing the land with us.

When it comes to questions concerning insects, we're lucky that our neighbor Gene is a top-notch entomologist with an encyclopedic memory. Over the years we have asked him to identify and talk about numerous odd-looking bugs, and he has always had a wealth of information to share with us. He really enjoys chatting about bugs, and I especially take pleasure in hearing all the fancy entomological terminology within his explanations.

Gene's expertise was especially useful to us in late May through early June of 1998. The calm of this particular spring was lost to the deafening song of a colossal chorus of joyous cicadas. These periodic intruders of solitude have performed for us before, but we had never experienced their "heavy metal" repertoire. Why was their incessant piercing cadence so intense, and were we in danger of permanent hearing loss? These and other natural questions were posed to our favorite neighbor and resident creepy-crawly specialist.

Gene told us that there are different species of periodical cicadas, some with a 13-year life cycle and others with a 17-year life cycle. In various years, groups of cicadas called "broods" emerge in northern and southern ranges. Missouri, in the spring of 1998, hosted a massive assembly of both the 13-year and 17-year cicadas at the same time. "Enjoy this cicada jamboree," said Gene, "because this is the first and last time in your lives that you will see and hear this particular species of 13-year and 17-year cicadas." We took Gene's advice and enjoyed one of nature's most infrequent events.

2.1 A FEW MATHEMATICAL QUESTIONS CONCERNING THE PERIODICAL CICADAS

1. When was the last time this particular species of 13-year and 17-year cicadas appeared in Missouri, and when will they appear again?
2. Why do you think the cicadas' cycles are a prime number of years long?

Our computations show that this particular species of 13-year and 17-year cicadas appear together on a 221-year cycle. Do your calculations agree with ours, and how did you arrive at your answer? The cicadas certainly benefit from this lengthy rotation rate, since the different-cycle cicadas compete for food less frequently. In the next several sections, we investigate the underlying mathematical concepts present in these questions and explore related topics.

2.2 CLASSROOM CONNECTIONS: MULTIPLES AND DIVISORS

In this section, we work on some problems involving common factors and multiples from a sixth grade unit in *Connected Mathematics* called *Prime Time* (see Figures 2.2.1 to 2.2.4 on pp. 65–69). In fact, Problem 4.2 (Looking at Locust Cycles) returns us to our cyclic cicada companions (nice alliteration). This work serves as a stepping-stone to a more expansive study of these and related arithmetic properties of the

Section 2.3   Reflections on Classroom Connections: Multiples and Divisors   65

FIGURE 2.1.1 Periodical cicadas in Missouri.

integers. Our development illuminates the central ideas present within this middle-grade unit and connects them to some fundamental notions of elementary number theory.

2.3 REFLECTIONS ON CLASSROOM CONNECTIONS: MULTIPLES AND DIVISORS

What were the fundamental mathematical notions you encountered while working on the selected problems from *Prime Time*? The concepts of multiples, divisors/factors, common multiples, common divisors, least common multiples, greatest common divisors, and prime numbers were the central ideas present among these problems. Let's look at each of the problems and while discussing them, we will formalize some definitions and terminology. This formalization is of the utmost importance, since the validity of your work can only be verified when tested against precise definitions or other known valid deductions. In particular, when presenting an argument to convince yourself or others that your conclusions are accurate and your reasoning is correct, it is imperative that all involved agree upon the exact meaning of the concepts involved. The power and utility of mathematics is built upon such precision!

Multiples, Common Multiples, and the Least Common Multiple: The Ferris Wheel Problem and the Cicadas Problem

Ferris Wheel Problem.   Recall from Problem 4.1 of *Prime Time* that you and your little sister are about to take a ride on two different Ferris wheels. The large Ferris wheel makes one revolution in 60 seconds, and the small Ferris wheel makes one revolution in 20 seconds. After both Ferris wheels start at the same time, when is the

Common Factors and Multiples

There are many things in the world that happen over and over again in set cycles. Sometimes we want to know when two things with different cycles will happen at the same time. Knowing about factors and multiples can help you to solve such problems.

Let's start by comparing the multiples of 20 and 30.

- The multiples of 20 are 20, 40, 60, 80, 100, 120, . . .
- The multiples of 30 are 30, 60, 90, 120, 150, 180, . . .

The numbers 60, 120, 180, 240, . . . are multiples of both 20 and 30. We call these numbers **common multiples** of 20 and 30.

Now let's compare the factors of 12 and 30.

- The factors of 12 are 1, 2, 3, 4, 6, and 12.
- The factors of 30 are 1, 2, 3, 5, 6, 10, 15, and 30.

The numbers 1, 2, 3, and 6 are factors of both 12 and 30. We call these numbers **common factors** of 12 and 30.

4.1 Riding Ferris Wheels

One of the most popular rides at a carnival or amusement park is the Ferris wheel.

Did you know?

The largest Ferris Wheel was built for the World's Columbian Exposition in Chicago in 1893. The wheel could carry 2160 people in its 36 passenger cars. Can you figure out how many people could ride in each car?

36 Prime Time

Reproduced from page 36 of *Prime Time* in *Connected Mathematics*.

FIGURE 2.2.1

> **Problem 4.1**
>
> You and your little sister go to a carnival that has both a large and a small Ferris wheel. You get on the large Ferris wheel at the same time your sister gets on the small Ferris wheel. The rides begin as soon as you are both buckled into your seats. Determine the number of seconds that will pass before you and your sister are both at the bottom again
>
> **A.** if the large wheel makes one revolution in 60 seconds and the small wheel makes one revolution in 20 seconds.
>
> **B.** if the large wheel makes one revolution in 50 seconds and the small wheel makes one revolution in 30 seconds.
>
> **C.** if the large wheel makes one revolution in 10 seconds and the small wheel makes one revolution in 7 seconds.

■ **Problem 4.1 Follow-Up**
For parts A–C in Problem 4.1, determine the number of times each Ferris wheel goes around before you and your sister are both on the ground again.

Investigation 4: Common Factors and Multiples 37

Reproduced from page 37 of *Prime Time* in *Connected Mathematics*.

FIGURE 2.2.2

first time that both you and your sister will be at the bottom (the original starting position)? Since 60 is a multiple of 20 (60 = 3 • 20), the first time (after the start) both you and your sister will be at the bottom is 60 seconds after the start of the wheels, and your sister will have made three revolutions compared to your one.

How would we approach this problem if the large wheel makes one revolution in 50 seconds and the small Ferris wheel makes one revolution in 30 seconds? Why is the answer not as apparent as it was in the previous situation? A simple strategy to solve this problem might be to list the multiples of 50 and 30, and then choose from the lists the least positive of the common multiples. Since 50 • 30 = 1,500, and 1,500 is always a common multiple of 50 and 30 (why?), there is no need to inspect multiples greater than 1,500.

Some multiples of 50 ≤ 1,500 are: 0, 50, 100, 150, 200, 250, 300, 350, 400, 450, ...

Some multiples of 30 ≤ 1,500 are: 0, 30, 60, 90, 120, 150, 180, 210, 240, 270, 300, ...

Inspecting the lists, we see that both Ferris wheels reach bottom together after 150 seconds have elapsed from the starting time and that the smaller wheel will have made five complete revolutions compared to the three complete revolutions of the

Looking at Locust Cycles

Cicadas spend most of their lives underground. Some cicadas—commonly called 13-year locusts—come above ground every 13 years, while others—called 17-year locusts—come out every 17 years.

> **Problem 4.2**
>
> Stephan's grandfather told him about a terrible year when the cicadas were so numerous that they ate all the crops on his farm. Stephan conjectured that both 13-year and 17-year locusts came out that year. Assume Stephan's conjecture is correct.
>
> **A.** How many years pass between the years when both 13-year and 17-year locusts are out at the same time? Explain how you got your answer.
>
> **B.** Suppose there were 12-year, 14-year, and 16-year locusts, and they all came out this year. How many years will it be before they all come out together again? Explain how you got your answer.

■ **Problem 4.2 Follow-Up**

For parts A and B of Problem 4.2, tell whether the answer is less than, greater than, or equal to the product of the locust cycles.

Reproduced from page 38 of *Prime Time* in *Connected Mathematics*.

FIGURE 2.2.3

large wheel. List a few other times when both wheels will be at the bottom position together.

Cicadas Problem. When will the particular species of 13-year and 17-year cicadas that appeared in Missouri in 1998 appear together again (compare to Problem 4.2 of *Prime Time*)? As in the Ferris wheel problem, we can list the multiples of 13 and the multiples of 17 that are less than or equal to $13 \cdot 17 = 221$, and then choose the least positive of their common multiples.

Some multiples of $13 \leq 221$ are: 0, 13, 26, 39, 52, 65, 78, 91, 104, 117, 130, 143, 156, 169, 182, 195, 208, 221, ...

Some multiples $17 \leq 221$ are: 0, 17, 34, 51, 68, 85, 102, 119, 136, 153, 170, 187, 204, 221, ...

By inspecting the lists, we see that this particular species of 13-year and 17-year cicadas will reappear in the year 2219 (i.e., 221 years after 1998). Note that in this case, the least common positive multiple is the product of 13 and 17, whereas the least common positive multiple of 30 and 50 in the second Ferris wheel problem is $150 < 30 \cdot 50 = 1{,}500$.

Section 2.3   Reflections on Classroom Connections: Multiples and Divisors   69

4.3 Planning a Picnic

Common factors and common multiples can be used to figure out how many people can share things equally.

> **Problem 4.3**
>
> Miriam's uncle donated 120 cans of juice and 90 packs of cheese crackers for the school picnic. Each student is to receive the same number of cans of juice and the same number of packs of crackers.
>
> What is the largest number of students that can come to the picnic and share the food equally? How many cans of juice and how many packs of crackers will each student receive? Explain how you got your answers.

■ **Problem 4.3 Follow-Up**

If Miriam's uncle eats two packs of crackers before he sends the supplies to the school, what is the largest number of students that can come to the picnic and share the food equally? How many cans of juice and how many packs of crackers will each receive?

Investigation 4: Common Factors and Multiples

Reproduced from page 39 of *Prime Time* in *Connected Mathematics*.

FIGURE 2.2.4

Although the method of listing multiples worked quite nicely in the previous problems, this method would be difficult and inefficient if the numbers within the problem were not as "user friendly." For example, what if the large Ferris wheel made a complete revolution in 5,173 seconds and the little wheel made a complete revolution in 3,959 seconds? The listing method in this case is quite cumbersome but eventually leads to an answer. A primary goal of this chapter is to develop more efficient methods for solving such problems and to further study these methods in more general settings.

Divisors, Common Divisors, and the Greatest Common Divisor: The Picnic Problem

Planning a Picnic Problem. Remember from Problem 4.3 of *Prime Time* that Miriam's uncle donated 120 cans of juice and 90 packs of cheese crackers for the school picnic. In the name of fairness, Miriam decided that each student is to receive the same number of cans of juice and the same number of packs of cheese crackers. Given these constraints, she wants to determine the largest number of students who can come to the picnic and share the food equally.

Betty, one of Miriam's friends, suggests that 90 students is the maximum number who could come to the picnic and share the food equally—one can of juice per student and one package of cheese crackers per student. Do you agree with Betty, and if not, why?

Miriam tells Betty that the students must equally share ALL the food her uncle donated. With this important clarification in mind, Miriam and Betty set out to solve their problem.

How would you proceed to find a solution to this problem? Since we are looking for the largest number of students who divide evenly into 90 and 120, let's list all the divisors (factors) of 90 and all the divisors of 120 (that are ≤ 90) and then choose the largest common one. This listing strategy is similar to the one we used in the Ferris wheel and cicadas problems.

Divisors of 90 are: 1, 2, 3, 5, 6, 9, 10, 15, 18, 30, 45, 90.
Divisors of 120 (that are ≤ 90) are: 1, 2, 3, 4, 5, 6, 8, 10, 12, 15, 20, 24, 30, 40, 60.

Inspection of the lists show that exactly 30 students can come to the picnic and equally share all of the food; each one can have four cans of juice and three packages of cheese crackers.

The method we used to find the solution to the "Planning a Picnic problem" was similar to the method employed in the "Ferris wheel and Cicadas problems." As we commented earlier, these methods are easy to understand but are not flexible enough to handle problems involving large numbers and more wide-ranging situations.

2.4 MULTIPLES AND DIVISORS

To begin, let's use our understanding of the specific numerical examples we've looked at and write some clear definitions that precisely capture the meanings of the concepts present.

The Universe. All of the work in this chapter takes place within the set of

$$\text{integers: } \mathbf{Z} = \{\ldots, -3, -2, -1, 0, 1, 2, 3, \ldots\},$$

and often times we restrict discussions to the

$$\text{nonzero integers: } (\mathbf{Z} - \{\mathbf{0}\}) = \{z \in \mathbf{Z}: z \neq 0\},$$

the

$$\text{positive integers (natural numbers): } \mathbf{Z}^+ = \{z \in \mathbf{Z}: z > 0\},$$

or the

$$\text{non-negative integers (whole numbers): } \mathbf{Z}^+ \cup \{\mathbf{0}\} = \{0, 1, 2, 3, \ldots\}.$$

As you will see, the concepts we concentrate on are meaningful within the set of integers but are less so within other sets, such as the set of rational numbers.

Multiple. An integer b is called a **multiple** of an integer a if $b = ca$, for some integer c.

For example, 10, 55, 0, -35, and 5 are all multiples of 5, whereas 101 is not a multiple of 5. Notice that 0 is a multiple of every integer, whereas 0 is the only multiple of 0.

The number line graphically illustrates the concept of multiples of an integer. For example, suppose a is some positive integer. We can represent the multiples of a on the number line in the following way:

Question. How would you draw the line if a was negative?

Divisor (Factor). A *nonzero* integer a is called a **divisor (factor)** of an integer b if $b = ac$, for some integer c. In this situation, b is said to be **divisible** by a. In case a is a divisor of b, we write $a|b$ and read this as *a* **divides** *b*.

This notation can be thought of as shorthand for the usual long division notation $a\overline{)b}$ (removing the horizontal bar), where a "goes into" b evenly. (In Section 3.1, we study the mathematics underlying long division—the division algorithm.)

Convention. Whenever the notation $a|b$ is used, it is always implicit that a is *nonzero*.

Be careful not to confuse this notation with the fraction notation a/b. The relation $a|b$ can be expressed by using several different terms. For example,

$$a|b \Leftrightarrow a \text{ \textbf{divides} } b \Leftrightarrow a \text{ is a \textbf{factor} of } b \Leftrightarrow b \text{ is a \textbf{multiple} of}$$
$$a \Leftrightarrow a \text{ is a \textbf{divisor} of } b.$$

We use these terms interchangeably throughout the text, as is the case in the mathematical literature.

The definitions we have just considered involve integers only. Would it be reasonable to extend our definitions to other sets of numbers, such as the rational numbers? For example, what are the rational factors of 1/2? Well, 2 is a factor of 1/2 since $1/2 = 2 \cdot (1/4)$, and 3/2 is a factor of 1/2 since $1/2 = (3/2) \cdot (1/3)$. Actually, every nonzero rational number is a factor of 1/2. (Why?) So, in the case of the rational numbers, it is not meaningful to consider these concepts.

Classroom Discussion. Since we have just made some precise definitions, it is appropriate to consider a variety of properties related to these concepts. The discovery and formalization of such properties often arise through observations or questions connected to concrete numerical computations. For example, notice that 9 is a factor of 18, and 9 is a factor of 27. Can one conclude that 9 is also a factor of 18 + 27? How would middle-school students approach this question? A common response is add the numbers and see that 9 divides into 45 evenly. This computational process does provide an answer and certainly can be applied to any specific example. However, does this method contribute to a clear understanding of the underlying idea(s) present, and, thus, can this method be applied to similar problems involving different numbers?

One way of looking at this example is to write $18 = 9 \cdot 2$ and $27 = 9 \cdot 3$, so $18 + 27 = (9 \cdot 2) + (9 \cdot 3) = 9 \cdot (2 + 3)$. It is now apparent that 9 is a factor of $18 + 27$. The advantage of this approach is that it illuminates the problem's core idea and naturally sets the stage for expressing that idea in more general terms. This in turn allows us to reach similar conclusions to analogous problems without treating each such problem as a singular entity.

Let's rephrase this problem in general terms using variables rather than explicit integers. Let a, b, and c be integers such that a is a factor of b and c, i.e., $a|b$ and $a|c$. We assert that $a|(b + c)$ and justify this by constructing an argument modeled after the previous numerical example. In particular, $b = ax$ and $c = ay$ for some integers x and y. (Why?) So $b + c = ax + ay = a(x + y)$, and thus $a|(b + c)$.

It is important to note that the converse of the previous result is not valid. For example, 9 is a factor of $63 = 22 + 41$, but 9 is not a factor of 22 and 41. ◆

The following divisibility properties can be justified in a similar manner.

Elementary divisibility properties of integers ($a, b, c,$ and d are integers):

1. If $a|b$ and $b|a$, then $a = \pm b$.
2. If $a|b$, then $a|bx$ for any integer x.
3. If $a|b$ and $a|c$, then $a|(bx \pm cy)$ for any integers x and y.
4. If $a|b$ and $a|(b \pm c)$, then $a|c$.
5. If $a|b$ and $b|c$, then $a|c$ (transitivity of "divides").
6. If $a|b$ and $c|d$, then $ac|bd$.

Property 4. We consider Statement 4 at this point and leave the proofs of the other properties for the Exercises or for classroom discussion.

Justification of Property 4. Write $b = ax$ and $(b + c) = ay$ (the subtraction case is handled similarly). Thus $c = (b + c) - b = ay - ax = a(y - x)$, so $a|c$.

EXERCISES 2.4

1. Write detailed arguments justifying properties 1, 2, 3, 5, and 6 (break those problems involving $\pm$ into two separate cases).
2. In each part, indicate whether the given statement concerning integers is true or false. If it is true, provide a sound argument demonstrating the truth of the statement or indicate how the statement follows from some other known statement(s). If it is false, provide a concrete counterexample with a complete explanation of why your example shows the statement is false.
    a. If $27a + 56b = c$, then $7|c$.
    b. If $51a - 801b = c$, then $3|c$.
    c. If $26a + 64b = c$, then $8|c$.
    d. If $15|(3a - 2b)$, then $3|(12a^2 - 8ab)$.
    e. If $5|(a - 45b)$, then $5|a$.
    f. If $12|(2a + 4b)$, then $2|a$.
    g. If $a|(b + c)$, then $a|b$ or $a|c$.
    h. If $a|bc$, then $a|b$ or $a|c$.
    i. If $7|ab$, then $7|a$ or $7|b$ (further discussed in Section 2.6.2).
    j. If $a|c$ and $b|c$, then $ab|c$.
    k. If $a|b$, then $a^2|b^2$.
    l. If $a^2|b^2$, then $a|b$ (further discussed in Section 2.6.9).
    m. $2^{35}|196^{20}$.
3. For positive integers a and n, show that $(a - 1)|(a^n - 1)$. Direction: Use induction to show that $(a^n - 1) = (a - 1)(a^{n-1} + a^{n-2} + \ldots + a^1 + 1)$.
4. Let $a_1, a_2, \ldots, a_s$ be integers greater than 1. Show that $(a_1 a_2 \cdots a_s) + 1$ is not divisible by a_i for each $1 \leq i \leq s$. Direction: Use an indirect argument.

2.5 LEAST COMMON MULTIPLE AND GREATEST COMMON DIVISOR

Let's continue to build on our understanding of the classroom examples we've looked at and formulate further precise definitions for the concepts we've encountered.

Common Multiple. An integer m is called a **common multiple** of integers a and b if m is both a multiple of a and a multiple of b (i.e., if $m = ax$ and $m = by$ for some integers x and y).

An uninteresting situation occurs if a or b equals zero, since then $m = 0$ is the only common multiple of a and b. For this reason, we shall restrict our attention to nonzero integers when we consider questions concerning common multiples.

Question. For nonzero integers a and b, is there always a nonzero positive common multiple of a and b (0 is always a common multiple), and if so, is there a smallest positive common multiple? (Why have we restricted this question to the positive integers?)

The answer to the first part is not difficult, since ab and $-(ab)$ are both common multiples of a and b, and one of these products is always positive. (Why?) Once we know there is at least one positive common multiple, then can we simply conclude there is a smallest positive common multiple? Intuitively this seems clear, but there is a subtle property of the positive integers that justifies this conclusion. This property arises as an axiom in the axiomatic development of the natural numbers (positive integers) and is logically equivalent to the Principle of Mathematical Induction (cf., 1.8.1).

2.5.1 The Well-Ordering Principle

Every nonempty set of positive integers has a least element.

Least Common Multiple. For nonzero integers a and b, the **least common multiple**, denoted **lcm(a, b)**, is the smallest *positive* common multiple of a and b.

We can restate this definition using the divisor notation.

2.5.2 Restating the Least Common Multiple Definition

Let a and b be nonzero integers. Then, $\text{lcm}(a, b) = m$ provided:

1. m is a positive integer;
2. $a|m$ and $b|m$;
3. If $a|c$ and $b|c$, where c is a positive integer, then $m \leq c$.

Common Divisor. A nonzero integer d is called a **common divisor** of integers a and b if d is both a divisor of a and a divisor of b (i.e., if $a = dx$ and $b = dy$ for some integers x and y).

For example, 3 is a common divisor of 198 and 735. What are the other common divisors, and which of these is the largest? In determining your answers, was it necessary to find all divisors of 198 and 735 and then choose the largest of their common divisors, or was it possible to reach the correct solution by focusing on some special divisors of a and b?

In Sections 2.6 and 2.7, we investigate a systematic method to find the set of positive divisors of a nonzero positive integer a and indicate some concrete and theoretical implications of this approach.

Greatest Common Divisor. For integers a and b that are not both 0, the **greatest common divisor (or greatest common factor)**, denoted **gcd(a, b)** (or **gcf(a, b)**), is the largest positive common divisor of a and b.

The inclusion of the word *positive* in the definition of $\gcd(a, b)$ is redundant, since the largest of the common divisors of a and b is necessarily positive. However, it is usually stated as part of the definition to accentuate this fact. Note that if $a = b = 0$, then all nonzero integers are common divisors of a and b, so we always insist that not both a and b are 0. With this restriction on a and b, it follows that $\gcd(a, b)$ always exists. (Why?) We have previously discovered that $\gcd(198, 735) = 3$ and shall soon consider a general framework for analyzing and computing such problems.

As in the lcm case, we can define the greatest common divisor using the language and notation of divisibility.

2.5.3 Restating the Greatest Common Divisor Definition

Let a and b be integers, not both zero. Then, $\gcd(a, b) = d$ provided:

1. d is a positive integer;
2. $d|a$ and $d|b$;
3. if $g|a$ and $g|b$, where g is a positive integer, then $g \leq d$.

If g is a common divisor of a and b, then g also divides any common multiple c of a and b. (Why?) Therefore,

2.5.4
$$\gcd(a, b) \mid \text{lcm}(a, b)$$

Convention. Since $\text{lcm}(a, b) = \text{lcm}(|a|, |b|)$ (why?) for nonzero integers a and b, and $\gcd(a, b) = \gcd(|a|, |b|)$ (why?) for integers a and b, one of which is nonzero, we usually frame our discussions concerning lcm and gcd within the set of positive integers.

Recall that the "Ferris wheel problem" amounted to determining (in our new notation) $\text{lcm}(20, 60)$, and another part of that problem involved finding $\text{lcm}(30, 50)$. The "Cicadas problem" was a search for $\text{lcm}(13, 17)$, and the "Planning a Picnic problem" dealt with determining $\gcd(90, 120)$. The conclusions we reached were: $\text{lcm}(20, 60) = 60$, $\text{lcm}(50, 30) = 150$, $\text{lcm}(13, 17) = 221$, and $\gcd(90, 120) = 30$. We arrived at all of these answers by listing multiples or divisors and then searching for the smallest of the common multiples or the largest of the common divisors. In the first two examples, the least common multiple was smaller than the product of the two given numbers, and in the third example, the least common multiple equaled the product of the two numbers. Why is this so, and what properties of the positive integers govern these outcomes? The theorem we explore in the next section provides a means to address this and other related questions and is a powerful tool for studying questions about the integers.

2.6 THE FUNDAMENTAL THEOREM OF ARITHMETIC

When students first learn positive integer multiplication, they quickly realize that the process can be "somewhat" reversed, and a positive integer can be expressed as a product of some of its factors. The process is called **factoring**, and the resulting representation is called a **factorization** of the number. For example, multiplying 2 and 9 results in 18, but this is not the only way to arrive at 18 through integer multiplication.

Some positive integers can be factored as a product of two or more smaller positive integers, while others cannot. We have noted that $18 = 2 \cdot 9 = 2 \cdot 3 \cdot 3$, whereas the only way (except for the order of the factors) to write 5 as a product of factors is $5 = 1 \cdot 5$. These observations lead to the following definitions:

76   Chapter 2   Arithmetic and Algebra of the Integers

Prime and composite numbers.   Note that an integer $n > 1$ either has exactly two distinct positive factors or more than two distinct positive factors. (Why?) If an integer $n > 1$ has *exactly two* distinct positive factors, it is called a **prime number** (or just **prime**). If it has more than two distinct positive factors, it is called a **composite number** (or just **composite**). The integer 1 is neither prime nor composite.

Let's see how the integers, with $1 < n \leq 50$, break up into primes and composites. It's good enough to find the primes in this range.

Prime numbers p with $1 < p \leq 50$: 2, 3, 5, 7, 11, 13, 17, 19, 23, 29, 31, 37, 41, 43, 47.

We soon discuss some methods to determine prime numbers, but first let's see that primes serve as the primitive (hence the term *prime*) "multiplicative" building blocks of the integers.

Prime factorization.   A composite number $n > 1$ can always be written as the product of two smaller positive integers, i.e., $n = ab$, where $1 < a < n$ and $1 < b < n$. For example, $540 = 10 \cdot 54$. If both a and b are prime, then n is a product of prime numbers. If either a or b is not prime, then one or both can further be factored into a product of smaller positive integers, e.g., $10 = 2 \cdot 5$ and $54 = 6 \cdot 9$ and so $540 = (2 \cdot 5) \cdot (6 \cdot 9)$. This process leads to, in a finite number of steps (why?), a factorization of n into prime numbers. In particular, a prime factorization of 540 is: $540 = 2 \cdot 2 \cdot 3 \cdot 3 \cdot 3$.

Summarizing these observations gives:

Each integer $n > 1$ is either a prime number or a product of prime numbers.

Prime factorization is of fundamental importance to the study of the integers, and we use it continually throughout our work.

To determine a prime factorization of an integer $n > 1$, it is sometimes convenient to use a tree diagram (called a **factor tree**) to help keep track of the successive factorizations leading to a prime factorization of n. The following examples illustrate this method.

EXAMPLES

1. Find a prime factorization of 252. Let's start with an obvious initial factorization of 252, namely $252 = 2 \cdot 126$, and then continue to factor the resulting composite factors (presuming there are some to be factored) until a prime factorization is reached. A factor tree is useful to record the successive factorizations.

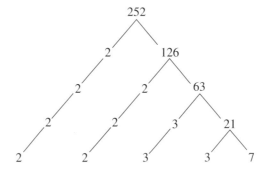

Section 2.6   The Fundamental Theorem of Arithmetic

We can extend this tree no farther, since the last discovered factors are all prime numbers. Therefore, a prime factorization of 252 is: $252 = 2 \cdot 2 \cdot 3 \cdot 3 \cdot 7$.

There are several different ways that we can factor 252. For example, what happens if we choose a different initial factorization, such as $252 = 14 \cdot 18$?

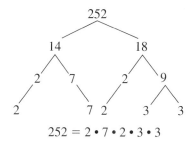

$$252 = 2 \cdot 7 \cdot 2 \cdot 3 \cdot 3$$

Notice that this factor tree is different from the previous one, but they both lead to the same prime factorization of 252 (except for the order in which the factors are multiplied). Are all prime factorizations of 252 the same, except for the order of the factors? This seems to be true for this particular example (more checking is necessary), but is it valid in general? We consider this question in more detail after another numerical example.

2. Let's look at $n = 2{,}057$. What is a good strategy for finding an initial factorization or showing that one does not exist (i.e., in the case n is prime)? To find a prime factor, we could start test divisions with the prime 3 and continue testing successive primes until we find a prime divisor d of n or show that n itself is a prime.

Before analyzing the example $n = 2{,}057$, let's think about factoring in the general case.

Classroom Discussion.   What method would you use to see if an integer $n > 1$ is prime or not? As mentioned earlier, one approach might be to look at all prime numbers less than the given number n and determine if any of them are factors of n (starting with the smallest and working toward n). Although this method always works in principle, it is not feasible in practice (even with the help of a computer). In fact, it is extremely difficult to determine if a given number is prime!

Let's consider a way to physically reduce the number of divisions needed to determine if a positive integer n is prime (this still does not make it practical for large numbers). As noted before, a composite number $n > 1$ can be expressed as $n = ab$ for some integers a and b greater than 1 and less than n. It follows that a and b cannot both be greater than $\sqrt{n}$ (why?), so a composite number n always has a prime divisor d with $1 < d \leq \sqrt{n}$. (Why?) These observations can be summarized and equivalently stated as a prime number test. ◆

2.6.1   Prime Number Test

If an integer $n > 1$ has no prime divisor d for $1 < d \leq \sqrt{n}$, then n is prime. (Why?)

Since 2 is the only even prime number, the remaining prime numbers are all odd. Thus, if n is an odd integer greater than 1, the test divisors for n can be restricted to all odd primes d with $1 < d \leq \sqrt{n}$.

Back to Example 2. To determine a prime factorization for 2,057, let's examine test prime divisors d such that $2 < d \leq \sqrt{2{,}057} \approx 45.35$. Test divisions show that 11 is the smallest prime divisor of 2,057, and $2{,}057 = 11 \cdot 187$. The focus now shifts to 187, and we can apply the same strategy. Our analysis leads to the following prime factorization of 2,057:

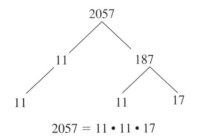

$$2057 = 11 \cdot 11 \cdot 17$$

Notice (with a little more checking), as in the previous example, that if the order of the factors is disregarded, then there is exactly one prime factorization of 2,057.

Question. We showed that each integer $n > 1$ always has a prime factorization, but is it conceivable that a given n might have different prime factorizations other than those obtained by reordering the primes in the given factorization?

The examples we have examined provide some evidence that prime factorizations are unique (except for the order of the factors), but this is certainly not sufficient to guarantee that all integers $n > 1$ have a unique prime factorization. Let's construct a proof that demonstrates uniqueness.

Uniqueness of Prime Factorization. If $n = p_1 p_2 \cdots p_s = q_1 q_2 \cdots q_t$, where the p_i and q_j are primes and $s \leq t$, then these two factorizations are identical except for the order of the prime factors.

Analysis. Note that $p_1 | q_1 q_2 \cdots q_t$. Does it follow that $p_1 | q_i$ for some $1 \leq i \leq t$? After looking at several specific examples, this conclusion seems reasonable, but formal justification of it is not obvious. For the time being, let's assume that this conclusion does follow and see where this takes us. If we can complete the proof using this assumption, then it would remain to establish the validity of this point.

If $p_1 | q_i$, then $p_1 = q_i$ (why?), and by relabeling, we can assume that $p_1 = q_1$. Thus, $p_1 p_2 \cdots p_s = p_1 q_2 \cdots q_t$, and so cancellation gives $p_2 \cdots p_s = q_2 \cdots q_t$. Repeating this argument and relabeling again if necessary gives that $p_2 = q_2$, and cancellation results in $p_3 \cdots p_s = q_3 \cdots q_t$. After s steps (relabeling if necessary at each step), this procedure leads to the equality $1 = q_{s+1} q_{s+2} \cdots q_t$, which is absurd unless $s = t$. (Why?)

Section 2.6   The Fundamental Theorem of Arithmetic   79

Therefore, the proof of the uniqueness of prime factorization is complete once we have verified the assumption concerning a prime dividing a product of integers, i.e.,

2.6.2 **If a prime p divides a product $q_1 q_2 \cdots q_s$ of positive integers, then $p|q_i$ for some $1 \leq i \leq s$**

We tackle this important result in Corollary 3.3.2, after we explore some new concepts.

The main ideas of this section can be summarized as one of the most important theorems in integer arithmetic.

2.6.3 **Fundamental Theorem of Arithmetic (FTA)**

Each integer $n > 1$ is either a prime number or can be represented as a product of prime numbers. This representation is unique except for the order of the primes involved.

By ordering the distinct prime factors of any integer $n > 1$ in increasing magnitude, the prime factorization of n can be written uniquely in the following **canonical form**:

$$n = p_1^{m_1} p_2^{m_2} \cdots p_s^{m_s},$$

where $p_1 < p_2 < \ldots < p_s$ and the m_i are positive integers. It is often convenient to use this canonical form when applying the Fundamental Theorem of Arithmetic to questions concerning the integers.

Excluding 1 as a Prime.   Can we define a prime number as an integer that is only divisible by 1 and itself? In this definition, the numbers we have called *prime* would still be prime, and in addition, the integer 1 would be also be prime. Using this definition, we could argue as before to show that each positive integer has a prime factorization, but unfortunately these factorizations are not unique. For example, $14 = 2 \cdot 7 = 1 \cdot 2 \cdot 7 = 1 \cdot 1 \cdot 2 \cdot 7$, etc. To avoid this unruly predicament, we have chosen to define primes for integers $n > 1$.

In Example 3.1.5 we consider a system of numbers that does not possess unique prime factorization.

Characterizing and Counting Divisors.   With the aid of the Fundamental Theorem of Arithmetic, it is possible to characterize all positive divisors of a positive integer n in terms of the prime divisors of n. For example, if d is a positive divisor of $81 = 3^4$, then $3^4 = dc$ for some integer c, and thus either $d = 1$ or d has one of the following prime factorizations (why?):

$$3^1, 3^2 = 9, 3^3 = 27, \text{ and } 3^4 = 81,$$

i.e., $d = 3^m$ for some integer $0 \leq m \leq 4$. Conversely, if $d = 3^m$ for some integer $0 \leq m \leq 4$, then $d|81$, so we have completely characterized the positive divisors d of 81.

Notice that the number of positive divisors of 81 is five, and, in particular, the number of positive divisors of $n = p^m$, where p is a prime and $m \geq 0$, is $m + 1$. (Why?)

A standard notation used to denote the **number of positive divisors of a positive integer n is $\tau(n)$**, thus, $\tau(p^m) = m + 1$ succinctly records the previous conclusion. This functional notation reflects the fact that the correspondence $n \to \tau(n)$ defines a function from $\mathbf{Z}^+$ into $\mathbf{Z}^+$.

Classroom Problems. Note that $\tau(n) = 2 \Leftrightarrow n$ is prime. (Why?) Characterize those positive integers n such that $\tau(n) = 3$ (see Exercise 5 of Section 3.2).

For an example involving more than one distinct prime factor, let's characterize the positive divisors of $n = 5^3 \cdot 7^8$ and also determine $\tau(n)$. First note, by arguing as done previously, that $d | (5^3 \cdot 7^8)$ if and only if $d = 5^{m_1} \cdot 7^{m_2}$, where $0 \le m_1 \le 3$ and $0 \le m_2 \le 8$. (Why?) Secondly, since $\tau(5^3) = 4$ and $\tau(7^8) = 9$, then by the Fundamental Counting Principle (1.9.2), $\tau(5^3 \cdot 7^8) = 4 \cdot 9 = 36$. ◆

In general terms, for positive integers d and $n = p_1{}^{m_1} p_2{}^{m_2} \cdots p_s{}^{m_s}$, where n is in its canonical prime factorization.

2.6.4 $d | n$ if and only if $d = p_1{}^{n_1} p_2{}^{n_2} \cdots p_s{}^{n_s}$, where $0 \le n_i \le m_i$ for $1 \le i \le s$, thus there are $m_i + 1$ choices for each exponent n_i. Therefore,

2.6.5
$$\tau(n) = \tau(p_1{}^{m_1} p_2{}^{m_2} \cdots p_s{}^{m_s}) = (m_1 + 1)(m_2 + 1) \cdots (m_s + 1)$$
$$= \tau(p_1^{m_1}) \tau(p_2^{m_2}) \cdots \tau(p_s^{m_s}).$$

The following statement is an immediate consequence of 2.6.5 (why?):

2.6.6 If $\gcd(a, b) = 1$, then $\tau(ab) = \tau(a)\tau(b)$.

2.6.7 Question

Is the converse of 2.6.6 true?
Although we could analyze this question now, it is easier to build an argument using some of the ideas of the next section. See Classroom Discussion 2.7.9 for a detailed response to this question.

Classroom Discussion: Squares and Their Prime Factorizations. In Exercises 2.4 k and l, we encountered the statements: if $a | b$, then $a^2 | b^2$ and the converse: if $a^2 | b^2$ then $a | b$. While it was straightforward to establish the first statement, it was not obvious how to find a proof or discover a counterexample for the converse. Let's use the Fundamental Theorem of Arithmetic to help us analyze this problem in a deeper way.

If $a^2 | b^2$, then $b^2 = a^2 c$, for some integer c. One conclusion that we can draw is that $b = a \sqrt{c}$. (Why?) However, it does not necessarily follow that $a | b$, since $\sqrt{c}$ might not be an integer. Of course, if we could prove that c is a perfect square, then we would be done. A point critical to reaching this end is that $a^2 c$ is a perfect square (why?), in fact, this is where the Fundamental Theorem of Arithmetic enters the picture. The next step is to specify a precise definition of *perfect square*.

A positive integer n is called a **square** (or a **perfect square**) if there is a positive integer a so that $n = a^2$ (in geometric terms, n is the area of some square having sides of positive integer length a units). For example, 21,609 is a square since $21{,}609 = 147^2$. The integer a can be found by several different methods. Prime factorization is one approach that gives a and, more importantly, leads to a general characterization of squares. Notice that $21{,}609 = 3^2 \cdot 7^4 = (3^1 \cdot 7^2)^2 = 147^2$.

More generally, if each prime factor of a positive integer n appears an even number of times in the prime factorization of n—i.e., the exponents in the canonical prime factorization are even integers (multiples of 2), as in $n = p_1^{2m_1} p_2^{2m_2} \cdots p_s^{2m_s}$—then $n = (p_1^{m_1} p_2^{m_2} \cdots p_s^{m_s})^2$ is a square. Conversely, if n is a square, then $n = a^2$ for some integer a having canonical prime factorization $a = p_1^{m_1} p_2^{m_2} \cdots p_s^{m_s}$. Therefore,

$$n = a^2 = (p_1^{m_1} p_2^{m_2} \cdots p_s^{m_s})^2 = p_1^{2m_1} p_2^{2m_2} \cdots p_s^{2m_s},$$

so the exponents in the canonical prime factorization of n are even integers (i.e., each prime factor of n appears an even number of times in the prime factorization of n).

2.6.8 Conclusion about Squares

An integer $n > 1$ is a square if and only if the exponents in the canonical prime factorization of n are even integers (i.e., each prime factor in the prime factorization of n appears an even number of times).

The characterization of the prime factorization of squares is useful for determining squares, and can be used to show that an integer is not a square. (Why?) For example, 49,392 is not a square since $49{,}392 = 2^4 \cdot 3^2 \cdot 7^3$.

Let's apply our structural knowledge of the prime factorization of a square to the problem under consideration.

2.6.9 Observation

If $b^2 = a^2 c$, then $a | b$.

Analysis. To analyze the square $a^2 c$ in more depth, it is useful to express it in terms of the canonical prime factorizations of a^2 and c:

$$a^2 c = (p_1^{2m_1} p_2^{2m_2} \cdots p_s^{2m_s})(q_1^{n_1} q_2^{n_2} \cdots q_t^{n_t}).$$

Note that this is a prime factorization of $a^2 c$, but it is not necessarily the canonical prime factorization of this product, even though $p_1 < p_2 < \ldots < p_s$ and $q_1 < q_2 < \ldots < q_t$. (Why?) Thus we cannot immediately apply 2.6.8 to the given prime factorization of $a^2 c$ and then conclude that each n_i is an even integer. However, with a bit more reasoning, the desired conclusion can be achieved.

It is good enough to show that n_1, the exponent of q_1, is even, and then similar reasoning can be applied to the other exponents n_i. Since either $q_1 \neq p_j$ for *all* $1 \leq j \leq s$ or $q_1 = p_j$ for *some* $1 \leq j \leq s$, then in the canonical prime factorization of $a^2 c$, the exponent of q_1 is either equal to n_1 or $2m_j + n_1$. (Why?) Thus, as q_1

must appear an even number of times in the prime factorization of the square a^2c, it follows that n_1 is even or $2m_j + n_1$ is even, which again implies that n_1 is even. (Why?) Finally, we have shown that c is a square, and this completes the analysis.

In the next section, we focus on using the Fundamental Theorem of Arithmetic to deepen our understanding of lcm and gcd. First some exercises.

EXERCISES 2.6

1. Determine whether each of the following integers is prime or composite. If composite, find its prime factorization and keep track of the factorization steps with a factor tree.

   a. 1,274
   b. 7,921
   c. 6,561
   d. 3,229
   e. 11,111
   f. 65,537

2. Two positive integers a and b are said to be **relatively prime** if they have no common prime factor. Show that a and b are relatively prime if and only if $\gcd(a, b) = 1$.

3. Explain how the following statements follow from the FTA:
   a. $d|13^{13} \Leftrightarrow d = 13^m$ for some integer $0 \le m \le 13$, and compute $\tau(13^{13})$.
   b. $d|(2^5 \cdot 3^6 \cdot 17^4) \Leftrightarrow d = 2^{m_1} \cdot 3^{m_2} \cdot 17^{m_3}$, where $0 \le m_1 \le 5, 0 \le m_2 \le 6$, and $0 \le m_3 \le 4$. Also, calculate $\tau(2^5 \cdot 3^6 \cdot 17^4)$.
   c. If $11^m = 23^n$, then $m = n = 0$.
   d. If at least one of the non-negative integers m_1, m_2, n_1, n_2, n_3 is positive, then $21^{m_1} \cdot 29^{m_2} \ne 7^{n_1} \cdot 19^{n_2} \cdot 23^{n_3}$.

4. In each part, indicate whether the given statement concerning integers is true or false. If it is true, provide a sound argument demonstrating the statement's truth or indicate how the statement follows from some other known statement(s). If it is false, provide a concrete counterexample with a complete explanation of why your example shows the statement is false.
   a. If $5|7a$, then $5|a$.
   b. If $3|a^4b^6$, then $3|a$ or $3|b$.
   c. If $6|25a$, then $6|a$.
   d. If $6|ab$ and 6 does not divide a, then $6|b$.
   e. $45|637^{25}$.
   f. If $15|24a$, then $5|a$.
   g. If $3|(a^2 - 9)$, then $3|a$.
   h. If $8|(24a + 35b)$, then $8|b$.

5. A positive integer $n > 1$ is an mth **power** if there is a positive integer a such that $n = a^m$. Show that an integer $n > 1$ is an mth power if and only if each of its distinct prime factors occur in multiples of m.

6. Prove the general statement: If $a^n|b^n$, then $a|b$, where n is any positive integer.

2.7 REVISITING THE LCM AND GCD

The Fundamental Theorem of Arithmetic (FTA) is a powerful tool for studying the arithmetic and algebra of the integers. In this section, we explore the FTA's utility in questions concerning the lcm and gcd of integers. Let's revisit some of the classroom problems we first encountered in Section 2.2 and see how we can employ the FTA in these particular problems.

In the Ferris wheel problem, we determined, by a listing method, that lcm(30, 50) = 150. What other methods could we use to find the correct answer?

Characterizing Common Multiples

Analysis. First note that if c is *any* positive common multiple of 30 and 50 (i.e., c is a positive integer such that $30|c$ and $50|c$), then we can write $c = 30x = 2 \cdot 3 \cdot 5x$ and $c = 50y = 2 \cdot 5^2 y$, for some integers x and y. So by the FTA, c must contain the factors 2, 3, and 5^2 (why?), thus, c can be expressed as $c = 2 \cdot 3 \cdot 5^2 z = 150z$, for some integer z. Therefore, any positive common multiple c of 30 and 50 is also a multiple of 150, so the smallest common multiple of 30 and 50 must be 150. (Why?) Notice that we have actually shown that lcm(30, 50)|c (i.e., c is a multiple of lcm(30, 50)), which is considerably stronger than lcm(30, 50) $\leq c$.

Let's conduct a similar analysis for the cicadas problem, where we previously concluded, by the listing technique, that lcm(13, 17) = 221.

Analysis. As in the previous analysis, if c is *any* positive common multiple of 13 and 17, then we can express $c = 13x$ and $c = 17y$, for some integers x and y. Using the FTA, it follows that $c = 13 \cdot 17z = 221z$ for some integer z, which means that any positive common multiple of 13 and 17 is also a multiple of 221. Consequently, lcm(13, 17) = 221. (Why?) As in the previous example, our reasoning has led to the stronger conclusion that lcm(13, 17)|c (rather than just lcm(13, 17) $\leq c$), where c is any common multiple of 13 and 17.

The previous considerations give rise to the following natural question.

Question. If c is any common positive multiple of positive integers a and b, does it always follow that c is a multiple of lcm(a, b), i.e., lcm(a, b)|c?

To investigate this question, let's abstract the arguments used in the previous problems to determine a general scheme for finding lcm(a, b) in terms of the prime factorizations of a and b.

Computing lcm(*a*, *b*) Using Unique Prime Factorization. Using the previous examples as a guide we can start by assuming that c is any common multiple of given positive integers a and b. The next stage in the numerical examples was to write c in terms of the prime factors of a and b, and then use FTA to express c as a multiple of the product of the maximal prime powers present in the respective prime factorizations of a and b.

For example, when $a = 2 \cdot 3 \cdot 5$ and $b = 2 \cdot 5^2$, we saw that $c = 2^1 \cdot 3^1 \cdot 5^2 z$ for some integer z. To establish the analogue of this step in the general setting, we introduce a slight modification of the canonical form factorizations for a and b (called the **modified canonical form**). Although a and b may have different prime factors, it is always possible to write a and b in terms of all of the distinct prime factors from both a and b. In particular, if

$$a = 2^3 \cdot 5^2 \cdot 11^4 \cdot 17^1 \text{ and } b = 3^1 \cdot 5^3 \cdot 7^2 \cdot 17^3 \cdot 29^1,$$

84 Chapter 2 Arithmetic and Algebra of the Integers

then by introducing 0 exponents, one has the **modified canonical** form

$$a = 2^3 \cdot 3^0 \cdot 5^2 \cdot 7^0 \cdot 11^4 \cdot 17^1 \cdot 29^0 \text{ and}$$
$$b = 2^0 \cdot 3^1 \cdot 5^3 \cdot 7^2 \cdot 11^0 \cdot 17^3 \cdot 29^1.$$

Continuing with the discussion, write

$$a = p_1^{m_1} p_2^{m_2} \cdots p_s^{m_s} \text{ and } b = p_1^{n_1} p_2^{n_2} \cdots p_s^{n_s},$$

where $p_1 < p_2 < \ldots < p_s$ are primes and m_i, n_i are non-negative integers, thus,

$$c = ax = p_1^{m_1} p_2^{m_2} \cdots p_s^{m_s} x \text{ and } c = by = p_1^{n_1} p_2^{n_2} \cdots p_s^{n_s} y$$

(c is a common multiple of a and b). Applying the FTA gives

$$c = p_1^{\max\{m_1,n_1\}} p_2^{\max\{m_2,n_2\}} \cdots p_s^{\max\{m_s,n_s\}} z,$$

for some integer z, therefore,

2.7.1 $$\text{lcm}(a, b) = p_1^{\max\{m_1,n_1\}} p_2^{\max\{m_2,n_2\}} \cdots p_s^{\max\{m_s,n_s\}}.$$

Finally, notice that we have also demonstrated that:

2.7.2 Any common multiple c of a and b is always a multiple of lcm(a, b)

Up to this point, all we knew was that $\text{lcm}(a, b) \leq c$ for any common multiple c of a and b, but now we know the stronger statement $\text{lcm}(a, b) | c$.

Applying formula 2.7.1 to the integers $a = 2^3 \cdot 3^0 \cdot 5^2 \cdot 7^0 \cdot 11^4 \cdot 17^1 \cdot 29^0$ and $b = 2^0 \cdot 3^1 \cdot 5^3 \cdot 7^2 \cdot 11^0 \cdot 17^3 \cdot 29^1$ previously listed gives

$$\text{lcm}(a, b) = 2^3 \cdot 3^1 \cdot 5^3 \cdot 7^2 \cdot 11^4 \cdot 17^3 \cdot 29^1.$$

Observe that Definition 2.5.2 of lcm can now be stated in an equivalent form.

2.7.3 Reformulating the Least Common Multiple Definition

Let a and b be positive integers. Then, $\text{lcm}(a, b) = m$ provided:

1. m is a positive integer;
2. $a|m$ and $b|m$;
3. If $a|c$ and $b|c$, where c is a positive integer, then $m|c$.

Now that we have developed a formula for the lcm of two positive integers in terms of their prime factorizations, let's conduct a similar analysis for the gcd of two positive integers. Recall from the Planning a Picnic problem in Section 2.2 that we concluded, by a listing method, that $\gcd(90, 120) = 30$.

Characterizing Common Divisors: Analysis. Notice that if g is *any* common divisor of 90 and 120, then $90 = 2^1 \cdot 3^2 \cdot 5^1 = gv$ and $120 = 2^3 \cdot 3^1 \cdot 5^1 = gw$, for some integers v and w. By the FTA, the only possibilities for g are: $1, 2, 3, 5, 2 \cdot 3, 3 \cdot 5, 2 \cdot 3 \cdot 5$ (why?), i.e., $g = 2^{m_1} \cdot 3^{m_2} \cdot 5^{m_3}$, where $0 \le m_1 \le 1, 0 \le m_2 \le 1$, and $0 \le m_3 \le 1$. The largest divisor from this set of possible divisors is $2^1 \cdot 3^1 \cdot 5^1$, so $\gcd(90, 120) = 30$. It is interesting to note that any divisor g of 90 and 120 is also a divisor of $\gcd(90, 120)$. We will show that this phenomenon is true in general.

Before moving to a general discussion, let's consider another example: $\gcd(378, 4{,}851)$.

Analysis. As in the prior example, if g is *any* common divisor of 378 and 4,851, then $378 = 2^1 \cdot 3^3 \cdot 7^1 = gv$ and $4{,}851 = 3^2 \cdot 7^2 \cdot 11^1 = gw$, for some integers v and w. We know by the FTA that $g = 2^{m_1} \cdot 3^{m_2} \cdot 7^{m_3} \cdot 11^{m_4}$, where $m_1 = 0, 0 \le m_2 \le 2$, $0 \le m_3 \le 1$, and $m_4 = 0$ (why?), so the greatest of the common divisor of a and b is $2^0 \cdot 3^2 \cdot 7^1 \cdot 11^0$, i.e., $\gcd(378, 4{,}851) = 2^0 \cdot 3^2 \cdot 7^1 \cdot 11^0$.

The stage is now set to develop a general formula for determining $\gcd(a, b)$ in terms of the prime factorizations of a and b.

Computing $\gcd(a, b)$ Using Unique Prime Factorization. The examples we have just considered, in combination with the formula development for lcm, will guide our gcd work.

For positive integers a and b, write the modified canonical form

$$a = p_1^{m_1} p_2^{m_2} \cdots p_s^{m_s} \text{ and } b = p_1^{n_1} p_2^{n_2} \cdots p_s^{n_s},$$

where $p_1 < p_2 < \cdots < p_s$ are primes and m_i, n_i are non-negative integers. If g is any common divisor of a and b, then

$$a = p_1^{m_1} p_2^{m_2} \cdots p_s^{m_s} = gv \text{ and } b = p_1^{n_1} p_2^{n_2} \cdots p_s^{n_s} = gw,$$

for some integers v and w. Thus, by employing the FTA, we can conclude that

$$g = p_1^{r_1} p_2^{r_2} \cdots p_s^{r_s},$$

where $0 \le r_1 \le \min\{m_1, n_1\}, 0 \le r_2 \le \min\{m_2, n_2\}, \ldots, 0 \le r_s \le \min\{m_s, n_s\}$. Therefore,

2.7.4
$$\gcd(a, b) = p_1^{\min\{m_1, n_1\}} p_2^{\min\{m_2, n_2\}} \cdots p_s^{\min\{m_s, n_s\}}.$$

An important consequence of our analysis is:

2.7.5 **Any common divisor g of a and b must also be a divisor of $\gcd(a, b)$**

The greatest common divisor definition 2.5.3, in combination with the previous comments, lead to the next equivalent definition. (Why is it equivalent?)

2.7.6 **Reformulating the Greatest Common Divisor Definition**

Let a and b be positive integers. Then, $\gcd(a, b) = d$ provided:

1. d is a positive integer;

2. $d|a$ and $d|b$;
3. if $g|a$ and $g|b$, where g is a positive integer, then $g|d$.

Definitions 2.7.3 and 2.7.6 do not rely on the order relation $\leq$ and are meaningful in certain algebraic structures that do not possess order relations among their elements.

2.7.7 Classroom Discussion

In the "Looking at Locust Cycles" problem 4.2 B from *Connected Mathematics*, you were asked to determine how many years it would take for the 12-year, 14-year, and 16-year locusts to appear again in the same year. In mathematical terms, the problem amounts to finding lcm(12, 14, 16).

Although our definitions of lcm and gcd were formulated in terms of two positive integers, it is easy to generalize those definitions to several positive integers. For example, given positive integers a, b, and c, then lcm$(a, b, c) = m$ provided:

1. m is a positive integer;
2. $a|m$, $b|m$, and $c|m$;
3. if $a|n$, $b|n$, and $c|n$ (where n is a positive integer), then $m|n$.

Using the same ideas as in the two positive integer cases, lcm(a, b, c) and gcd(a, b, c) can be analogously expressed in terms of the prime factors of a, b, and c. For example, when

$$a = 12 = 2^2 \cdot 3^1 \cdot 7^0, b = 14 = 2^1 \cdot 3^0 \cdot 7^1, \text{ and } c = 16 = 2^4 \cdot 3^0 \cdot 7^0,$$

$$\text{lcm}(12, 14, 16) = 2^{\max\{2,1,4\}} \cdot 3^{\max\{1,0,0\}} \cdot 7^{\max\{0,1,0\}} = 2^4 \cdot 3^1 \cdot 7^1 = 336. \quad \blacklozenge$$

2.7.8 Classroom Problems

1. Extend definitions 2.7.3 and 2.7.6 to define lcm$(a_1, \ldots, a_n)$ and gcd$(a_1, \ldots, a_n)$ for positive integers $a_1, \ldots, a_n$, and express both quantities in terms of the prime factors of $a_1, \ldots, a_n$.
2. **a.** Show that lcm$(a, b, c) = $ lcm(lcm$(a, b), c) = $ lcm$(a,$ lcm$(b, c))$.
   b. Show that gcd$(a, b, c) = $ gcd(gcd$(a, b), c) = $ gcd$(a,$ gcd$(b, c))$.

Let's tackle part of 2a and show that lcm$(a, b, c) = $ lcm(lcm$(a, b), c)$ by using the prime factor representation of the appropriate quantities. The remaining parts are handled in a similar fashion.

First, represent a, b, and c in modified canonical form:

$$a = p_1^{m_1} p_2^{m_2} \cdots p_s^{m_s}, b = p_1^{n_1} p_2^{n_2} \cdots p_s^{n_s}, c = p_1^{k_1} p_2^{k_2} \cdots p_s^{k_s}.$$

The desired equality follows from an application of the appropriate prime factor representations of lcm. Namely,

$$\text{lcm}(\text{lcm}(a, b), c) = \text{lcm}(p_1^{\max\{m_1, n_1\}} p_2^{\max\{m_2, n_2\}} \cdots p_s^{\max\{m_s, n_s\}}, c)$$

$$p_1^{\max\{\max\{m_1,n_1\},k_1\}} p_2^{\max\{\max\{m_2,n_2\},k_2\}} \cdots p_s^{\max\{\max\{m_s,n_s\},k_s\}}$$
$$p_1^{\max\{m_1,n_1,k_1\}} p_2^{\max\{m_2,n_2,k_2\}} \cdots p_s^{\max\{m_s,n_s,k_s\}} = \text{lcm}(a,b,c). \;\blacklozenge$$

Our work thus far on lcm and gcd has illuminated the significance of the Fundamental Theorem of Arithmetic in questions concerning multiples and divisors. The previous formulae for lcm(a, b) and gcd(a, b) depend upon finding the prime factorizations of both a and b, and this makes their application impractical for use with "large" numbers. On the other hand, we soon see that this prime factorization method is quite useful for general deductions involving lcm and gcd.

Classroom Connection. Before considering some applications, let's read through Heidi's prime factorization method for finding lcm and gcd as presented in *Prime Time* from *Connected Mathematics* (Figure 2.7.1 on p. 88), then below we discuss the substance of her method.

Our comprehensive discussion on lcm and gcd shows that Heidi's methods always lead to correct answers and precisely details the mathematics underlying her methods. This fundamental understanding supports both computation and theory and will aid us in further investigations.

Question. How would you develop and present Heidi's method to middle-grade students so that they arrive at a deeper understanding of the essential mathematics?

2.7.9 Classroom Discussion

The ideas of this section help us answer Question 2.6.7 in an efficient manner. In particular, we show that

if $\tau(ab) = \tau(a)\tau(b)$, then gcd($a$, b) = 1 (a and b are positive integers).

Analysis. We'll establish the converse indirectly by assuming that $\tau(ab) = \tau(a)\tau(b)$ and gcd(a, b) = $d > 1$ and then deriving a contradiction from these assumptions. Start by writing the prime factorizations of a and b in their modified canonical form:

$$a = p_1^{m_1} p_2^{m_2} \cdots p_s^{m_s} \text{ and } b = p_1^{n_1} p_2^{n_2} \cdots p_s^{n_s}.$$

We know from 2.6.5 that

$$\tau(a)\tau(b) = [(m_1 + 1) \cdots (m_s + 1)][(n_1 + 1) \cdots (n_s + 1)]$$
$$= (m_1 + 1)(n_1 + 1) \cdots (m_s + 1)(n_s + 1)$$
$$= (m_1 n_1 + m_1 + n_1 + 1) \cdots (m_s n_s + m_s + n_s + 1), \text{ and}$$
$$\tau(ab) = (m_1 + n_1 + 1) \cdots (m_s + n_s + 1).$$

Since gcd(a, b) > 1, there exists $m_i > 0$ and $n_i > 0$ for some $1 \leq i \leq s$ (why?), so $\tau(ab) < \tau(a)\tau(b)$. (Why?) This contradiction completes the proof. $\blacklozenge$

Using Prime Factorizations

In Investigation 4, you found common multiples and common factors of numbers by comparing lists of their multiples and factors. In this problem, you will explore a method for finding the *greatest common factor* and the *least common multiple* of two numbers by using their prime factorizations.

Heidi says she can find the greatest common factor and the least common multiple of a pair of numbers by using their prime factorizations. The **prime factorization** of a number is a string of factors made up only of primes. Below are the prime factorizations of 24 and 60.

$$24 = 2 \times 2 \times 2 \times 3 \qquad 60 = 2 \times 2 \times 3 \times 5$$

Heidi claims that the greatest common factor of two numbers is the product of the longest string of prime factors that the numbers have in common. For example, the longest string of factors that 24 and 60 have in common is $2 \times 2 \times 3$.

$$24 = 2 \times \underline{2 \times 2 \times 3} \qquad 60 = \underline{2 \times 2 \times 3} \times 5$$

According to Heidi's method, the greatest common factor of 24 and 60 is $2 \times 2 \times 3$, or 12.

Heidi claims that the least common multiple of two numbers is the product of the shortest string that contains the prime factorizations of both numbers. The shortest string that contains the prime factorizations of 24 *and* 60 is $2 \times 2 \times 2 \times 3 \times 5$.

Contains the prime factorization of 24   Contains the prime factorization of 60
$\underline{2 \times 2 \times 2 \times 3} \times 5$ $2 \times \underline{2 \times 2 \times 3 \times 5}$

According to Heidi's method, the least common multiple of 24 and 60 is $2 \times 2 \times 2 \times 3 \times 5$, or 120.

Problem 5.3

A. Try using Heidi's methods to find the greatest common factor and least common multiple of 48 and 72 and of 30 and 54.

B. Are Heidi's methods correct? Explain your thinking. If you think Heidi is wrong, revise her methods so they are correct.

---

50 Prime Time

Reproduced from page 50 of *Prime Time* in *Connected Mathematics*.

FIGURE 2.7.1

2.8 RELATIONS AND RESULTS CONCERNING LCM AND GCD

In the previous section, we saw how the prime factorizations of two positive integers can be used to determine the lcm and gcd of those integers. Our work culminated in the following formulae:

$$\mathbf{lcm}(a, b) = p_1^{\max\{m_1,n_1\}} p_2^{\max\{m_2,n_2\}} \cdots p_s^{\max\{m_s,n_s\}}$$

$$\mathbf{gcd}(a, b) = p_1^{\min\{m_1,n_1\}} p_2^{\min\{m_2,n_2\}} \cdots p_s^{\min\{m_s,n_s\}}.$$

An immediate bonus derived from these representations was that $\mathrm{lcm}(a,b) | c$ for any common multiple c of a and b, and $g | \gcd(a,b)$ for any common divisor g of a and b.

Are there further interesting facets of lcm and gcd that can be uncovered with the employment of these formulae? Answers to this question further deepen our understanding of lcm and gcd.

Linking lcm and gcd. Recall that Heidi computed $\mathrm{lcm}(24, 60)$ and $\gcd(24, 60)$ by using the prime factorizations of 24 and 60. In our notation, she concluded that

$$24 = 2^3 \cdot 3^1 \cdot 5^0 \text{ and } 60 = 2^2 \cdot 3^1 \cdot 5^1,$$

$$\mathrm{lcm}(24, 60) = 2^{\max\{3,2\}} \cdot 3^{\max\{1,1\}} \cdot 5^{\max\{0,1\}},$$

$$\gcd(24, 60) = 2^{\min\{3,2\}} \cdot 3^{\min\{1,1\}} \cdot 5^{\min\{0,1\}}.$$

We have chosen not to simplify these expressions, since we intend to generalize this computation to $\mathrm{lcm}(a, b)$ and $\gcd(a, b)$ for arbitrary positive integers a and b (sometimes simplification tends to obscure the general nature of the computation). Notice that

$$\mathrm{lcm}(24, 60) \cdot \gcd(24, 60)$$
$$= (2^{\max\{3,2\}} \cdot 3^{\max\{1,1\}} \cdot 5^{\max\{0,1\}}) \cdot (2^{\min\{3,2\}} \cdot 3^{\min\{1,1\}} \cdot 5^{\min\{0,1\}})$$
$$= 2^{\max\{3,2\}+\min\{3,2\}} \cdot 3^{\max\{1,1\}+\min\{1,1\}} \cdot 5^{\max\{0,1\}+\min\{0,1\}}$$
$$= 2^5 \cdot 3^2 \cdot 5^1 = 24 \cdot 60.$$

It is now easy to construct a general argument modeled after this example. The prime factorizations of positive integers a and b in their modified canonical form and the $\mathrm{lcm}(a, b)$ and $\gcd(a, b)$ in terms of these forms gives rise to the following equalities:

$$ab = (p_1^{m_1} p_2^{m_2} \cdots p_s^{m_s})(p_1^{n_1} p_2^{n_2} \cdots p_s^{n_s})$$
$$= p_1^{m_1+n_1} p_2^{m_2+n_2} \cdots p_s^{m_s+n_s}, \text{ and}$$

$$\mathrm{lcm}(a, b) \cdot \gcd(a, b) = (p_1^{\max\{m_1,n_1\}} p_2^{\max\{m_2,n_2\}} \cdots p_s^{\max\{m_s,n_s\}})$$
$$\cdot (p_1^{\min\{m_1,n_1\}} p_2^{\min\{m_2,n_2\}} \cdots p_s^{\min\{m_s,n_s\}})$$

$$= p_1^{\max\{m_1,n_1\}+\min\{m_1,n_1\}} p_2^{\max\{m_2,n_2\}+\min\{m_2,n_2\}}$$
$$\cdots p_s^{\max\{m_s,n_s\}+\min\{m_s,n_s\}}.$$

Since $\max\{m_i, n_i\} + \min\{m_i, n_i\} = m_i + n_i$ for each $1 \leq i \leq s$ (why?), then the elegant and useful relationship between $\text{lcm}(a, b)$ and $\gcd(a, b)$ is expressed as:

2.8.1
$$\text{lcm}(a, b) \cdot \gcd(a, b) = ab.$$

An important upshot of this relationship is that computing either $\text{lcm}(a, b)$ or $\gcd(a, b)$ leads to the computation of the other through simple division. Another useful consequence of this linkage occurs when a and b are relatively prime (have no prime factors in common or equivalently $\gcd(a, b) = 1$). Applying relationship 2.8.1, we see that

2.8.2
$$\text{lcm}(a, b) = ab \Leftrightarrow \gcd(a, b) = 1.$$

Factoring a Common Divisor out of the lcm and the gcd. The following computation leads to the correct answer for $\text{lcm}(24, 60)$, but is this process valid in general?

$$\text{lcm}(24, 60) = \text{lcm}(12 \cdot 2, 12 \cdot 5) = 12 \cdot \text{lcm}(2, 5) = 12 \cdot 10 = 120.$$

Let's formulate and establish validity for a general statement that has the previous calculation as a consequence.

Observation. If a, b, c are positive integers, then

2.8.3
$$\text{lcm}(ca, cb) = c \cdot \text{lcm}(a, b).$$

Justification. As in previous analyses, express $\text{lcm}(ca, cb)$ and $c \cdot \text{lcm}(a, b)$ in terms of the prime factorizations of $a, b,$ and c in their modified canonical forms:

$$a = p_1^{m_1} p_2^{m_2} \cdots p_s^{m_s}, b = p_1^{n_1} p_2^{n_2} \cdots p_s^{n_s}, c = p_1^{v_1} p_2^{v_2} \cdots p_s^{v_s};$$

$$c \cdot \text{lcm}(a,b) = (p_1^{v_1} p_2^{v_2} \cdots p_s^{v_s})(p_1^{\max\{m_1,n_1\}} p_2^{\max\{m_2,n_2\}} \cdots p_s^{\max\{m_s,n_s\}})$$
$$= p_1^{v_1+\max\{m_1,n_1\}} p_2^{v_2+\max\{m_2,n_2\}} \cdots p_s^{v_s+\max\{m_s,n_s\}}; \text{ and}$$

$$\text{lcm}(ca, cb) = p_1^{\max\{m_1+v_1, n_1+v_1\}} p_2^{\max\{m_2+v_2, n_2+v_2\}} \cdots p_s^{\max\{m_s+v_s, n_s+v_s\}}.$$

Note that
$$v_i + \max\{m_i, n_i\} = \max\{m_i + v_i, n_i + v_i\} \text{ for each } 1 \leq i \leq s,$$
since
$$m_i + v_i \geq n_i + v_i \Leftrightarrow m_i \geq n_i \text{ and likewise,}$$
$$m_i + v_i \leq n_i + v_i \Leftrightarrow m_i \leq n_i \text{ (explain).}$$

Therefore, $\text{lcm}(ca, cb) = c \cdot \text{lcm}(a, b)$.

The argument just given was constructed using the prime factorization formula for lcm. It is sometimes useful to develop alternate proofs that can possibly be utilized in other contexts. Let's consider such an alternate proof, which primarily relies on the definition of lcm (2.7.3). The nature of this argument is somewhat more sophisticated than the previous one, but the benefit is (as remarked after definition 2.7.6) that it can be used in certain algebraic structures that do not enjoy order relations between their elements.

Alternate Argument. Let $\mathrm{lcm}(ca, cb) = y$, so $y = cau = cbv$ for some positive integers u and v. Set $x = au$, and note that $x = au = bv$. To complete the proof, we show that $x = \mathrm{lcm}(a, b)$. (Why?) Since x is a common multiple of a and b, we must show (by definition 2.7.3) that any common multiple z of a and b is divisible by x. Observe that $ca|cz$ and $cb|cz$ (why?), so $y|cz$ (why?). Thus $cz = yw = cauw = cxw$ for some positive integer w, and hence $z = xw$ (i.e., $x|z$). Therefore, by definition 2.7.3, $x = \mathrm{lcm}(a, b)$, and finally, $\mathrm{lcm}(ca, cb) = c \cdot \mathrm{lcm}(a, b)$.

Using the prime factorization representation of $\gcd(a, b)$, and arguing as in the lcm case, it follows that

2.8.4
$$\gcd(ca, cb) = c \cdot \gcd(a, b).$$

The proof of this, and the analogous alternate argument, is left for the exercises. A consequence of 2.8.4, which is often useful in reducing other arguments to the relatively prime case, is as follows:

2.8.5   **If $\gcd(a, b) = d$, then $\gcd(a/d, b/d) = 1$.**

Justification. Write $a = dx$ and $b = dy$ for some integers x and y and then apply 2.8.4 (write out the details).

2.8.6 Classroom Discussion

Our work thus far has vividly illustrated the value of the Fundamental Theorem of Arithmetic and its consequences in questions concerning lcm and gcd. Even though the unique prime factorization for large integers is often very difficult to determine, and thus not always practical as a tool to find lcm and gcd, it still provides an efficient means to analyze various questions concerning integers. In the next chapter, we explore an effective method (that does not rely on prime factorization) to find the lcm and gcd of positive integers. ◆

Let's use what we have learned to help investigate some interesting mathematical applications.

Classroom Problems.
Let $a \geq b$ be positive integers and $\gcd(a, b) = 1$.
Determine $g = \gcd(a + b, a - b)$.
A good first step is to look at some examples.

EXAMPLE

a	b	$\gcd(a, b) = 1$	$\gcd(a + b, a - b) = g$
3	2	1	$\gcd(5, 1) = 1$
5	3	1	$\gcd(8, 2) = 2$
8	3	1	$\gcd(11, 5) = 1$
15	11	1	$\gcd(26, 4) = 2$
275	189	1	$\gcd(464, 86) = 2$

Create some of your own examples and tabulate the results as in the table. Using the data collected, it is now possible to formulate an initial conjecture, and if necessary, modifications can be made later.

Conjecture. If $\gcd(a, b) = 1$, where $a \geq b$, then $g = \gcd(a + b, a - b) = 1$ or 2.

Analysis. In an effort to validate this conjecture, we attempt to construct a proof of the proposed statement. In several of the previous deductions, we employed a prime factorization approach to reach the desired conclusions. However, this method may not always be directly applicable in the study of related problems. In particular, the prime factorizations of a and b are not, in general, related to the prime factorizations of $a + b$ and $a - b$ (e.g., $a = 5$, $b = 3$ and $a + b = 8 = 2^3$, $a - b = 2$), so it unnecessarily complicates matters to use the prime factorizations of a and b to determine g. Of course, this does not rule out utilizing valuable deductions made from employing prime factorizations.

Where, then, is a good place to start the proof? Since we are trying to show that $g = 1$ or 2, we write down what we know about g. Since $g|(a + b)$ and $g|(a - b)$, it follows that that $g|2a$ and $g|2b$ (why?), so $g|\gcd(2a, 2b)$ (why?). Furthermore, formula 2.8.4 gives $\gcd(2a, 2b) = 2 \cdot \gcd(a, b)$, so $g|2$. (Why?) Therefore, $g = 1$ or 2, and the argument is complete. ◆

What is a likely generalization of the previous conjecture?

Generalized Conjecture. If $\gcd(a, b) = d$, where $a \geq b$, then $g = \gcd(a + b, a - b) = d$ or $2d$.

Analysis. As before, examining several examples is always a good first step. Secondly, it is sensible to try to derive this new conjecture from the original one, rather than retracing the steps of the original argument.

An application of result 2.8.5 allows us to reduce to the relatively prime case. In particular, if $\gcd(a, b) = d$, with $a \geq b$, then $\gcd(a/d, b/d) = 1$. Thus,

$$\gcd(a/d + b/d, a/d - b/d) = 1 \text{ or } 2 \text{ (why?)},$$

so $d \cdot \gcd(a/d + b/d, a/d - b/d) = d$ or $2d$.

Therefore, $\gcd(a + b, a - b) = d$ or $2d$ (why?), and the generalized conjecture is verified.

Our next application concerns the Fibonacci sequence.

Adjacent Terms of the Fibonacci Sequence. Recall that the Fibonacci sequence is defined recursively by: $F_0 = F_1 = 1$ and $F_n = F_{n-1} + F_{n-2}$, where $n \geq 2$. In Section 1.11, we studied many interesting properties of this sequence, and there is still much to learn about its nature. For example, is it true that the adjacent members of the Fibonacci sequence 1, 1, 2, 3, 5, 8, 13, 21, 34, 55, 89, 144, ... are relatively prime? Let's construct an inductive proof showing that this is true.

2.8.7 Observation

If F_n and F_{n+1} are successive terms of the Fibonacci sequence, then $\gcd(F_n, F_{n+1}) = 1$.

Proof **(by Induction).** Since the statement is obviously true when $n = 0$ (basis step), assume that it is true for $n = k$, i.e., $\gcd(F_k, F_{k+1}) = 1$ (inductive hypothesis). We wish to show that $\gcd(F_{k+1}, F_{k+2}) = 1$, then by induction, the observation will be true for all $n \geq 0$.

Set $\gcd(F_{k+1}, F_{k+2}) = d$, and note that $d | F_k$, since $F_k = F_{k+2} - F_{k+1}$. Therefore, $d | \gcd(F_k, F_{k+1})$ (why?), so $d = 1$. This completes the proof. □

Question. What can be said about the gcd of any two members of the Fibonacci sequence? Try several examples and formulate a conjecture. We further comment on this question in Remark 3.2.3.

We end this section with an interesting observation concerning certain polynomials and rational roots.

2.8.8 Observation

Let $f(x) = x^n + a_{n-1}x^{n-1} + \ldots + a_1 x^1 + a_0$ be a polynomial with integer coefficients. If $f(u) = 0$, where $u = a/b$ ($a, b \neq 0$ integers), then u is an integer.

Analysis. There exist integers c and d so that $a/b = c/d$, where $\gcd(c, d) = 1$. (Why?) Since $(c/d)^n + a_{n-1}(c/d)^{n-1} + \ldots + a_1(c/d)^1 + a_0 = 0$, it follows that
$$c^n + d a_{n-1} c^{n-1} + \ldots + d^{n-1} a_1 c + d^n a_0 = 0 \text{ and}$$
$$c^n = -d(a_{n-1} c^{n-1} + \ldots + d^{n-2} a_1 c + d^{n-1} a_0).$$

Thus, $d | c^n$, so $d = \pm 1$, since $\gcd(c^n, d) = 1$. (Why?) Finally, $u = a/b = \pm c$, and we have shown that u is an integer.

EXERCISES 2.8

1. Find $\text{lcm}(a, b)$ and $\gcd(a, b)$ by using the prime factorization method (represent your answers in modified canonical form).
   a. $a = 2^1 \cdot 3^2 \cdot 5^3 \cdot 7^2 \quad b = 2^4 \cdot 5^2 \cdot 7^3 \cdot 13^2$
   b. $a = 1{,}232 \quad\quad\quad\quad b = 4{,}840$
   c. $a = 1{,}984 \quad\quad\quad\quad b = 2{,}003$
   d. $a = 1{,}001 \quad\quad\quad\quad b = 100{,}001$
   e. $a = 17{,}296 \quad\quad\quad b = 18{,}416$

2. If a is any positive integer, show that
   a. $\gcd(a, a+1) = 1$ (consecutive positive integers are relatively prime)
   b. $\gcd(3a+7, 3a+8) = 1$
   c. $\gcd(30a+14, 10a+4) = 2$
   d. $\gcd(14a+11, 4a+3) = 1$
   e. $\gcd(12a+4, 28a+8) = 4$
3. a. Do there exist positive integers a and b so that $\gcd(a,b) = 4$ and $\text{lcm}(a,b) = 6$? Explain your answer.
   b. Do there exist positive integers a and b so that $\gcd(a,b) = 4$ and $\text{lcm}(a,b) = 8$? Explain your answer.
   c. Given positive integers d and m, prove that $d|m \Leftrightarrow$ there exist positive integers a and b so that $\gcd(a,b) = d$ and $\text{lcm}(a,b) = m$ (cf., 2.5.4).
4. For positive integers a and b, characterize when $\text{lcm}(a,b) = \gcd(a,b)$. Justify your characterization with a detailed argument.
5. a. Given positive integers a and b, show that if $1 = ax + by$ for some integers x and y, then $\gcd(a,b) = 1$.
   b. Can part a be generalized to the following statement: if $d = ax + by$ for some integers x and y, then $\gcd(a,b) = d$? If true, demonstrate it, if false, provide a counterexample.
6. a. For positive integers a and b, show that $\gcd(a,b) = \gcd(a, a+b)$. Direction: Let $d = \gcd(a,b)$. Using definition 2.7.6 show that $d = \gcd(a, a+b)$.
   b. Can part a be generalized to $\gcd(a,b) = \gcd(a, ax+b)$, where x is any positive integer? If this extension is valid, prove it, if it is false, provide a counterexample.
7. Given positive integers a and b, show that if $\gcd(a,b) = d$, then $\gcd(a^2, b^2) = d^2$. Generalize for any positive integer n.
8. Justify each of the following statements for positive integers a, b, and c:
   a. If $\gcd(a,b) = 1$, then $\gcd(a^2, b^2) = 1$. Generalize for any positive integer n.
   b. If $\gcd(a,b) = 1$, then $\gcd(a+b, ab) = 1$. Direction: Let $d = \gcd(a+b, ab)$. Show that $d|a^2$ and $d|b^2$, and then use part a.
   c. If $\gcd(a,b) = 1$ and $\gcd(a,c) = 1$, then $\gcd(a, bc) = 1$.
   d. If $\gcd(a,b) = 1$ and ab is a square, then both a and b are squares.
   e. If $\gcd(a,b) = 1$ and $a|c$ and $b|c$, then $ab|c$.
9. Prove: For positive integers a, b, and c, if $\gcd(a,b) = 1$ and $c|(a+b)$, then $\gcd(c,a) = 1$ and $\gcd(c,b) = 1$. Direction: Let $d_1 = \gcd(c,a)$ and $d_2 = \gcd(c,b)$. Show that d_1 and d_2 are common divisors of a and b.
10. Prove that $\gcd(ca, cb) = c \cdot \gcd(a,b)$, where a, b, and c are positive integers (cf., 2.8.4). Give one proof using prime factorization, and give an alternate proof analogous to the lcm alternate argument presented in this section (cf., 2.8.3, alternate argument).

The Division Algorithm and the Euclidean Algorithm

CHAPTER 3

3.1 MEASURING INTEGER LENGTHS AND THE DIVISION ALGORITHM
3.2 THE EUCLIDEAN ALGORITHM
3.3 APPLICATIONS OF THE REPRESENTATION gcd(a, b) = $ax + by$
3.4 PLACE VALUE
3.5 PRIME THOUGHTS

Thus far we have learned a great deal about the lcm and gcd, but we still do not have an effective means for computing these quantities for large integers. The next discussion helps set the stage for Euclid's algorithm, which is a powerful method for finding gcd (and hence lcm) that does not directly depend upon determining prime factorizations of the integers involved.

3.1 MEASURING INTEGER LENGTHS AND THE DIVISION ALGORITHM

Which integer lengths L inches can be measured with 3-inch and 5-inch unmarked straightedges?

3 inch

5 inch

Certainly, lengths that are positive multiples of 3 inches or 5 inches (inclusive "or") or sums of such multiples can be measured. Expressing this algebraically, lengths $L = 3x + 5y$ inches, where x and y are non-negative integers, can be determined. For example, lengths of 51 inches ($x = 17, y = 0$), 105 inches ($x = 0, y = 21$), and 57 inches ($x = 4, y = 9$) can be laid out. Note that this description does not account for all integer lengths that can be constructed using 3-inch and 5-inch unmarked straightedges. In particular, 1, 2, and 7 are not of this form (x and y are non-negative integers), but it is possible to measure lengths of 1 inch, 2 inches, and 7 inches.

One Inch. Mark 6 inches and then mark 5 inches of those 6 inches: $1 = 3 \cdot 2 + 5 \cdot (-1)$. Notice that 1 inch can also be measured by: $1 = 3 \cdot 7 + 5 \cdot (-4)$. (Can you find other ways to measure 1 inch?)

Two Inches. Mark 5 inches and then mark 3 inches of those 5 inches: $2 = 3 \cdot (-1) + 5 \cdot 1$.

Alternate Measurement. $2 = 3 \cdot 4 + 5 \cdot (-2)$.

Seven inches. Mark 10 inches and then mark 3 inches of those 10 inches: $7 = 3 \cdot (-1) + 5 \cdot 2$.

Alternate Measurement. $7 = 3 \cdot 14 + 5 \cdot (-7)$.

It was not difficult to determine whether some small integer lengths can be measured with 3-inch and 5-inch unmarked straightedges. However, larger integer lengths, such as 9,382 inches, or more arbitrary integer lengths, might be more challenging. In the alternate measurements mentioned earlier, we used the 1-inch measurement to find other ways to measure 2 inches and 7 inches. These were not the most efficient approaches, but they accomplished the task. The equality

$$9{,}382 = 3 \cdot 18{,}764 + 5 \cdot (-9{,}382)$$

demonstrates that a length of 9,382 inches can be measured, and the equality

$$s = 3 \cdot 2s + 5 \cdot (-1)s$$

shows that any integer length s can be measured with 3-inch and 5-inch unmarked straightedges.

Conclusion. All integer lengths of L inches can be measured with 3-inch and 5-inch unmarked straightedges, since 1 inch can be measured that way.

Question. Which integer lengths of L inches can be measured with 14-inch and 25-inch unmarked straightedges? In algebraic terms, for which positive integers L do there exist integers x and y so that

$$L = 14x + 25y?$$

What methods, other than trial and error, are there to solve such equations? Up to this point, we have not developed any procedures for analyzing equations of this type. Our work in this section leads to a simple and efficient method for solving these equations. Using this method, we determined that $1 = 14 \cdot 9 + 25 \cdot (-5)$, and therefore all integer lengths can be measured with a 14-inch and a 25-inch unmarked straightedge. Can you find other solutions to the equation $1 = 14x + 25y$?

Question. Which integer lengths of L inches can be measured with 12-inch and 15-inch unmarked straightedges (i.e., for which positive integers L are there integers x and y so that $L = 12x + 15y$)?

We know that this equality cannot occur for $L = 1$ since 3 is a factor of $12x + 15y$ but not a factor of 1. Therefore, not all integer lengths of L inches can be measured with 12-inch and 15-inch unmarked straightedges, but those that can be measured this way *must* be a multiple of 3. (Why?) Reiterating symbolically, if there exists integers x and y so that $L = 12x + 15y$, then $L = 3z$ for some positive integer z. We have not yet established the converse (state the converse), but this can be accomplished easily since $3 = 12 \cdot (-1) + 15 \cdot 1$, and thus $3z = 12(-z) + 15z$ for any positive integer z.

Conclusion. The integer lengths (in inches) that can be measured with 12-inch and 15-inch unmarked straightedges are $L = 3z$ inches, where z is any positive integer.

Using reasoning similar to that employed in the previous example, it follows that:

3.1.1 General Conclusion.
If a length of L inches can be measured with unmarked straightedges of integer lengths a inches and b inches, then L is a multiple of $d = \gcd(a, b)$. (Why?)

This conclusion can be stated solely as a statement about integers. Namely: If a and b are positive integers and $L = ax + by$ for some integers x and y, then $d = \gcd(a, b)$ is a divisor of L.

We are now prepared to state a general question that includes all the previous ones.

3.1.2 Question (Converse of General Conclusion).
Which integer lengths of L inches that are multiples of $d = \gcd(a, b)$ can be measured with a-inch and b-inch unmarked straightedges, where a and b are positive integers?

Stated exclusively in terms of integers, one has: if a and b are positive integers and L is a positive integer divisible by $d = \gcd(a, b)$, then do there exist integers x and y such that $L = ax + by$?

Our work in the remainder of this section leads to an affirmative answer to this question and provides an algorithm (the Euclidean algorithm) for finding specific solutions. A first step in proceeding to this end involves an analysis of the mathematics underlying long division.

Integer Division. Our studies have clearly illustrated that divisibility is an important concept in the study of the integers, and it plays a significant role in the subsequent chapters. We know that if a nonzero integer a divides an integer b, then $b = qa$ for some integer q. In geometric terms, b falls on a point on the number line that is a multiple of a (for example, assume $a > 0$ and $q = 3$).

If a does not divide b, then b resides (strictly) between some consecutive multiples of a. In either case, this leads to the next question.

Question. Given integers $a > 0$ and b, is it always possible to determine **the largest** multiple of a that is **less than or equal to** b? (The case of $a < 0$ is considered in the exercises.)

A look at the number line helps illustrate this question. Suppose b lies between $3a$ and $4a$ and $b \neq 4a$.

In this example, $3a$ is the largest multiple of a that is less than or equal to b, and the inequality $3a \leq b < 4a$, or equivalently $0 \leq b - 3a < a$, precisely captures this conclusion. (Why?) In summary, b can be written as

$$b = 3a + r, \text{ where } r = b - 3a \text{ and } 0 \leq r < a.$$

Let's consider an example when b is negative, say b lies between $-5a$ and $-6a$ and $b \neq -5a$.

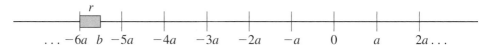

For this example, $-6a$ is the largest multiple of a that is less than or equal to b, and thus $-6a \leq b < -5a$, which is the same as $0 \leq b - (-6a) < a$. As in the other example,

$$b = -6a + r, \text{ where } r = b - (-6a) \text{ and } 0 \leq r < a.$$

With these examples in mind, the previous question can be paraphrased as follows:
  Equivalent questions (explain why these questions are equivalent):

1. Given integers $a > 0$ and b, does there exist an integer q such that $qa \leq b < (q + 1)a$ or equivalently, $0 \leq b - qa < a$?
2. Given integers $a > 0$ and b, do there exist integers q and r so that $b = qa + r$ and $0 \leq r < a$?

The determination of integers q and r so that $b = qa + r$ and $0 \leq r < a$ is called **division of b by a** with **quotient** q and **remainder** r and is denoted by $b \div a$ or $a\overline{)b}$.

Section 3.1   Measuring Integer Lengths and the Division Algorithm

In the first number line example given previously, the quotient was $q = 3$ with remainder $r = b - 3a$, in the second example, the quotient was $q = -6$ with remainder $r = b - (-6a)$.

The geometric reasoning previously employed certainly makes it believable that the equivalent questions have an affirmative answer in general. In fact, using these ideas, we construct a rigorous proof that demonstrates their truth. Part of the proof shows that if such a q and r exist with $b = qa + r$ and $0 \leq r < a$, then they are the only integers satisfying these conditions (which was clear in the geometric models). After the theorem is precisely stated and the proof is complete, we then illustrate how the standard long division procedure for finding q and r utilizes this result.

3.1.3 Theorem: Division Algorithm. If $a > 0$ and b are integers, then there exist unique integers q and r such that $b = qa + r$, where $0 \leq r < a$.

Proof **(Existence).** In this part of the proof, we show that q and r exist with the specified properties. We noted earlier that it suffices to find an integer q so that $0 \leq b - qa < a$, so we prove that some member of the set $R = \{b - xa : x \in \mathbf{Z} \text{ and } b - xa \geq 0\}$ is less than a.

Before we can make meaningful assertions about the elements of R, we must demonstrate that $R \neq \emptyset$. Since $a \geq 1$, then $|b|a \geq |b|$, so $b + |b|a \geq b + |b| \geq 0$. Therefore, by choosing $x = -|b|$, we see that $b - (-|b|)a \in R$. (Why?)

If $0 \in R$, then $0 = b - qa$ for some integer q, so $b = qa$. Thus, in this case, a quotient q and a remainder $r = 0 < a$ have been determined.

If $0 \notin R$, then R is a non-empty set of positive integers, and by the Well-Ordering Principle 2.5.1, there is a least element $r \in R$. Accordingly, there exists an integer q so that $r = b - qa \geq 0$.

To complete this part of the proof, we show that $r < a$. Suppose that this is not the case and that $r \geq a$. Thus, $r - a = (b - qa) - a = b - (q + 1)a \geq 0$, and so $r - a \in R$. Hence, $r - a < r$ since $a > 0$, and this contradicts the fact that r is the least element of R. Therefore, $r < a$, so $b = qa + r$, where $0 < r < a$, which completes the existence part of the proof.

(Uniqueness). Suppose that $b = qa + r = \bar{q}a + \bar{r}$, where $0 \leq r < a$ and $0 \leq \bar{r} < a$. We wish to show that $q = \bar{q}$ and $r = \bar{r}$. Note that $a(q - \bar{q}) = r - \bar{r}$, and since the absolute value of a product is equal to the product of the absolute values (justify this), it follows that

$$a|q - \bar{q}| = |r - \bar{r}|.$$

Moreover, $|r - \bar{r}| < a$, since $0 \leq r < a$ and $0 \leq \bar{r} < a$ (why?), and thus $a|q - \bar{q}| < a$. Hence, $0 \leq |q - \bar{q}| < 1$, which implies that $|q - \bar{q}| = 0$. Therefore, $q = \bar{q}$ and consequently $r = \bar{r}$. (Why?) We have established the proof's existence and uniqueness parts, and the proof is complete.

Classroom Problems.   In each of the following divisions, find the unique quotient and remainder, and represent the division on the number line:

a. $3\overline{)9}$ c. $3\overline{)8}$ e. $8\overline{)3}$ ◆

b. $3\overline{)-9}$ d. $3\overline{)-8}$

Long Division. The standard long division algorithm that we all learned in grade school is an efficient division procedure that utilizes the division algorithm. A specific example illustrates this point. Consider the division of 2,301 by 7:

$$\begin{array}{r} 328 \\ 7\overline{)2{,}301} \\ -21 \\ \hline 20 \\ -14 \\ \hline 61 \\ -56 \\ \hline 5 \end{array} \qquad 2{,}301 = 328 \cdot 7 + 5$$

Long division steps	Division algorithm steps
1. How many times does 7 go into 23? This question is a short form of the question, How many hundreds of 7s are there in 2,301? $2{,}301 - 300 \cdot 7 = 201$.	$23 = 3 \cdot 7 + 2$
2. How many times does 7 go into 20? This question is a short form of the question, How many tens of 7s are there in 201? $201 - 20 \cdot 7 = 61$.	$20 = 2 \cdot 7 + 6$
3. How many times does 7 go into 61? $61 - 8 \cdot 7 = 5$.	$61 = 8 \cdot 7 + 5$

Conclusion: $2{,}301 = (300 + 20 + 8) \cdot 7 + 5$.

Question. How can the long division procedure be applied when $a > 0$ and $b < 0$?

For example, how would you divide $-2{,}301$ by 7 using long division? One approach that leads to a correct solution is to first divide 2,301 by 7 using long division and obtain (as previously) $2{,}301 = 328 \cdot 7 + 5$. Multiplying both sides of the equation by -1 results in

$$-2{,}301 = -328 \cdot 7 + (-5),$$

but -328 and -5 are not the unique quotient and the remainder guaranteed by the division algorithm 3.1.3. (Why?) The problem is that we have not found the largest multiple of 7 that is less than or equal to $-2{,}301$, since -5 can further be divided by 7, yielding a quotient of -1 and a remainder of 2, that is, $-5 = (-1) \cdot 7 + 2$.

Thus, taking into account this "extra division step" and substituting back into the previous equation gives

$$-2{,}301 = -328 \cdot 7 + (-5) = -328 \cdot 7 + [(-1) \cdot 7 + 2] = -329 \cdot 7 + 2,$$

which produces the correct quotient and remainder.

Summary: Dividing a Negative Integer by a Positive Integer. We express the process employed in the previous example in general terms. Namely, suppose a solution to the problem $a\overline{)-b}$ is sought, where a and b are positive integers. Using long division, compute $a\overline{)b}$, and then write $b = qa + r$, where $0 \leq r < a$. Thus, $-b = (-q)a + (-r)$, and dividing $-r$ by a yields $-r = (-1)a + (a - r)$, where $0 \leq (a - r) < a$. Finally,

$$-b = (-q)a + (-r) = (-q)a + [(-1)a + (a - r)]$$
$$= (-q - 1)a + (a - r),$$

so the division $a\overline{)-b}$ has the unique quotient $-q - 1$ and the remainder $a - r$.

The division algorithm is useful for computational purposes and is an indispensable theoretical tool that we call upon frequently throughout our work.

An elementary application concerns even and odd integers.

Even and Odd Integers. Recall that an integer b is called **even** if it is divisible by 2, which means $b = 2q$ for some integer q. Note that 0 is even, since $0 = 2 \cdot 0$. If an integer b is not even, then it is called **odd**; in this case, $b = 2q + 1$ for some integer q. (Why?)

3.1.4 Classroom Problems. In questions 1 through 6, determine whether the specified computations yield even or odd integers, and then give arguments to justify your answers. Let a and b be even integers, and c and d be odd integers. Direction: Break statements involving $\pm$ into two cases.

1. even $\pm$ even $= a \pm b = ?$
2. odd $\pm$ odd $= c \pm d = ?$
3. even $\pm$ odd $= a \pm c = ?$
4. even $\cdot$ even $= a \cdot b = ?$
5. odd $\cdot$ odd $= c \cdot d = ?$
6. even $\cdot$ odd $= a \cdot c = ?$ ◆

3.1.5 Example: An Arithmetic Structure Lacking Prime Factorization. The Fundamental Theorem of Arithmetic (FTA) guarantees that any positive integer greater than 1 has a unique prime factorization. It is interesting to observe that there are other arithmetic systems of numbers (sets of numbers endowed with arithmetic operations and axioms governing those operations) for which this result does not hold.

To see such an example, let E be the set of all positive even integers. How would a prime p be defined in the set E?

Recall that we defined a positive integer p to be prime if it has exactly two distinct positive integer factors. To remain consistent with this integer definition of prime, and since each even integer $p > 2$ either has exactly two distinct factors in E or more than two distinct factors in E (why?), it is reasonable to say an even integer $p > 2$ is **prime in E** if it has exactly two distinct factors in E. Under this definition, 4 is a prime number in E, and $6, 10, 14, \ldots, 2(2m + 1)$, with $m \geq 1$ an integer, are also primes in E. (Why?) Moreover, notice that $8, 12, 16, \ldots, 4m$, with $m \geq 2$ an integer, are all composites in E. (Why?)

To see that a prime factorization in E need not be unique, consider the following prime factorizations in E:

$$540 = 18 \cdot 30 = 10 \cdot 54,$$

where 18, 30, 10, and 54 are all prime in E.

3.2 THE EUCLIDEAN ALGORITHM

Our studies thus far have focused on several important divisibility aspects of the integers, and in this direction we have learned many interesting properties and relationships. In particular, we have extensively investigated lcm(a, b) and gcd(a, b) for positive integers a and b, and through this work have expanded our understanding of these concepts. One key realization was that, although prime factorization is a powerful tool, it is not a practical way to compute lcm and gcd. This kind of dilemma is common in mathematics, and the endeavor to discover new feasible methods contributes to the vitality of the subject.

The Euclidean algorithm (in Euclid's *Elements*, Book VII, Proposition VII.2) is an efficient computational process for computing gcd(a, b) that relies on repeated applications of the division algorithm and does not depend upon finding the explicit prime factorizations of the positive integers a and b. An additional benefit to this method is that it provides a means for determining integers x and y so that gcd$(a, b) = ax + by$, which in turn leads to a complete answer to the questions concerning measurement of varying lengths with given unmarked straightedges.

The following elementary observation is at the heart of the Euclidean algorithm.

3.2.1 Observation. Let $a > 0$ and b be integers. If $b = qa + r$, where q and r are integers, then gcd$(a, b) = $ gcd(a, r).

Justification. Let $d = $ gcd(a, b). Using Definition 2.7.6, we show that $d = $ gcd(a, r). First note that since d is a common divisor of a and b, then $d | (b - qa) = r$, so d is a common divisor of a and r. Next notice that if g is any common divisor of a and r, then $g | (qa + r) = b$, and thus $g | gcd(a, b) = d$. (Why?) Therefore, gcd$(a, b) = d = $ gcd(a, r).

It may not seem apparent how this observation, in combination with repeated applications of the division algorithm, can be used to determine gcd, but a few specific examples help illustrate this point. The process described in these examples is referred to as the **Euclidean algorithm**.

a. Illustration of the Euclidean algorithm: $\gcd(324, 392) = ?$

Divisions	GCD conclusions
1. Find q and r in the division $324\overline{)392}$ $392 = 1 \cdot 324 + 68;\ 0 \leq 68 < 324$	$\gcd(324, 392) = \gcd(324, 68)$
2. Find q and r in the division $68\overline{)324}$ $324 = 4 \cdot 68 + 52;\ 0 \leq 52 < 68$	$\gcd(324, 68) = \gcd(68, 52)$
3. Find q and r in the division $52\overline{)68}$ $68 = 1 \cdot 52 + 16;\ 0 \leq 16 < 52$	$\gcd(68, 52) = \gcd(52, 16)$
4. Find q and r in the division $16\overline{)52}$ $52 = 3 \cdot 16 + 4;\ 0 \leq 4 < 16$	$\gcd(52, 16) = \gcd(16, 4)$
5. Find q and r in the division $4\overline{)16}$ $16 = 4 \cdot 4 + 0;\ 0 \leq 0 < 4$	$\gcd(16, 4) = \gcd(4, 0) = 4$

Justification a. Repeatedly applying Observation 3.2.1 to the previous divisions and linking the results shows that the *last nonzero remainder* obtained in this process is the greatest common divisor of the original given integers 324 and 392:

$$\gcd(324, 392) = \gcd(324, 68) = \gcd(68, 52) = \gcd(52, 16)$$
$$= \gcd(16, 4) = \gcd(4, 0) = 4.$$

The algorithm always stops in a finite number of steps since the remainders obtained in each division stage get progressively smaller until the remainder of 0 is reached. In this example, the decreasing remainder sequence is: $68 > 52 > 16 > 4 > 0$.

The collection of equations that just demonstrated that $\gcd(324, 392) = 4$ can also be used to find integers x and y so that $4 = 324x + 392y$. To see how this works, first list the equations in the order they appeared. Next, starting with the equation that has the last nonzero remainder (the one preceding the last), sequentially back solve the equations in terms of the remainders until 4 is expressed as a linear combination of 324 and 392.

$\gcd(324, 392) = 4$	$4 = 324x + 392y$
$392 = 1 \cdot 324 + 68$	$4 = 52 - 3 \cdot 16$
$324 = 4 \cdot 68 + 52$	$= 52 - 3 \cdot (68 - 1 \cdot 52)$
$68 = 1 \cdot 52 + 16$	$= 4 \cdot 52 - 3 \cdot 68$
$52 = 3 \cdot 16 + 4$	$= 4 \cdot (324 - 4 \cdot 68) - 3 \cdot 68$
$16 = 4 \cdot 4 + 0$	$= 4 \cdot 324 - 19 \cdot 68$
	$= 4 \cdot 324 - 19 \cdot (392 - 1 \cdot 324)$
	$= 23 \cdot 324 - 19 \cdot 392$

Conclusion. For $x = 23$ and $y = -19$, $\gcd(324, 392) = 4 = 324 \cdot (23) + 392 \cdot (-19)$.

Let's use the Euclidean algorithm to find $\gcd(675, 14{,}787)$ and this time simply indicate each stage by listing the resulting equation and the corresponding remainder inequality.

b. Illustration of the Euclidean algorithm: $\gcd(675, 14787) = ?$

<u>Divisions</u>

$$14{,}787 = 21 \cdot 675 + 612 \qquad 0 \leq 612 < 675$$
$$675 = 1 \cdot 612 + 63 \qquad 0 \leq 63 < 612$$
$$612 = 9 \cdot 63 + 45 \qquad 0 \leq 45 < 63$$
$$63 = 1 \cdot 45 + 18 \qquad 0 \leq 18 < 45$$
$$45 = 2 \cdot 18 + 9 \qquad 0 \leq 9 < 18$$
$$18 = 2 \cdot 9 + 0 \qquad 0 \leq 0 < 9$$

$$\gcd(675, 14{,}787) = \gcd(675, 612) = \gcd(612, 63) = \gcd(63, 45) = \gcd(45, 18)$$
$$= \gcd(18, 9) = \gcd(9, 0) = 9.$$

<u>Expressing the gcd as a linear combination: $9 = 675x + 14{,}787y$</u>

$$9 = 45 - 2 \cdot 18$$
$$= 45 - 2 \cdot (63 - 1 \cdot 45)$$
$$= 3 \cdot 45 - 2 \cdot 63$$
$$= 3 \cdot (612 - 9 \cdot 63) - 2 \cdot 63$$
$$= 3 \cdot 612 - 29 \cdot 63$$
$$= 3 \cdot 612 - 29 \cdot (675 - 1 \cdot 612)$$
$$= 32 \cdot 612 - 29 \cdot 675$$
$$= 32 \cdot (14{,}787 - 21 \cdot 675) - 29 \cdot 675$$
$$= 32 \cdot 14{,}787 - 701 \cdot 675.$$

Choosing $x = -701$ and $y = 32$, $\gcd(675, 14{,}787) = 9 = 675 \cdot (-701) + 14{,}787 \cdot (32)$.

Although we have not given a general proof of the Euclidean algorithm and the consequence that $\gcd(a, b)$ can be written as a linear combination of a and b, developing rigorous arguments modeled after the examples we considered is a straightforward exercise.

An Answer to the General Question Concerning Measurements of Lengths. We are now prepared to answer Question 3.1.2 regarding which lengths of L inches can be measured with unmarked straightedges of lengths a inches and b inches. As noticed (General Conclusion 3.1.1), any such L must be divisible by $d = \gcd(a, b)$. Applying the Euclidean algorithm to given positive integers $a \le b$, it follows that there exist integers x and y such that $d = ax + by$. If L is some positive integer divisible by d, then $L = dz$, for some positive integer z. Therefore, $L = dz = axz + byz$, so any length of $L = dz$ inches (z any positive integer) can be measured with unmarked straightedges of lengths a inches and b inches.

3.2.2 Summary. The measurement work we just completed can be summarized as a general statement about integers. Let a, b, and L be positive integers, and $d = \gcd(a,b)$. Then,

$$L = ax + by \text{ for some integers } x \text{ and } y \text{ if and only if } d|L,$$

so, $\gcd(a,b)$ is the *least positive* member of the set $S = \{ax + by : x \text{ and } y \text{ integers}\}$.

In the next section, we determine *all* possible ways to measure lengths of L inches with unmarked straightedges of lengths a inches and b inches, i.e., we'll find all solutions x and y to the equation $L = ax + by$, where $\gcd(a,b)|L$.

Historical Note. The equation $L = ax + by$ is an example of a Diophantine equation. In his work *Arithmetica*, the third-century Greek mathematician Diophantus, began the detailed study of equations having integer coefficients and integer solutions and because of this initial work, such equations are termed **Diophantine equations**. The Diophantine equations we have worked with thus far are linear (exponent one) with two variables, whereas in Chapter 5 we'll study quadratic Diophantine equations in three variables.

3.2.3 Remark. In Observation 2.8.7, we saw that consecutive terms of the Fibonacci sequence are relatively prime. In addition to this result, it can be shown using the Euclidean algorithm and some other technical results, that the greatest common divisor of any two terms of the Fibonacci sequence is also a term of the Fibonacci sequence. The Fibonacci sequence is truly a mathematical wonderland!

EXERCISES 3.2

1. In each division, find the unique quotient and remainder.
    a. $29 \overline{)35{,}784}$
    b. $757 \overline{)525}$
    c. $38 \overline{)-47{,}893}$
    d. $91 \overline{)-52}$
2. For an integer a, show that a is even if and only if a^2 is even, and a is odd if and only if a^2 is odd.
3. If a is an integer, then
    a. a^2 is divisible by 4 or has a remainder of 1 when divided by 4. Direction: Consider two cases: $a = 2q$ and $a = 2q + 1$;
    b. a^2 is divisible by 3 or has a remainder of 1 when divided by 3. Direction: Consider three cases: $a = 3q$, $a = 3q + 1$, and $a = 3q + 2$.

4. a. Given three consecutive integers a, $a + 1$, and $a + 2$, show that exactly one of the them is divisible by 3. Direction: Same hint as in 3b.
   b. Given integers a, $a + 2$, $a + 4$, show that exactly one of them is divisible by 3.
5. In Section 2.6, we determined the number of positive divisors of a positive integer n to be $\tau(n) = \tau(p_1^{m_1} p_2^{m_2} \cdots p_s^{m_s}) = (m_1 + 1)(m_2 + 1) \cdots (m_s + 1)$. Show that $\tau(n)$ is odd $\Leftrightarrow n$ is a square (Conclusion about squares 2.6.8).
6. Prove the **generalized division algorithm**: If $a \neq 0$ and b are integers, then there exist unique integers q and r such that $b = aq + r$, where $0 \leq r < |a|$. Direction: Apply the division algorithm with $|a|$ and b.
7. Using the Euclidean algorithm, find the gcd of the given integers, and then write it as a linear combination of those integers.
   a. $a = 87$ $b = 111$
   b. $a = 585$ $b = 2{,}480$
   c. $a = 496$ $b = 8{,}128$
   d. $a = 10{,}285$ $b = 89{,}523$
8. What lengths of L inches can be measured with unmarked straightedges of lengths 20 inches and 72 inches?
9. Given a collection of unmarked straightedges of lengths 2^n inches for each integer $n \geq 0$, what lengths of L inches can you measure if you can use only straightedges of lengths 2^n inches at most one time in the measurement of L and if straightedge measurements can only be added? (In Section 3.4, we revisit this question in a general setting.) For example, 20 inches can be measured, since $20 = 2^4 + 2^2$, but it cannot be accomplished as $20 = 2^3 + 2^3 + 2^2$, since this is not a permitted measurement under the established rules. Does your conclusion apply if the unmarked straightedges were of length 3^n inches rather than 2^n? If not, what additional condition(s) are needed?
10. **The 7–11 Casino Problem.** At the 7–11 Riverboat Casino, all chips are worth $7 or $11 dollars, and the dealers only pay out or give change in $7 and/or $11 dollar chips. Explain how it is possible to place a $344 bet. Suppose you win that bet at 3-to-1 odds. Explain how the dealer pays out in $7 and $11 chips.
11. *Die Hard 3* **Problem.** In the movie *Die Hard 3*, Bruce Willis and Samuel L. Jackson try to keep a bomb from exploding by figuring out how to determine exactly 4 gallons of water using one unmarked 3-gallon jug, one 5-gallon jug, and lots of water from a fountain. They managed to solve the problem just in the nick of time.
    a. How would you solve this problem? (Explain your reasoning.) Express your solution as $4 = 3x + 5y$ for some integers x and y, and explain how this equation encodes the measuring procedure you used for determining exactly 4 gallons of water.
    b. Could 4 gallons be determined if the unmarked jugs held 6 gallons and 9 gallons? (Explain your reasoning.)

3.3 APPLICATIONS OF THE REPRESENTATION gcd(a, b) = ax + by

In the last section we saw how we could use the Euclidean algorithm to find integers x and y so that $\gcd(a, b) = ax + by$. This important representation provided an answer to a general question concerning the measurement of lengths using unmarked straightedges of given lengths, and it is especially useful in studying gcd-related problems.

Recall that our proof of the uniqueness part of the FTA depended on the following result: If a prime p divides a product $q_1 q_2 \cdots q_s$ of primes, then $p|q_i$ for some $1 \leq i \leq s$ (see result 2.6.2). This proposition can be deduced from a more general result referred to as Euclid's lemma. For our purposes, assume a, b, and c are positive integers.

3.3.1 Euclid's Lemma. If $a|bc$ and $\gcd(a, b) = 1$, then $a|c$.

Proof. The hypotheses give us $bc = as$ for some integer s, and since $\gcd(a, b) = 1$, there exist integers x and y so that $1 = ax + by$. To show that $c = at$ for some integer t, we multiply the equation by c and substitute as for cb to obtain

$$c = cax + cby = acx + asy = a(cx + sy), \text{ so } a|c.$$

The key ingredient in unique prime factorization occurs when $a = p$ is a prime number.

3.3.2 Corollary. If $p|bc$ where p is a prime number, then $p|b$ or $p|c$.

Proof **(Indirect).** Assume b and c are not divisible by p, so, in particular, $\gcd(p, b) = 1$. (Why?) The assumption $p|bc$, in conjunction with Euclid's lemma, gives $p|c$, which is a contradiction.

With regards to result 2.6.2 that was used to prove uniqueness in the FTA, it follows by an induction argument that if a prime divides a finite product of integers, then it must divide at least one of them (see Exercises 3.3, 1).

Constructing Proofs Using $\gcd(a, b) = ax + by$. Many of the deductions we have previously made were based on various prime factorization arguments. Using the representation $\gcd(a, b) = ax + by$, it is often possible to develop arguments that do not depend upon prime factorization. Let's consider a few such results at this time and examine some others in Exercises 3.3.

1. In problem 10 of Exercises 2.8, you had to prove that $\gcd(ca, cb) = c \cdot \gcd(a, b)$, where a, b, and c are positive integers, by giving one proof using prime factorization and an alternate proof analogous to the lcm alternate argument presented in Section 2.8. We now give a third demonstration of this fact, using the linear combination expression for gcd.

For positive integers a and b, let $d = \gcd(a, b)$. We use Definition 2.7.6 to show that $\gcd(ca, cb) = cd$, where c is a positive integer.

Justification. Since cd is a common divisor of ca and cb (why?), it suffices to show that if g is any common divisor of ca and cb, then $g|d$. Writing $d = ax + by$ for some integers x and y and then multiplying the equality by c gives $cd = cax + cby$, which in turn implies that $g|d$.

2. In problem 8e of Exercises 2.8, you had to show that if $\gcd(a,b) = 1$ and $a|c$ and $b|c$, then $ab|c$. The next argument employs the linear combination expression for gcd to justify this statement.

Justification. We know that $c = as = bt$ for some integers s and t and that $1 = ax + by$ for some integers x and y. As before, multiplying the equality by c and suitably substituting for c gives $c = cax + cby = (bt)ax + (as)by = ab(tx + sy)$, so $ab|c$.

Solving the equation $ax + by = L$: Finding all ways to measure lengths with unmarked straightedges. The principal conclusion reached in our measurement work was that a length of L inches can be measured with unmarked straightedges of lengths a inches and b inches if and only if $\gcd(a,b)|L$. In addition, we observed that it is possible to measure a given length in more than one way. For example, three different approaches for measuring a 1-inch length with $a = 3$ and $b = 5$ are

$$1 = 3 \cdot 2 + 5 \cdot (-1), 1 = 3 \cdot 7 + 5 \cdot (-4), \text{ and } 1 = 3 \cdot (-8) + 5 \cdot 5.$$

Two ways to measure a 3-inch length with $a = 12$ and $b = 15$ are

$$3 = 12 \cdot (-1) + 15 \cdot 1 \text{ and } 3 = 12 \cdot 4 + 15 \cdot (-3).$$

These examples prompt the next problem (for our purposes, we state this problem in positive integers).

3.3.3 Classroom Discussion. In this analysis, we find all integer solutions to the Diophantine equation $ax + by = L$, where $a, b,$ and L are positive integers and $\gcd(a,b)|L$.

Analysis. Since $\gcd(a,b)|L$, we know that the equation $ax + by = L$ has at least one integer solution, say $x = x_0$ and $y = y_0$. We show that there are infinitely many such solutions, which can all be expressed in terms of the particular solution $x = x_0$ and $y = y_0$.

Suppose $x = u$ and $y = v$ is any integer solution to the equation $ax + by = L$. Thus,

$$ax_0 + by_0 = au + bv, \text{ so } a(x_0 - u) = b(v - y_0).$$

Using the fact that $\gcd(a,b) = d$ implies that $\gcd(a/d, b/d) = 1$ (see result 2.8.5), we can simplify matters further by "dividing out the greatest common divisor d" to get

$$(a/d)(x_0 - u) = (b/d)(v - y_0).$$

Notice that $(a/d)|(b/d)(v - y_0)$, and since $\gcd(a/d, b/d) = 1$, it follows from Euclid's lemma 3.3.1 that $(a/d)|(v - y_0)$. Hence, $v - y_0 = (a/d)t$ for some integer t, and thus substitution and cancellation gives

$$(a/d)(x_0 - u) = (b/d)(v - y_0) = (b/d)(a/d)t, \text{ so } (x_0 - u) = (b/d)t.$$

Solving for u and v leads to the expressions

$$u = x_0 - (b/d)t \text{ and } v = y_0 + (a/d)t.$$

Therefore, if (u, v) is a solution of the equation $ax + by = L$, then it can be represented in terms of the particular solution $x = x_0$ and $y = y_0$.

Conversely, for any integer t, $x = x_0 - (b/d)t$ and $y = y_0 + (a/d)t$ is a solution to the equation $ax + by = L$, since

$$a(x_0 - (b/d)t) + b(y_0 + (a/d)t) = a\,x_0 + b\,y_0 = L.$$

Therefore, if $\gcd(a,b)|L$, then the equation $ax + by = L$ has infinitely many solutions given by

$$x = x_0 - (b/d)t \text{ and } y = y_0 + (a/d)t, \qquad 3.3.4$$

where $x = x_0$ and $y = y_0$ is a particular solution and t is an arbitrary integer. ◆

3.3.5 General Coefficients. The results in Summary 3.2.2 and result 3.3.4 were stated with the assumption that $a, b,$ and L are all positive integers, since this was the context in which we were working. However, both results remain valid (using similar proofs) under the conditions that a, b, and L are integers and that a and b are not both zero.

Geometric Remark. The previous problem and solution has an interesting geometric interpretation. The graph of the given equation $L = ax + by$ (a, b, and L are positive integers) is a line in the plane, and we have shown that if this line passes through one point (x_0, y_0) where both coordinates are integers, then it passes through infinitely many points having integer coordinates. Points in the plane having integer coordinates are called **lattice points**.

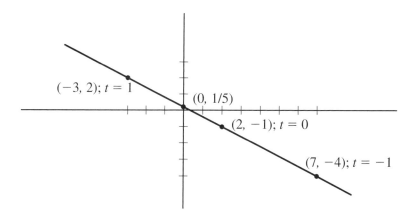

FIGURE 3.3.1 Graph of the line $y = -\frac{3}{5}x + \frac{1}{5}$

For example, let's consider the graph of the linear equation $1 = 3x + 5y$, which emerged in our study of measurements with unmarked straightedges. We initially observed the solution $(2, -1)$ and have subsequently seen that the line passes through infinitely many lattice points whose integer coordinates are given by the equations $x = 2 - 5t$ and $y = -1 + 3t$, where t is an arbitrary integer.

The study of algebraic equations with coefficients and solutions restricted to the integers (Diophantine equations) is an incredibly rich and complex subject with an extensive history, and we have only considered a very small part of this important area. Research mathematicians continue to work on problems in this field, and their new discoveries continue to deepen our understanding of this interesting subject area.

EXERCISES 3.3

1. Let p be a prime number and $a_1, a_2, \ldots, a_s$ positive integers. Show by induction that if $p | a_1 a_2 \cdots a_s$, then $p | a_i$ for some $1 \leq i \leq s$.
2. Let n be a positive integer and p a prime number.
   a. Prove: $p | n! \Leftrightarrow p \leq n$. Direction: ($\Rightarrow$). Use Exercise 1 above to handle this part of the proof.
   b. Prove: If $p | (n! - 1)$, then $p > n$. Direction: Indirect proof. Determine $\gcd(n! - 1, n!)$ (consecutive integers), and then use part a to reach a contradiction.
   c. Prove: For any positive integer $n > 2$, there is a prime p such that $n < p < n!$ Direction: Use part b to show that any prime divisor of $n! - 1$ will do the job.
3. Note that by Summary 3.2.2, $\gcd(a, b) = 1 \Leftrightarrow 1 = ax + by$ for some integers x and y. (Explain.) Use this fact to construct proofs for the following statements:
   a. If $\gcd(a, b) = 1$, then $\gcd(a + b, ab) = 1$ (see Exercise 2.8, 8b).
   b. If $\gcd(a, b) = 1$ and $\gcd(a, c) = 1$, then $\gcd(a, bc) = 1$ (see Exercise 2.8, 8c). Direction: Write $1 = ax + by$ and $1 = az + cw$ and then multiply the equations.
4. Determine which of the following Diophantine equations have an integer solution. For those that do, find all their integer solutions.
   a. $21x + 75y = 42$ c. $147x + 185y = 6$
   b. $32x + 92y = 82$ d. $68x + 289y = 51$

3.4 PLACE VALUE

As you know, the decimal system (Hindu–Arabic system) is a simple but ingenious numeration system based on powers of 10. The **place value** of a digit is dependent upon its position in a given number (the ten symbols 0, 1, 2, 3, 4, 5, 6, 7, 8, 9 are called *digits*). For example, in the number 5,247, the digit 5 has place value "thousands," the digit 2 has place value "hundreds," the digit 4 has place value "tens," and the digit 7 has place value "ones."

A digit's **value** in a number is determined by multiplying the digit by its place value. For instance, the value of 2 in the number 5,247 is $2 \cdot 100 = 200$. A number can be uniquely represented as the sum of the values of the digits, and this is called the number's **expanded form**. For example, the expanded form of the integer 5,247 is:

$$5{,}247 = (5 \cdot 10^3) + (2 \cdot 10^2) + (4 \cdot 10^1) + (7 \cdot 10^0).$$

The discovery of the decimal system is attributed to the Hindus and Arabs, and its development spanned a thousand-year period of time (circa 300 BC–AD 700). It was not until AD 1100 that the Hindu–Arabic system reached Europe, and it took several hundred years after that time before it was extensively used. The advance of typeset printing in the 1400s led to the publication of mathematics textbooks that popularized the decimal system. The most prominent advantages of this system over other earlier systems are the simplicity of representing numbers and the relative ease (compared to systems such as Roman numerals) of performing arithmetic calculations with them.

In this brief section, we investigate numeration systems similar to the decimal system. A question in problem 9 of Exercise 3.2 is relevant to this work.

Question. Given a collection of unmarked straightedges of lengths 2^n inches for each integer $n \geq 0$, what lengths of L inches can be measured if you can use only straightedges of lengths 2^n inches at most one time in the measurement of L and if straightedge measurements can only be added?

It was noted that 20 inches can be measured this way, since $20 = 2^4 + 2^2$, in fact, all positive integer lengths of L inches can be measured by this method.

In an effort to provide justification for the previous general statement, we consider a few approaches. The first method was given by one of my former students (Susan) as part of a homework assignment.

Susan's Representation Method. I will show how to write a particular number in the form described in the question, and you will see that my method works for any positive integer. For example, let $L = 157$. My process can best be described in stages.

Stage 1. Find the largest power of 2 that is less than or equal to 157.

By calculating increasingly higher powers of 2, we find out that $2^7 = 128$ is the largest power of 2 that is less than or equal to 157.

Stage 2. Find the largest power of 2 that is less than or equal to the difference $157 - 128 = 29$.

In this case, $2^4 = 16$ is the largest power of 2 that is less than or equal to 29.

Stage 3. Find the largest power of 2 that is less than or equal to the difference $29 - 16 = 13$.

Here we see that $2^3 = 8$ is the largest power of 2 that is less than or equal to 13.

Stage 4. Find the largest power of 2 that is less than or equal to the difference $13 - 8 = 5$.

In this case, $2^2 = 4$ is the largest power of 2 that is less than or equal to 5.

Stage 5. Find the largest power of 2 that is less than or equal to the difference $5 - 4 = 1$. This is the final stage, since $2^0 = 1$ is the largest power of 2 that is less than or equal to 1.

Summary. Combining the five stages gives

$$157 = 128 + 16 + 8 + 4 + 1 = 2^7 + 2^4 + 2^3 + 2^2 + 2^0.$$

As you can see, I have described a general procedure for measuring any positive integer length of L inches as a sum of distinct powers of 2.

Although Susan did not give a formal proof showing that all positive integers can be represented (uniquely) as a sum of distinct powers of 2, her process does lead to the desired representation and can easily be applied to any positive integer (as long as we can do the arithmetic). We revisit Susan's method later in this section and show how it can be applied to other related problems.

Alternate Representation Method. Let's consider an approach that reaches the desired representation through a path different from Susan's route. As in her method, we solve the problem for the length $L = 157$ inches. Our procedure may not be the quickest way to solve the problem for a particular number, but it will always work for any positive integer L, and it extends easily to a general proof.

Analysis. We start by dividing 157 by 2 and get a quotient of 78 with a remainder of 1. In terms of the division algorithm,

$$157 = 78 \cdot 2 + 1.$$

Since the quotient 78 is greater than or equal to 2, we can divide it by 2 and obtain the quotient 39 and the remainder 0. Therefore,

$$78 = 39 \cdot 2 + 0.$$

This process can be repeated as long as the newly obtained quotients are greater than or equal to 2. Proceeding as above leads to the following equalities:

$$39 = 19 \cdot 2 + 1$$
$$19 = 9 \cdot 2 + 1$$
$$9 = 4 \cdot 2 + 1$$
$$4 = 2 \cdot 2 + 0$$
$$2 = 1 \cdot 2 + 0$$

The process stops at this point, since the quotient 1 is less than 2. The sought-after representation can now be determined by combining (through substitution) the previous equalities. In particular,

$$157 = 78 \cdot 2 + 1 = 39 \cdot 2^2 + 1 = (19 \cdot 2 + 1) \cdot 2^2 + 1 = 19 \cdot 2^3 + 2^2 + 1$$
$$= (9 \cdot 2 + 1) \cdot 2^3 + 2^2 + 1 = 9 \cdot 2^4 + 2^3 + 2^2 + 1$$
$$= (4 \cdot 2 + 1) \cdot 2^4 + 2^3 + 2^2 + 1$$

$$= 4 \cdot 2^5 + 2^4 + 2^3 + 2^2 + 1 = (2 \cdot 2) \cdot 2^5 + 2^4 + 2^3 + 2^2 + 1$$
$$= 2 \cdot 2^6 + 2^4 + 2^3 + 2^2 + 1 = 2^7 + 2^4 + 2^3 + 2^2 + 2^0.$$

Therefore, the number 157 has been represented in the desired form. Notice that both methods led to the same representation, and this uniqueness property is true in general.

As in Susan's method, this single example analysis does not constitute a proof. However, we now show that it is straightforward to construct an argument based on the ideas of this second approach. We can state the result purely in terms of integers.

3.4.1 Theorem. If L is a positive integer, then there exists some integer $n \geq 0$ such that L can be expressed uniquely in the form

$$L = 2^n + r_{n-1}2^{n-1} + \ldots + r_2 2^2 + r_1 2^1 + r_0 2^0,$$

where $r_i \in \{0, 1\}$ for each i.

***Proof* (Existence).** First we establish that such a representation always exists. The proof of this part mirrors the previous analysis.

We may as well assume that $L > 2$, otherwise there is nothing to prove. As in the specific example, we can divide L by 2 (division algorithm) and get an integer quotient $q_0 > 0$ (since $L > 2$) and an integer remainder r_0, such that

$$L = q_0 2 + r_0 \text{ and } 0 \leq r_0 < 2 \text{ (i.e., } r_0 \in \{0, 1\}\text{).}$$

If $q_0 = 1$, then we are done (why?), so let's assume that $q_0 \geq 2$ and divide q_0 by 2. Thus, there exist integers q_1 and r_1, such that

$$q_0 = q_1 2 + r_1 \text{ and } 0 \leq r_1 < 2.$$

By replacing q_0 with $q_1 2 + r_1$ in the first equality, we may rewrite L as

$$L = (q_1 2 + r_1)2 + r_0 = q_1 2^2 + r_1 2 + r_0 2^0.$$

As before, if $q_1 = 1$ ($q_1 > 0$, since $q_0 \geq 2$), then our task is complete, so we'll suppose that $q_1 \geq 2$. Hence,

$$q_1 = q_2 2 + r_2 \text{ and } 0 \leq r_2 < 2.$$

Therefore,
$$L = q_1 2^2 + r_1 2 + r_0 = (q_2 2 + r_2) 2^2 + r_1 2 + r_0$$
$$= q_2 2^3 + r_2 2^2 + r_1 2^1 + r_0 2^0.$$

Note that $L > q_0 > q_1 > q_2 > 0$, and thus this process must ultimately lead to a quotient $q_{n-1} = 1$. (Why?) Therefore, as in the earlier cases, L can be written as

$$L = 2^n + r_{n-1} 2^{n-1} + \ldots + r_2 2^2 + r_1 2^1 + r_0 2^0,$$

and the existence part of the proof is complete.

114   Chapter 3   The Division Algorithm and the Euclidean Algorithm

(Uniqueness). Suppose some positive integer L has two representations as distinct powers of 2. Thus

$$L = 2^n + r_{n-1}2^{n-1} + \ldots + r_2 2^2 + r_1 2^1 + r_0 2^0$$
$$= 2^m + s_{m-1}2^{m-1} + \ldots + s_2 2^2 + s_1 2^1 + s_0 2^0,$$

where $r_i, s_i \in \{0, 1\}$ for each i. Since $n \geq m$ or $m \geq n$, we lose no generality in assuming that $n \geq m$. We show that $n = m$ and that $r_i = s_i$ for $i = 0, 1, 2, \ldots, n - 1$.

Collecting the terms having 2 as a factor results in the equality

$$(2^n + r_{n-1}2^{n-1} + \ldots + r_2 2^2 + r_1 2^1)$$
$$- (2^m + s_{m-1}2^{m-1} + \ldots + s_2 2^2 + s_1 2^1) = s_0 - r_0,$$

so $2 | (s_0 - r_0)$, which implies that $s_0 = r_0$. (Why?) Therefore,

$$(2^n + r_{n-1}2^{n-1} + \ldots + r_2 2^2 + r_1 2^1)$$
$$= (2^m + s_{m-1}2^{m-1} + \ldots + s_2 2^2 + s_1 2^1),$$

and dividing each side of the equality by 2 gives

$$(2^{n-1} + r_{n-1}2^{n-2} + \ldots + r_2 2^1 + r_1 2^0)$$
$$= (2^{m-1} + s_{m-1}2^{m-2} + \ldots + s_2 2^1 + s_1 2^0).$$

As above, $2 | (s_1 - r_1)$, and it follows that $s_1 = r_1$. Thus,

$$(2^{n-1} + r_{n-1}2^{n-2} + \ldots + r_2 2^1) = (2^{m-1} + s_{m-1}2^{m-2} + \ldots + s_2 2^1);$$

again dividing each side of the equality by 2 yields

$$(2^{n-2} + r_{n-1}2^{n-3} + \ldots + r_2 2^0) = (2^{m-2} + s_{m-1}2^{m-3} + \ldots + s_2 2^0).$$

Repeating this process shows that $s_i = r_i$ for $i = 0, 1, 2, \ldots, m - 1$, so subtracting the quantity $(r_{m-1}2^{m-1} + \ldots + r_2 2^2 + r_1 2^1 + r_0 2^0)$ from each side of the following (hypothesized) representations for L,

$$L = 2^n + r_{n-1}2^{n-1} + \ldots + r_m 2^m$$
$$+ (r_{m-1}2^{m-1} + \ldots + r_2 2^2 + r_1 2^1 + r_0 2^0)$$
$$= 2^m + s_{m-1}2^{m-1} + \ldots + s_2 2^2 + s_1 2^1 + s_0 2^0,$$

results in the equality

$$2^n + r_{n-1}2^{n-1} + \ldots + r_m 2^m = 2^m.$$

Dividing each side of this equality by 2^m results in the equality

$$2^{n-m} + r_{n-1}2^{(n-1)-m} + \ldots + r_m 2^0 = 1.$$

Therefore, $r_{n-1} = r_{n-2} = \ldots = r_m = 0$ and $n = m$ (why?), so we have demonstrated that any positive integer can be uniquely expressed as a sum of distinct powers of 2.

Binary Place Value Notation. We have just demonstrated that any positive integer L can be uniquely expressed in the form

$$L = 2^n + r_{n-1}2^{n-1} + \ldots + r_2 2^2 + r_1 2^1 + r_0 2^0,$$

where $r_i \in \{0,1\}$ for each i.

This representation is called the **base 2 expanded form of** L, and since it is unique, L can also be written using the digits 0 and 1 in the **binary** (base 2) **place value** notation

$$L = (1r_{n-1}r_{n-2}\ldots r_2 r_1 r_0)_2.$$

Recall that $157 = 2^7 + 0 \cdot 2^6 + 0 \cdot 2^5 + 2^4 + 2^3 + 2^2 + 0 \cdot 2^1 + 2^0$, so

$$157 = (10011101)_2.$$

This base 2 numeral is read: (one, zero, zero, one, one, one, zero, one) base 2.

Classroom Problems. Convert the following integers into base 2 numerals:

a. 42
b. 63
c. 238 ◆

As we have seen, even relatively small numbers represented in the binary number system require long strings of the digits 0 and 1, so this system is impractical for everyday arithmetic. However, just the opposite is true for computer systems.

The digits 1 and 0 of the binary number system represent two state changes (i.e., whether a switch [electrical, magnetic, optical, etc.] is either on or off). Information in the form of binary code can be stored on a computer's hard disk through a sequence of such state changes. Let's consider a simple magic trick that illustrates this fundamental idea of computer science.

A Base 2 Magic Trick. To construct this trick so that you can perform it properly, get five blank 8 × 10 inch (or larger) sheets of paper or (even better) poster board (which we refer to as **cards**) and a marker with which to label the cards with certain numbers. Next, convert each of the integers 1 through 31 into base 2 numerals.

$1 = (1)_2$	$9 = (1001)_2$	$17 = (10001)_2$	$25 = (11001)_2$
$2 = (10)_2$	$10 = (1010)_2$	$18 = (10010)_2$	$26 = (11010)_2$
$3 = (11)_2$	$11 = (1011)_2$	$19 = (10011)_2$	$27 = (11011)_2$
$4 = (100)_2$	$12 = (1100)_2$	$20 = (10100)_2$	$28 = (11100)_2$
$5 = (101)_2$	$13 = (1101)_2$	$21 = (10101)_2$	$29 = (11101)_2$
$6 = (110)_2$	$14 = (1110)_2$	$22 = (10110)_2$	$30 = (11110)_2$
$7 = (111)_2$	$15 = (1111)_2$	$23 = (10111)_2$	$31 = (11111)_2$
$8 = (1000)_2$	$16 = (10000)_2$	$24 = (11000)_2$	

We are now ready to prepare each of the cards. The order in which these cards appear during the trick is of the utmost importance.

Ones Place Card. To start with, write on one of the five cards all those integers from 1 to 31 having a 1 in the ones place in their base 2 representation. In particular, this card will contain the integers 1, 3, 5, 7, 9, 11, 13, 15, 17, 19, 21, 23, 25, 27, 29, 31 (display the numbers on the card in evenly spaced rows or columns). To remind the magician (that's you) of this card's defining property, label the card's reverse side with an inconspicuous 1 on the upper right-hand corner.

Twos Place Card. On one of the remaining four blank cards, write those integers from 1 to 31 having a 1 in the twos place of their base 2 representation (2, 3, 6, 7, 10, 11, 14, 15, 18, 19, 22, 23, 26, 27, 30, 31), mark the reverse side of this card with a 2 in the upper right-hand corner.

Fours Place Card. On the next blank card, write those integers from 1 to 31 having a 1 in the fours place of their base 2 representation (4, 5, 6, 7, 12, 13, 14, 15, 20, 21, 22, 23, 28, 29, 30, 31), label the back of this card with a 4 in the upper right-hand corner.

Eights Place Card. For the next blank card, write the integers from 1 to 31 having a 1 in the eights place of their base 2 representation (8, 9, 10, 11, 12, 13, 14, 15, 24, 25, 26, 27, 28, 29, 30, 31), label the back of this card with an 8 in the upper right-hand corner.

Sixteenths Place Card. On the final blank card, write the integers from 1 to 31 having a 1 in the sixteenths place of their base 2 representation (16, 17, 18, 19, 20, 21, 22, 23, 24, 25, 26, 27, 28, 29, 30, 31), mark the back of this card with a 16 in the upper right-hand corner.

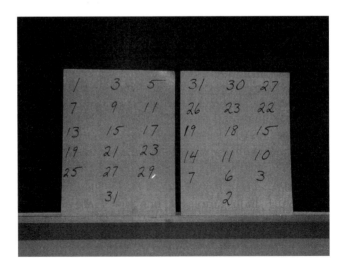

FIGURE 3.4.1 Ones place card and twos place card

FIGURE 3.4.2 Fours place card and eights place card

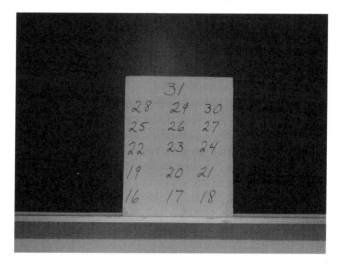

FIGURE 3.4.3 Sixteenths place card

Important Point. Note that each integer from 1 to 31 lies on one and only one ordered subset of these cards, since these integers have unique base 2 representations. Thus, if you know what cards a number is on, then you know the number.

Performing the Trick. Ask your students or friends to choose a whole number L from 1 to 31 and have them write it on a sheet of paper without you seeing it. Your task is to determine L from five yes or no questions that they will respond to. In terms of computers, these yes or no questions relate to a switch being on or off (i.e., current flowing or not flowing).

To add a little "shtick" to your performance, tell the audience that sometimes this trick does not work and you hope that they will not laugh at your efforts.

Show the first card (with the 1 on the back upper right-hand corner) to the viewers and ask them if their number is on this card. They need only respond yes or no to this question and to the other four questions that follow.

If their response is yes, then they have told you that their number L has a 1 in the ones place in its base 2 representation, if they answer no, then L has a 0 in the ones place in its base 2 notation. You need to remember their responses to this question and the four questions to follow, since they tell you their number in base 2 and you convert it in your head to its decimal (base 10) representation.

Next you show the second card (with the 2 in the back upper right-hand corner) and ask if it contains their number L. If it does, then L has a 1 in the twos place, if it does not, then L has a 0 in the twos place. At this stage, you have learned two base 2 digits of their number L, and you can secretly convert (in your head) this information to base 10 numerals.

For example, if they responded 0 to the first question and 1 to the second question, then you know that

$$L = (\_\_\_10)_2 = r_4 2^4 + r_3 2^3 + r_2 2^2 + 1 \cdot 2^1 + 0 \cdot r^0 = \_ + \_ + \_ + 2 + 0.$$

The trick proceeds in the same fashion through the remaining three cards, and after each response is given, you compute part of L from the information given. Once you know all five responses, then L will be completely determined. At this point of your magic routine, you can announce their number L (in base 10) and sheepishly ask if you are correct. When they say yes, you can "scold" them by saying that they should have chosen a more difficult number and should try again.

Let's rehearse the trick before taking it on stage. Suppose the number chosen is $L = 23$. The ordered responses to the five questions would be 1, 1, 1, 0, 1, and thus

$$L = (10111)_2 = 1 \cdot 2^4 + 0 \cdot 2^3 + 1 \cdot 2^2 + 1 \cdot 2^1 + 1 \cdot 2^0$$
$$= 16 + 0 + 4 + 2 + 1 = 23.$$

To make the trick flow smoothly, keep a running total in your head so that after the fifth response you can immediately announce the number that they secretly chose.

Base b Numeration Systems. We started this section by reviewing some terminology and notation concerning the decimal (base 10) system. Familiarity with this system paved the way for our exploration of the binary (base 2) numeration system.

Our base 2 investigations were stimulated by a measurement question raised in problem 9 of Exercises 3.2. In fact, we showed that all positive integer lengths of L inches can be measured with straightedges of lengths 2^n inches, where given straightedges of lengths 2^n inches can be used at most one time in the measurement of L and straightedge measurements can only be added.

In problem 9 of Exercises 3.2, it was also asked whether an analogous conclusion could be reached if the unmarked straightedges were of length 3^n inches rather than 2^n, and if not, what additional condition(s) were needed? Clearly not all integer lengths can be measured in this manner (e.g., 2 inches can't be measured this way), however, it is possible to handle all positive integer lengths with a slightly modified measurement scheme. In particular, the proof of Theorem 3.4.1 can be adapted (without much difficulty) to show that any positive integer L can be uniquely expressed as:

$$L = r_n 3^n + r_{n-1} 3^{n-1} + \ldots + r_2 3^2 + r_1 3^1 + r_0 3^0,$$

where $r_i \in \{0, 1, 2\}$ for each i and $r_n \neq 0$.

The above sum is called the **base 3 expanded form of L**, and in analogy to the binary system, L can also be written using the digits 0, 1, and 2 in the **ternary** (base 3) **place value** notation

$$L = (r_n r_{n-1} r_{n-2} \ldots r_2 r_1 r_0)_3.$$

Example. Let's find the unique base 3 place value representation of the integer $L = 157$. We'll proceed in a fashion similar to the base 2 analyses.

Analysis. We approach this task as we did the alternate method preceding Theorem 3.4.1 (which is the same method used in the proof of Theorem 3.4.1).

Dividing 157 by 3 yields a quotient of 52 with reminder 1, i.e.,

$$157 = 52 \cdot 3 + 1.$$

Since the quotient 52 is greater than or equal to 3, we can divide it by 3 and obtain the quotient 17 and the remainder 1. Therefore,

$$52 = 17 \cdot 3 + 1.$$

By replacing 52 with $17 \cdot 3 + 1$ in the first equality, L can be rewritten as

$$L = 157 = (17 \cdot 3 + 1) \cdot 3 + 1 = 17 \cdot 3^2 + 1 \cdot 3^1 + 1 \cdot 3^0.$$

Since the quotient 17 is greater than or equal to 3, we divide it by 3 to get

$$17 = 5 \cdot 3 + 2.$$

Substituting $5 \cdot 3 + 2$ for 17 in the latest expression for L yields

$$L = 157 = (5 \cdot 3 + 2) \cdot 3^2 + 3 + 1 = 5 \cdot 3^3 + 2 \cdot 3^2 + 1 \cdot 3^1 + 1 \cdot 3^0.$$

Again, as the quotient 5 is greater than or equal to 3, dividing it by 3 gives

$$5 = 1 \cdot 3 + 2,$$

and replacing 5 with $1 \cdot 3 + 2$ in the evolving form of L results in

$$L = 157 = (1 \cdot 3 + 2) \cdot 3^3 + 2 \cdot 3^2 + 3 + 1$$
$$= 1 \cdot 3^4 + 2 \cdot 3^3 + 2 \cdot 3^2 + 1 \cdot 3^1 + 1 \cdot 3^0.$$

Since the previous quotient 1 is less than 3, the process is complete, and we have found the unique base 3 expanded form of $L = 157$. Therefore, the base 3 place value representation of L is

$$L = 157 = (12211)_3.$$

Classroom Problems. Convert the following integers into base three numerals:

a. 65
b. 182
c. 909 ◆

The decimal system (respectively, binary system, ternary system) are examples of base 10 (respectively, base 2, base 3) place value numeration systems and in fact, for any positive integer $b \geq 2$, there exists a base b place value numeration system. More specifically, Theorem 3.4.1 and its proof can be extended (in a straightforward manner) to the following more general result.

3.4.2 Theorem. Let b be an integer with $b \geq 2$. If L is a positive integer, then there exists some integer $n \geq 0$ such that L can be expressed uniquely in the form

$$L = r_n b^n + r_{n-1} b^{n-1} + \ldots + r_2 b^2 + r_1 b^1 + r_0 b^0,$$

where $r_i \in \{0, 1, 2, \ldots, b-1\}$ for each i and $r_n \neq 0$.
The **base b place value** notation for L is given by

$$L = (r_n r_{n-1} \ldots r_2 r_1 r_0)_b.$$

Important Notational Conventions.

1. It is standard practice to delete the base b subscript in the base 10 place value notation for an integer L, i.e., $L = r_n r_{n-1} \ldots r_2 r_1 r_0$.
2. If $b > 10$, then in order to avoid ambiguities in base b place value representations, it is necessary to create new digits corresponding to integers greater than 9. For example, in base 13 we could introduce digits $a = 10$, $c = 11$, and $d = 12$, so that the complete set of thirteen available digits to represent an integer in base 13 would be 0, 1, 2, 3, 4, 5, 6, 7, 8, 9, a, c, and d. With this convention, the base 13 place value representation of 130 would be

$$130 = a \cdot 13^1 + 0 \cdot 13^0 = (a0)_{13}.$$

Notice the undesirable situation that would result if the numeral 10 was used instead of the newly created digit a. Under this scenario, the base 13 place value notation for 130 would be $(100)_{13}$. However, $(100)_{13}$ also equals 169, since $13^2 + 0 \cdot 13^1 + 0 \cdot 13^0 = 169$, and this circumstance is mathematically unacceptable.

Algorithmic Methods for Converting Base 10 Numerals into Other Bases. Susan's representation method and the alternate representation method were two procedures used to express integers in base 2 place value form. Both of these methods can also determine other base b place value representations. With a bit of streamlining, these approaches can be expressed as algorithms.

Susan's Representation Method: Focusing on Remainders. By reviewing and slightly modifying Susan's process for expressing any integer as a sum of distinct powers of 2, we can develop a general procedure for converting integers into base b place value notation.

To illustrate this variation of Susan's method (which we'll call Susan's base b algorithm), let's (again) write $L = 157$ in its base 2 place value form.

Susan's base b algorithm

Divide 157 by $2^7 = 128$, the largest power of 2 that is less than or equal to 157.	$128\overline{)157}$ 1 $\underline{128}$ 29
Divide the previous remainder 29 by the $2^6 = 64$ (Susan did not include this step, since she was not specifically looking for the unique base 2 place value representation of 157).	$64\overline{)29}$ 0 $\underline{0}$ 29
Divide the previous remainder 29 by $2^5 = 32$.	$32\overline{)29}$ 0 $\underline{0}$ 29
Divide the previous remainder 29 by $2^4 = 16$.	$16\overline{)29}$ 1 $\underline{16}$ 13
Divide the previous remainder 13 by $2^3 = 8$.	$8\overline{)13}$ 1 $\underline{8}$ 5
Divide the previous remainder 5 by $2^2 = 4$.	$4\overline{)5}$ 1 $\underline{4}$ 1
Divide the previous remainder 1 by $2^1 = 2$.	$2\overline{)1}$ 0 $\underline{0}$ 1
Divide the previous remainder 1 by $2^0 = 1$.	$1\overline{)1}$ 1 $\underline{1}$ 0
The quotients from the sequence of divisions give the base 2 representation.	$157 = (10011101)_2$

122 Chapter 3 The Division Algorithm and the Euclidean Algorithm

Classroom Problems. Using Susan's algorithm, perform the following conversions:

a. Convert 532 into base 3 place value notation.
b. Convert 30,314 into base 8 place value notation.
c. Convert 6,565 into base 11 place value notation (use the letter a to represent the digit having value 10).

Analysis. We'll do part b and leave the others for you. To further abbreviate the procedure, only the relevant powers of 8 and the accompanying division steps are noted.

$8^4 = 4{,}096$	$8^3 = 512$	$8^2 = 64$	$8^1 = 8$	$8^0 = 1$
$4{,}096 \overline{)30{,}314}$ 7 $\underline{28{,}672}$ $1{,}642$	$512 \overline{)1{,}642}$ 3 $\underline{1{,}536}$ 106	$64 \overline{)106}$ 1 $\underline{64}$ 42	$8 \overline{)42}$ 5 $\underline{40}$ 2	$1 \overline{)2}$ 2 $\underline{2}$ 0

The quotients of the division steps give the base 8 representation $30{,}314 = (73152)_8$. ◆

Alternate Representation Method: Focusing on Quotients. Now that Susan's method has been streamlined, we will show how the alternate representation method can also be condensed. The calculations in Susan's algorithm involved remainders of successive divisions, whereas the calculations in the alternate base b algorithm will involve quotients of successive divisions.

Let's illustrate the alternate base b algorithm (which is nothing more than an abbreviated version of the alternate representation method) by expressing the integer 1,861 in base 5 place value notation.

Alternate base b algorithm

Divide 1,861 by 5.	$\begin{array}{r} 372 \\ 5\overline{)1{,}861} \\ \underline{15} \\ 36 \\ \underline{35} \\ 11 \\ \underline{10} \\ 1 \end{array}$	$1{,}861 = 372 \cdot 5 + 1$
Divide the quotient 372 of the previous division by 5.	$\begin{array}{r} 74 \\ 5\overline{)372} \\ \underline{35} \\ 22 \\ \underline{20} \\ 2 \end{array}$	$372 = 74 \cdot 5 + 2$ $1{,}861 = 372 \cdot 5 + 1$ $\phantom{1{,}861} = (74 \cdot 5 + 2) \cdot 5 + 1$ $\phantom{1{,}861} = 74 \cdot 5^2 + 2 \cdot 5 + 1$

Divide the quotient 74 of the previous division by 5.	$\begin{array}{r} 14 \\ 5\overline{)74} \\ \underline{5} \\ 24 \\ \underline{20} \\ 4 \end{array}$	$74 = 14 \cdot 5 + 4$ $\begin{aligned} 1{,}861 &= 74 \cdot 5^2 + 2 \cdot 5 + 1 \\ &= (14 \cdot 5 + 4) \cdot 5^2 \\ &\quad + 2 \cdot 5 + 1 \\ &= 14 \cdot 5^3 + 4 \cdot 5^2 \\ &\quad + 2 \cdot 5 + 1 \end{aligned}$
Divide the quotient 14 of the previous division by 5.	$\begin{array}{r} 2 \\ 5\overline{)14} \\ \underline{10} \\ 4 \end{array}$	$14 = 2 \cdot 5 + 4$ $\begin{aligned} 1{,}861 &= 14 \cdot 5^3 + 4 \cdot 5^2 \\ &\quad + 2 \cdot 5 + 1 \\ &= (2 \cdot 5 + 4) \cdot 5^3 + 4 \cdot 5^2 \\ &\quad + 2 \cdot 5 + 1 \\ &= 2 \cdot 5^4 + 4 \cdot 5^3 + 4 \cdot 5^2 \\ &\quad + 2 \cdot 5 + 1 \end{aligned}$
Divide the quotient 2 of the previous division by 5. The process is complete when the quotient is 0.	$\begin{array}{r} 0 \\ 5\overline{)2} \\ \underline{0} \\ 2 \end{array}$	$2 = 0 \cdot 5 + 2$: Replacing 2 with this expression does not change the representation below. $\begin{aligned} 1{,}861 &= 2 \cdot 5^4 + 4 \cdot 5^3 \\ &\quad + 4 \cdot 5^2 + 2 \cdot 5 + 1 \end{aligned}$
The remainders, in the reverse order of the divisions, give the base 5 form.	$1{,}861 = (24421)_5$	

In my high school American history class, we learned that the Civil War began in 1861. In fact, my history teacher taught us the mnemonic "The war begun in 61." I wonder what grade I would have received on my high school American history final exam if I had written down that the Civil War began in $(24421)_5$?

Classroom Problems. Using the alternate base b algorithm, perform the following conversions:

a. Convert 3,719 into base 4 place value notation.
b. Convert 22,875 into base 7 place value notation.
c. Convert 17,996 into base 12 place value notation (use the letter a to represent the digit having value 10 and the letter c to represent the digit having value 11; we skipped the letter b since we use it in the general terminology, "base b").

Analysis. We'll do c and leave the others for you. To further abridge the procedure, only the relevant division steps are noted.

$$\begin{array}{r} 1{,}499 \\ 12\overline{)17{,}996} \\ \underline{17{,}988} \\ 8 \end{array} \qquad \begin{array}{r} 124 \\ 12\overline{)1{,}499} \\ \underline{1{,}536} \\ 11 \end{array} \qquad \begin{array}{r} 10 \\ 12\overline{)124} \\ \underline{120} \\ 4 \end{array} \qquad \begin{array}{r} 0 \\ 12\overline{)10} \\ \underline{0} \\ 10 \end{array}$$

The remainders of the division steps (in reverse order) give the base 12 representation, $17{,}996 = (a4c8)_{12}$. ◆

Our limited investigation of base b numeration systems was primarily motivated by a measurement question in Exercises 3.2, 9, and this led us to the study of base b place value representations. We have seen that the decimal system is a special case of a base b numeration system, and in fact, the arithmetic of base b numeration systems is completely analogous to the arithmetic of the decimal system. To help illustrate this point, we'll consider a few base 3 arithmetic calculations.

To add and multiply arbitrary base 3 numerals, it is first important to understand base 3 addition and multiplication for the digits 0, 1, and 2. Since $0 = (0)_3$, $1 = (1)_3$, and $2 = (2)_3$, we will suppress the subscripts of the digits in the following base 3 addition and multiplication tables:

Base 3 addition table

+	0	1	2
0	0	1	2
1	1	2	$(10)_3$
2	2	$(10)_3$	$(11)_3$

Base 3 multiplication table

·	0	1	2
0	0	0	0
1	0	1	2
2	0	2	$(11)_3$

Base 3 addition problem

$(122021)_3$
$+(22102)_3$
$(221200)_3$

Base 3 addition check

$(122021)_3 = 466$
$+(22102)_3 = +227$
$(221200)_3 = 693$

Base 3 multiplication problem

$$(2102)_3$$
$$\times (22)_3$$
$$\overline{(11211)_3}$$
$$(11211)_3$$
$$\overline{(201021)_3}$$

Base 3 multiplication check

$$(2102)_3 = 65$$
$$\times (22)_3 = \times 8$$
$$\overline{(201021)_3 = 520}$$

Classroom Problems. Construct the addition and multiplication tables for base 5 arithmetic, and then use them to complete the following addition and multiplication problems. Check your results with base 10 arithmetic.

$$(432103)_5$$
$$+(133402)_5$$

$$(230412)_5$$
$$\times (423)_5$$ ◆

EXERCISES 3.4

1. First use Susan's base b algorithm to complete each of the following conversions, and then use the alternate base b algorithm to make the notational changes.
   a. Convert 2,004 into base 3 notation.
   b. Convert 30,235 into base 6 notation.
   c. Convert $(1100110011)_2$ into base 5 notation.
   d. Convert 276664 into base 13 notation (use $a = 10, c = 11$, and $d = 12$).
   e. Convert $(564a017)_{11}$ into base 9 notation.
   f. Convert your home phone number (area code included) into base 2 notation. (It would be really difficult to remember your phone number in base 2).
2. Construct the base 8 addition and multiplication tables and then complete the following arithmetic problems. Check your results in base 10 arithmetic.

   a. $(5072164)_8$
      $+(345117)_8$

   b. $(6052133)_8$
      $\times (2406)_8$

3. Construct the base 11 addition and multiplication tables (use $a = 10$) and then complete the following arithmetic problems. Check your results in base 10 arithmetic.

   a. $(6a892a)_{11}$
      $+(47a65)_{11}$

   b. $(a18a9)_{11}$
      $\times (23a)_{11}$

3.5 PRIME THOUGHTS

We have seen that the prime numbers are the "multiplicative" building blocks of the integers and have used this knowledge to broaden and deepen our understanding of the integers. Although we have considered many interesting and useful facts concerning prime numbers, there is still a voluminous collection of "prime" results available for mathematically inquisitive minds. Moreover, if the monumental existing literature does not satisfy the appetite of the hungriest students of mathematics, don't fret, since mathematicians are continually exploring and discovering new insights in the fascinating world of prime numbers.

In this section, we study a small sampling of some classical results concerning prime numbers.

Recall from Section 2.6 that we wanted to find a reasonable primality test for a given positive integer n. Rather than checking all positive integers d in the range $1 < d < n$ as potential divisors, we saw that it was good enough to check for possible *prime* divisors d in the range $1 < d \leq \sqrt{n}$. Although this test materially decreases the number of potential divisors, it still is a daunting task to test for and find large prime numbers. In fact, the speed of newly developed super computers is often gauged by their ability to find extremely large prime numbers.

News Alert. On February 18, 2005, Mersenne.org reported that Dr. Martin Nowak, an eye surgeon from Michelfeld, Germany, discovered the largest known prime number, $p = 2^{25,964,951} - 1$. This humongous integer has 7,816,230 digits; later in this section, we discuss an interesting collection of prime numbers (called Mersenne primes) having the same structure as this "monster" prime.

Panning for Primes with the Sieve of Eratosthenes. The year 1849 was the beginning of the great California gold rush, where hopeful people full of dreams traveled across America to California in search of gold. They stood side by side in rivers and streams and with their screened pans spent countless hours sifting through river sand for those rare shiny nuggets. Little did these 49ers know that some 2,000 years earlier the Greek mathematician Eratosthenes was also sifting for some rare gems, but of a much different variety. In fact, the sieve of his invention, which is described here, was created to pan the integers for the gems we call *primes*.

To illustrate the Sieve of Eratosthenes, we sift for all primes less than $n = 130$.

The method involves successive elimination of proper prime multiples $mp (m \geq 2)$, where $p \leq \sqrt{130} = 11.4$, culminating in a list of all primes less than $n = 130$. More specifically, starting with the first prime 2, cross out all multiples of 2 other than 2; then cross out all multiples of 3 other than 3 that you have not previously crossed out; keep doing this for all multiples of 5 other than 5, all multiples of 7 other than 7, and finally all multiples of 11 other than 11.

Conclusion. The crossed out integers are all the composites less than or equal to $n = 130$ (why?), and all the remaining uncrossed out numbers are the primes less than $n = 130$ (why?). In effect, the composites have slipped through the sieve, and only the primes remain.

Variations of the Sieve of Eratosthenes are used by present-day researchers to count the number of primes below a positive integer n that possess certain specific properties, such as pairs of primes differing by 2 (see "Twin Primes" discussed later in this chapter).

	2	3	~~4~~	5	~~6~~	7	~~8~~	~~9~~	~~10~~
11	~~12~~	13	~~14~~	~~15~~	~~16~~	17	~~18~~	19	~~20~~
~~21~~	~~22~~	23	~~24~~	~~25~~	~~26~~	~~27~~	~~28~~	29	~~30~~
31	~~32~~	~~33~~	~~34~~	~~35~~	~~36~~	37	~~38~~	~~39~~	~~40~~
41	~~42~~	43	~~44~~	~~45~~	~~46~~	47	~~48~~	~~49~~	~~50~~
~~51~~	~~52~~	53	~~54~~	~~55~~	~~56~~	~~57~~	~~58~~	59	~~60~~
61	~~62~~	~~63~~	~~64~~	~~65~~	~~66~~	67	~~68~~	~~69~~	~~70~~
71	~~72~~	73	~~74~~	~~75~~	~~76~~	~~77~~	~~78~~	79	~~80~~
~~81~~	~~82~~	83	~~84~~	~~85~~	~~86~~	87	~~88~~	89	~~90~~
91	~~92~~	~~93~~	~~94~~	~~95~~	~~96~~	97	~~98~~	~~99~~	~~100~~
101	~~102~~	103	~~104~~	~~105~~	~~106~~	107	~~108~~	109	~~110~~
~~111~~	~~112~~	113	~~114~~	~~115~~	~~116~~	~~117~~	~~118~~	119	~~120~~
~~121~~	~~122~~	~~123~~	~~124~~	~~125~~	~~126~~	127	~~128~~	~~129~~	~~130~~

FIGURE 3.5.1 The Sieve of Eratosthenes.

Even though primes are not that easy to locate, they are in great abundance. This naturally leads to the next question.

Question. Are there infinitely many prime numbers?

A Proof or Not a Proof. A student once claimed that the answer is obviously yes and offered a simple argument as proof. Since there are infinitely many positive integers, and each positive integer is a product of primes, then "it follows" that there are infinitely many prime numbers. Is this a complete argument, or is it flawed? (Explain.)

Evidence and intuition suggest that the previous question does have an affirmative answer, but what methods or reasoning can we use to show that the primes are truly infinite in abundance? The computer is an indispensable tool in the search for primes but is not helpful for settling the infinite question (why?). Maybe it is feasible to consider methods used to demonstrate the infinity of other sets of integers, such as the even integers or the Fibonacci numbers, and then try to adapt these ideas to prime numbers.

Firstly, how would we show that the even integers form an infinite set? One argument might be that each integer n can be matched up to an even integer $2n$, and since there are infinitely many integers, then there are infinitely many even integers. Put more succinctly, the function $f(n) = 2n$ defined from the integers into the integers, describes a one-to-one correspondence between the integers and the even integers.

$$\begin{array}{rl} \text{Integers} & \ldots\ -n\ \ldots\ -3\ -2\ -1\ 0\ 1\ 2\ 3\ \ldots\ n\ \ldots \\ & \updownarrow \qquad\quad \updownarrow\ \updownarrow\ \updownarrow\ \updownarrow\ \updownarrow\ \updownarrow\ \updownarrow \qquad \updownarrow \\ \text{Even integers} & \ldots\ -2n\ \ldots\ -6\ -4\ -2\ 0\ 2\ 4\ 6\ \ldots\ 2n\ \ldots \end{array}$$

Secondly, the infinitude of the Fibonacci numbers is a direct consequence of their recursive definition $a_1 = a_2 = 1$ and $a_n = a_{n-1} + a_{n-2}$, where $n \geq 3$ (why?).

There Are Infinitely Many Prime Numbers: Indirect Approach. Can similar "formula-type" arguments be developed for application to the set of prime numbers? Unfortunately, no useful prime-producing formula has ever been discovered, so it is not clear how to proceed with this kind of direct approach.

Analysis. Let us try an indirect attack by assuming that there are only finitely many distinct primes $p_1, p_2, \ldots, p_r$ and then showing that this assumption leads to a contradiction. We know by the Fundamental Theorem of Arithmetic that all integers $n > 1$ can be uniquely factored into a product of prime numbers, and by our assumption, the primes in any factorization must come from the list $p_1, p_2, \ldots, p_r$. If an integer $n > 1$ could be constructed so that it is not divisible by any of these primes, then we would have found a contradiction to the FTA.

1. Euclid, in Book IX of *Elements*, produced the desired contradiction by considering the integer $n = p_1 p_2 \cdots p_r + 1$ and noting that none of the p_i divide n. (Why?) (see Exercise 2.4, 4).
2. A similar idea is to form the integer $m = p_1 + (p_2 \cdots p_r)$, and again notice that m is not divisible by any of the p_i (why?). What are some other variations to this construction that would also lead to a contradiction?

Either of the constructions just considered lead to a contradiction of the FTA, thus our original assumption is false, and consequently there are infinitely many prime numbers.

Smell the Roses. Take a moment to reflect on the beauty of Euclid's proof. In its stunning simplicity, we can all appreciate the power of mathematical reasoning.

Gaps between Primes. Even though the prime numbers are infinite in extent, successive primes p_t and p_{t+1} may be separated by varying numbers of composites, and except for the pair of primes (2, 3), there is always an odd number $n = (p_{t+1} - p_t) - 1$ of composites between successive primes (why?). This observation leads to the following questions.

Questions.

1. How many consecutive composite numbers can be listed?
2. For each *odd* positive integer n, does there exist a pair of successive prime numbers p_t and p_{t+1} that are separated by exactly n consecutive composite numbers?

In terms of the second question, it might be helpful to look at some pairs of primes that have differing numbers of composites between them.

Some successive primes	n consecutive composites
3, 5	1
7, 11	3
31, 37	5
89, 97	7
113, 127	13

It is challenging to find examples of successive primes separated by an odd number n of consecutive composites. In fact, it is not known whether this can be accomplished for any odd integer n (Ribenboim, *The New Book of Prime Number Records*, p. 250).

Actually, much less is known. In particular, it is not yet known whether there exists distinct primes $p < q$ (not necessarily successive) so that $(q - p) = n$ for any even integer $n \geq 2$ (or equivalently so that $(q - p) - 1 = m$ for any odd integer $m \geq 1$). Question 2 asks for significantly more than this fairly innocent question, and it is remarkable—but not uncommon within the world of integers—how such simply stated natural questions can elude the grasp of generations of brilliant mathematicians.

Although Question 2 remains unanswered, it is straightforward to answer Question 1. Indeed, a clever use of "factorial" makes it quite easy to produce n consecutive composites for *any* integer $n \geq 1$. For example, if $n = 3$, the following list consists of three consecutive composites:

$$4! + 2 = 26, 4! + 3 = 27, \text{ and } 4! + 4 = 28.$$

Notice (from the previous chart) that these are not the smallest three consecutive composites, but nevertheless, they are three consecutive composites. For another example, say $n = 6$ consecutive composites, consider the list

$$7! + 2, 7! + 3, 7! + 4, 7! + 5, 7! + 6, \text{ and } 7! + 7,$$

and for $n \geq 1$ consecutive composites, the following general list works as well:

$$(n + 1)! + 2, (n + 1)! + 3, (n + 1)! + 3, \ldots, (n + 1)! + n, \text{ and}$$
$$(n + 1)! + (n + 1).$$

Twin Primes. While searching for successive primes p_t and p_{t+1} separated by exactly n consecutive composites, it becomes apparent that there is an abundance of pairs (p_t, p_{t+1}) separated by *one* composite, i.e., $p_{t+1} - p_t = 2$, and such pairs of primes are referred to as **twin primes**.

Some examples of twin primes are: (3, 5), (5, 7), (11, 13), (17, 19), (29, 31), (41, 43), and (59, 61).

130 Chapter 3 The Division Algorithm and the Euclidean Algorithm

The largest known twin primes to date are the pair

$$([(33{,}218{,}925 \cdot 2^{169{,}690}) - 1], [(33{,}218{,}925 \cdot 2^{169{,}690}) + 1]).$$

These mammoth primes were discovered in 2002, and each contains 51,090 digits. It has been conjectured that there are infinitely many pairs of twin primes (**twin prime conjecture**), but no one has been able to verify this or refute it. A solution to this conjecture would bring instant fame and (some) fortune to the solver, and someday (maybe not in my lifetime) a creative mathematical mind will "slay this dragon."

Prime Triples. The triple $(3, 5, 7)$ consists of three primes that successively differ by two and is referred to as a **prime triple**, i.e., a triple of primes of the form

$$(p, p + 2, p + 4).$$

Although the number of pairs of twin primes is not yet known, the number of prime triples can be determined.

A search for other prime triples must begin with twin primes, so let's tabulate some examples.

Twin primes	Prime triples	Yes	No
(3, 5)	(3, 5, 7)	√	
(5, 7)	(5, 7, 9)		√
(11, 13)	(11, 13, 15)		√
(17, 19)	(17, 19, 21)		√
(29, 31)	(29, 31, 33)		√

A pattern that has emerged from these few examples is that if $(p, p + 2)$ are twin primes and $p > 3$, then $3 \mid (p + 4)$. If we can show that this is true in general, then $(3, 5, 7)$ would be the only prime triple.

Analysis. To prove this statement, suppose that $(p, p + 2)$ is a pair of twin primes and that $p > 3$. By the division algorithm, either $p = 3q + 1$ or $p = 3q + 2$. However, since $p + 2$ is a prime, then $p \neq 3q + 1$ (why?). Therefore, $p + 4 = (3q + 2) + 4$ is divisible by 3 (why?), which is exactly what we wanted to prove.

Actually, a more general statement can be established. In fact, if a is any integer, then exactly one of the integers $a, a + 2, a + 4$ is divisible by 3 (see Exercises 3.2, 4b).

Polynomials Having Prime Values. The sporadic nature of the primes is part of their fascination, but it is also a reason why questions concerning primes are often difficult to answer. Life among the integers would be much simpler, but definitely not as interesting, if a useful formula for generating the nth prime was known.

Linear Polynomials. One thing is for sure, it is easy to see that a linear polynomial $f(x) = ax + b$ (a and b are positive integers) could never possess the property that $f(n)$ = the nth prime ($n \geq 1$). (Construct an argument.) In fact, a linear polynomial $f(x) = ax + b$ (a and b are positive integers) cannot solely have prime values for each positive integer n.

Analysis. To establish this assertion, suppose to the contrary, i.e., assume that $f(n) = an + b$ is prime for each positive integer n. By assumption, $f(1) = p = a \cdot 1 + b$, where p is some prime number, and so

$$f(1 + p) = (1 + p)a + b = a + b + pa = p + pa = p(1 + a), \quad \textbf{3.5.1}$$

which is a composite (note that $a > 0$ by assumption). This contradiction completes the argument.

Although a linear polynomial with integer coefficients cannot exclusively produce prime values for each positive integer n, it could take on infinitely many prime values. For example, $f(x) = x + 1$ produces infinitely many prime values when evaluated on the positive integers (why?), whereas, $f(x) = 6x + 4$ only gives composite numbers for positive integer values of n (why?).

More generally, for a linear polynomial $f(x) = ax + b$ (a and b are positive integers), if $\gcd(a, b) = d > 1$, then $f(n) = an + b$ is composite for each positive integer n (why?).

Question. Is the converse of the previous statement true, i.e., if $f(n) = an + b$ is composite for each integer $n \geq 1$, then must $\gcd(a, b) = d > 1$?

Analysis. Using an indirect approach, assume $f(n) = an + b$ is composite for each integer $n \geq 1$ and that $\gcd(a, b) = 1$. To reach a contradiction, it would suffice to show that there exists a positive integer n_0 so that $an_0 + b$ is prime. Reaching this conclusion involves mathematics beyond the scope of these materials, in fact, a famous theorem in number theory proved in 1837 by Dirichlet guarantees infinitely many primes of this form.

3.5.2 Theorem Dirichlet. If $\gcd(a, b) = 1$ (a and b positive integers), then $f(x) = ax + b$ produces infinitely many primes when evaluated on the positive integers, i.e., for positive integers a and b, the arithmetic sequence $a + b, 2a + b, 3a + b, \ldots, na + b, \ldots$ contains infinitely many primes (see Section 1.3).

It is interesting to note that to prove Dirichlet's theorem, it is good enough to show that the following statement is true.

3.5.3 If $\gcd(s, t) = 1$ for positive integers s and t, then there exists a positive integer n_0 so that $sn_0 + t$ is prime.

Analysis. We wish to show that the truth of Statement 3.5.3 implies the conclusion of Dirichlet's theorem. Suppose that Statement 3.5.3 is true, and let a and b be positive integers with $\gcd(a, b) = 1$. Thus there exists a positive integer n_0 so that $an_0 + b = p_0$ is prime. We wish to show that there are infinitely many primes of the form $an + b$ with n a positive integer. Since $\gcd(a, an_0 + b) = 1$ (why?), then again by Statement 3.5.3 there exists a positive integer n_1 such that $an_1 + (an_0 + b) = p_1$ is prime. Moreover, $p_1 = an_1 + (an_0 + b) = a(n_1 + n_0) + b$ and $p_0 \neq p_1$, so we have found another prime of the form $an + b$. This iterative process shows that if Statement 3.5.3 is true, then there are infinitely many primes of the form $an + b$, where n is a positive integer, and this is precisely the conclusion of Dirichlet's theorem.

Quadratic Polynomials. Can our work on linear polynomials be generalized to quadratic polynomials, i.e., does it follow that a quadratic polynomial $f(x) = ax^2 + bx + c$ (a, b, and c are positive integers) cannot have prime values for each positive integer n?

Classroom Problems. Determine the *smallest positive* integer n for which:

a. $f(n) = 2n^2 + 29$ is composite;
b. $f(n) = n^2 + n + 41$ is composite.

In both examples, each polynomial produces several primes, but neither exclusively generates only primes. In fact, an argument similar to the linear case shows that no quadratic, and more generally no polynomial with positive integer coefficients, can solely have prime values. We'll handle the quadratic analysis now and leave the general case for the exercises. ◆

Analysis. Proceeding indirectly as in the linear case, assume that $f(n) = an^2 + bn + c$ is prime for each integer $n \geq 1$ (a, b, and c are positive integers). In particular, $f(1) = p = a + b + c$, for some prime p, and then evaluating, expanding, and rearranging as in calculation (3.4.1) gives

$$f(1 + p) = a(1 + p)^2 + b(1 + p) + c = (a + b + c) + p(2 + b + p)$$
$$= p + p(2 + b + p),$$

which is composite and contradicts the assumption.

Infinitely Many Prime Values. Even though nonconstant polynomials with integer coefficients cannot exclusively have prime values, we saw that certain linear polynomials possess infinitely many prime values (Dirichlet's theorem 3.5.2). An analogous general result for higher-degree polynomials is not yet known, in particular, the polynomials $f(x) = 2x^2 + 29$ and $f(x) = x^2 + x + 41$ (considered in the previous Classroom Problem) might or might not have infinitely prime values.

Question. Does the polynomial $f(x) = x^2 + 1$ have infinitely many prime values? A table is useful to organize our findings.

x	$f(x) = x^2 + 1$	Prime
1	2	yes
2	5	yes
4	17	yes
6	37	yes
8	65	no
10	101	yes
12	145	no
14	197	yes
16	257	yes
18	325	no
20	401	yes

Classroom Problems. Complete four more rows of the table.

The evidence from the table indicates that $f(x) = x^2 + 1$ has many prime values, but it does not bring us any closer to knowing whether the given polynomial has infinitely many prime values. An indirect approach is certainly more feasible, but no one (to this date) has crafted an argument that settles the given question. Even more amazing is the general state of affairs. ◆

Stark Reality. The existence of a quadratic polynomial possessing infinitely many prime values has not yet been established, and current researchers continue to work on this and related problems.

Finitely Many Prime Values. There are some specific quadratic and higher-degree polynomials for which it is possible to determine all prime values they produce. For an interesting example, consider the quadratic polynomial $f(x) = x^2 - 1$. Notice if $x = 2$, then $f(2) = 3$ is a prime value of this polynomial.

Question. Are there positive integers other than $x = 2$ that produce prime values for the polynomial $f(x) = x^2 - 1$?

Analysis. To answer this question, let's suppose $f(a) = p$ is a prime number for some positive integer $x = a$ and see what deductions we can make from this assumption. We know that

$$f(a) = a^2 - 1 = (a-1)(a+1) = p,$$

and since p is prime and $a - 1 < a + 1$, then $a - 1 = 1$ (why?). Therefore, $x = a = 2$ is the only positive integer making $f(x)$ prime.

The Cubic Case. Notice that $f(2) = 2^3 - 1 = 7$ is prime. Is $x = 2$ the only positive integer making $f(x) = x^3 - 1$ prime?

Analysis. Proceeding as in the quadratic case, assume that $f(a) = p$ is a prime number for some positive integer $x = a$. Hence,

$$f(a) = a^3 - 1 = (a - 1)(a^2 + a + 1);$$

thus it follows (by an argument similar to the quadratic case) that $x = 2$ is the only positive integer making $f(x) = x^3 - 1$ prime. (Justify.)

The Quartic Case. Unlike the quadratic and cubic cases, the quartic polynomial $f(x) = x^4 - 1$ is not prime at $x = 2$, i.e., $f(2) = 2^4 - 1 = 15$. In fact, the polynomial $f(x) = x^4 - 1$ has no prime values for any positive integer x.

Analysis. Suppose that $f(a) = p$ is a prime number for some positive integer $x = a \neq 2$. Since

$$f(a) = (a^4 - 1) = (a - 1)(a^3 + a^2 + a^1 + 1),$$

it follows that $a = 2$ (justify), a contradiction. Therefore, the polynomial $f(x) = x^4 - 1$ has no prime values for any positive integer x.

3.5.4 Classroom Problems: The General Case. Prove. If $f(x) = x^n - 1$ is prime for some positive integer $x = a$, then $a = 2$. Direction: $a^n - 1 = (a - 1)(a^{n-1} + a^{n-2} + \ldots + a^2 + a^1 + 1)$. ◆

At this point of our work, we are naturally drawn to the next question.

Question. For which positive integers $n > 1$ is $2^n - 1$ is prime?

A table of some examples might help formulate a conjecture.

n	$2^n - 1$	Prime
2	3	yes
3	7	yes
4	15	no
5	31	yes
6	63	no
7	127	yes
8	255	no

By inspecting the data in the table, it is tempting to conjecture that $2^n - 1$ is prime if n is prime. However, further calculations reveal that $2^{11} - 1 = 2,047 = 23 \cdot 89$ is composite.

Another question that arises from the limited evidence considered is whether the converse of the previous conjecture is valid.

Question. If $2^n - 1$ is prime, does it follow that n is prime?

Analysis (Indirect). Suppose that $2^n - 1$ is prime and n is not prime, i.e., $n = ab$, where $1 < a < n$ and $1 < b < n$. Notice that

$$2^n - 1 = 2^{ab} - 1 = (2^a)^b - 1 = x^b - 1,$$

where $x = 2^a$. Since $x^b - 1$ is prime by assumption, then we have previously seen (Classroom problem 3.5.4) that $x = 2$. Therefore, $2 = 2^a$, so $a = 1$. This contradiction implies that n must be prime, and the analysis is now complete.

3.5.5 Summary. If $2^n - 1$ is prime, then n is prime.

Primes of the form $2^n - 1$ play a central role in the topic we consider next.

Perfect Numbers. Are some numbers better than others, and if so, what makes them so appealing? Many of us have lucky numbers or numbers that have special significance to our lives (e.g., my wedding anniversary is on 7/7/77). Sometimes our favorites are selected because they possess some interesting mathematical property, but oftentimes they are chosen for purely personal reasons. (A former student once told me that his favorite number was π because this was also his favorite food.)

The ancient Greek philosophers and mathematicians were fascinated with a special class of numbers that had both mathematical and spiritual meaning to them. They noticed that certain positive integers, such as the number 6, were equal to the sum of their *positive proper divisors*, i.e., all divisors of n other than n itself. Because of this astonishing property and certain divine associations, these integers were called **perfect**.

An equivalent definition is:

n is **perfect** if the sum of all positive divisors of n equals $2n$, i.e., $\sum_{d|n} d = 2n$.

Examples of perfect numbers that were known in ancient times are 6, 28, 496, and 8,128.

$$1 + 2 + 3 = 6$$
$$1 + 2 + 4 + 7 + 14 = 28$$
$$1 + 2 + 4 + 8 + 16 + 31 + 64 + 124 + 248 = 496.$$

Problem. Show that 8,128 is perfect.

A Perfect Surprise. We soon investigate perfect numbers in more depth, but first let's consider an interesting observation concerning perfect numbers. The positive divisors of the perfect number 6 are 1, 2, 3, and 6. We can also represent these divisors as $\frac{6}{1}, \frac{6}{2}, \frac{6}{3}$, and $\frac{6}{6}$. Thus,

$\frac{6}{1} + \frac{6}{2} + \frac{6}{3} + \frac{6}{6} = 12$, so multiplying both sides of the equality by $\frac{1}{6}$ gives

$$\frac{1}{1} + \frac{1}{2} + \frac{1}{3} + \frac{1}{6} = 2.$$

Similarly, the positive divisors of the perfect number 28 can be expressed as $\frac{28}{1}, \frac{28}{2}, \frac{28}{4}, \frac{28}{7}, \frac{28}{14}, \frac{28}{28}$. Therefore,

$$\frac{28}{1} + \frac{28}{2} + \frac{28}{4} + \frac{28}{7} + \frac{28}{14} + \frac{28}{28} = 56, \text{ and thus}$$

$$\frac{1}{1} + \frac{1}{2} + \frac{1}{4} + \frac{1}{7} + \frac{1}{14} + \frac{1}{28} = 2.$$

We could certainly make analogous computations for the perfect numbers 496 and 8,128, but instead let's use the previous examples to help construct an argument showing that the sum of the reciprocals of the positive divisors of a perfect number n will always be 2.

Given a perfect number n, denote the positive divisors of n by $d_1, d_2, \ldots, d_t$. We know that these divisors can also be represented as $\frac{n}{d_1}, \frac{n}{d_2}, \ldots, \frac{n}{d_t}$ (why?), so

$$\frac{n}{d_1} + \frac{n}{d_2} + \ldots + \frac{n}{d_t} = 2n, \text{ which in turn gives } \frac{1}{d_1} + \frac{1}{d_2} + \ldots + \frac{1}{d_t} = 2.$$

Summary: A Characterization of Perfect Numbers. A positive integer n is perfect if and only if the sum of the reciprocals of the positive divisors of n equals 2, i.e., $\sum_{d|n} \frac{1}{d} = 2$.

Consequence of the Reciprocal Characterization of Perfect Numbers. As we have seen before and will continue to see throughout these materials, expressing mathematical statements in equivalent forms is a fundamental tool in solving mathematical problems. Often a question that is not evident using one definition becomes more accessible using an equivalent definition.

For example, the next statement follows quite easily from the reciprocal characterization of perfect numbers but is not as transparent using the original definition of perfect number.

Observation. If g is a *proper* divisor of a perfect number n, then g is not perfect.

Analysis. To see why this observation is true, note that

$$\sum_{d|g} \frac{1}{d} < \sum_{d|n} \frac{1}{d} = 2,$$

since any divisor of g is a divisor of n and $g \neq n$. Therefore, g is not perfect.

Problem. Construct another proof of the same assertion, but this time use the original definition of perfect number.

The Structure of Even Perfect Numbers. Euclid, in *Elements* (Proposition 36 of Book IX), showed that if the prime factorization of an even positive integer has a very specific structure (described here in Theorem 3.5.6), then it is a perfect number.

A closer inspection of the prime factorizations of the perfect numbers we've already examined serve to motivate Euclid's result.

$$6 = 2 \cdot 3 = 2 \cdot (2^2 - 1) \qquad\qquad 496 = 2^4 \cdot 31 = 2^4 \cdot (2^5 - 1)$$
$$28 = 2^2 \cdot 7 = 2^2 \cdot (2^3 - 1) \qquad\qquad 8{,}128 = 2^6 \cdot 127 = 2^6 \cdot (2^7 - 1)$$

3.5.6 Theorem. If $p = 2^m - 1$ is a prime number for $m > 1$, then $n = 2^{m-1}p$ is a perfect number.

Proof. To show that $n = 2^{m-1}p$ is perfect, we need to verify that $\sum_{d|n} d = 2n$. With the aid of the Fundamental Theorem of Arithmetic, the divisors of n are

$$1, 2, 2^2, \ldots, 2^{m-1}, p, 2p, 2^2p, \ldots, 2^{m-1}p = n,$$

and their sum is

$$\sum_{d|n} d = 1 + 2 + 2^2 + \ldots + 2^{m-1} + p + 2p + 2^2p + \ldots + 2^{m-1}p$$
$$= (1 + 2 + 2^2 + \ldots + 2^{m-1}) + p(1 + 2 + 2^2 + \ldots + 2^{m-1})$$
$$= (1 + 2 + 2^2 + \ldots + 2^{m-1})(1 + p).$$

An application of the formula for the sum of a geometric progression (see Section 1.4) shows that $1 + 2 + 2^2 + \ldots + 2^{m-1} = (2^m - 1)$. Therefore,

$$\sum_{d|n} d = (1 + 2 + 2^2 + \ldots + 2^{m-1})(1 + p) = (2^m - 1)[1 + (2^m - 1)]$$
$$= (2^m - 1)2^m = 2(2^{m-1}p) = 2n,$$

thus $n = 2^{m-1}(2^m - 1)$ is perfect.

3.5.7 Question. Is the converse of Euclid's result valid? (State the converse.)

If so, this would give a complete characterization for even perfect numbers. The answer to this important question is revealed in the ensuing discussion (see Theorem 3.5.8).

Mersenne Primes. Euclid's result brought special significance to primes of the form $2^m - 1$, since each of these primes would produce perfect numbers. Furthermore, as $2^m - 1$ is prime only when m is prime (Summary 3.5.5) but not conversely (see next paragraph), the discovery of new perfect numbers is fundamentally linked to determining which primes m result in $2^m - 1$ being prime. Given these facts, it is unimaginable that the fifth perfect number was not discovered until the fifteenth or sixteenth century (it is not clear when and who first discovered the fifth perfect number). However, a few calculations using Roman numerals vividly demonstrates the arithmetical limitations inherent to this number system and the incredible difficulties encountered in calculations involving large numbers.

The discovery of the fifth perfect number is often attributed to Hudalrichus Regius, where in his 1536 publication *Utriusque Arithmetices* he showed that $2^{13} - 1 = 8{,}191$ is prime, and consequently $2^{12}(2^{13} - 1) = 33{,}550{,}336$ is perfect. He also demonstrated, through the factorization $2^{11} - 1 = 23 \cdot 89$, that it is possible for m to be prime while $2^m - 1$ is not prime.

It was not until the start of the seventeenth century that Pietro Cataldi verified, by means of a table of primes he constructed, that $2^{17} - 1 = 131{,}071$ is prime and that $2^{19} - 1 = 524{,}287$ is also prime. His accomplishments gave birth to the sixth and seventh perfect numbers,

$$2^{16}(2^{17} - 1) = 8{,}589{,}869{,}056 \text{ and } 2^{18}(2^{19} - 1) = 137{,}438{,}691{,}328.$$

In 1644, a French monk named Friar Marin Mersenne conjectured that $2^m - 1$ is prime for $m = 2, 3, 5, 7, 13, 17, 19, 31, 67, 127, 257$, and for no other primes $m \leq 257$. Although Mersenne was aware of much of the existing relevant work (e.g., Fermat had previously shown that both $2^{23} - 1$ and $2^{37} - 1$ are composite), it is still not entirely clear what led him to make his conjecture, since it was essentially impossible during that time period to accomplish such colossal computations. Whatever reasoning or intuition he might have used was absolutely astonishing, considering that his conjecture turned out to be correct in all but five cases. In recognition of Mersenne's work, primes of the form $2^m - 1$ are referred to as **Mersenne primes**.

The following table lists all 55 of the primes $m \leq 257$. Mersenne's prediction that $2^m - 1$ is prime for the 11 primes he listed **failed** for the primes $m = 67, 257$ but succeeded for the primes $m = 61, 89, 107$ that he excluded from his list.

2	3	5	7	11	13	17	19	23	29	31
37	41	43	47	53	59	61	67	71	73	79
83	89	97	101	103	107	109	113	127	131	137
139	149	151	157	163	167	173	179	181	191	193
197	199	211	223	227	229	233	239	241	251	257

Primes m $\leq$ 257

More than one hundred years passed before the brilliant mathematician Leonard Euler discovered the eighth perfect number. He showed that $2^{31} - 1 = 2{,}147{,}483{,}647$ is prime, and thus the number $2^{30}(2^{31} - 1) = 2{,}305{,}843{,}008{,}139{,}952{,}128$ is perfect. In addition to finding this gargantuan perfect number, he also crafted a proof demonstrating the converse of Euclid's result (Theorem 3.5.6), thus producing a complete characterization for even perfect numbers. Although we will not work through the details of Euler's theorem, it is worthwhile to record the stunning characterization of even perfect numbers resulting from the work of Euclid and Euler.

3.5.8 Theorem (Euclid, Euler): Characterizing Even Perfect Numbers.
An even number n is perfect if and only if $n = 2^{m-1}(2^m - 1)$, where $2^m - 1$ is prime.

The Silent Lecture. Mersenne's provocative conjecture attracted the interest of many distinguished researchers; in fact, it was not until the early part of the twentieth century that all cases were completely resolved. Most of the work on the conjecture involved immense computations, but some depended on deductive arguments. For example, Edouard Lucas in 1876 developed a means to show that $2^{67} - 1$ is composite without explicitly listing any of its (nontrivial) factors. It was not until 1903 that a concrete factorization was found. What makes this discovery worth mentioning is not the actual factorization, but rather the unusual way it was communicated to the mathematical community.

It was in the Fall of 1903 at a meeting of the American Mathematical Society that Frederick Cole was scheduled to present his paper, "On the Factorization of Large Numbers." When it was his turn to speak, he approached the blackboard, and without uttering so much as one single word, he started to write.

First he chalked the following computation,

$$2^{67} - 1 = 147,573,952,589,676,412,927 \quad \text{(check on calculator)}.$$

Next, still in total silence, he wrote two numbers down and then commenced to carefully multiply them together until he reached the answer. The product he computed was $(761,838,257,287) \cdot (193,707,721) = 147,573,952,589,676,412,927$ (check on calculator).

At this point of the **silent lecture**, he took his seat, and at that moment his efforts were rewarded by a rousing ovation from the audience. His computations were clearly worth a thousand words!

Largest Known Mersenne Prime (to This Date). At the beginning of this section, we remarked that the largest known prime is $p = 2^{25,964,951} - 1$, and as you can see, it happens to be a Mersenne prime.

Undiscovered Odd Perfect Numbers. The four perfect numbers of antiquity, and all the other perfect numbers discovered to this date, are even numbers. There are only thirty-nine perfect numbers currently known, and no one thus far has provided arguments demonstrating the existence of an odd perfect number or determined whether there are infinitely many perfect numbers. It is known that if an odd perfect number n were to exist, then $n > 10^{300}$, its prime factorization would contain at least eight distinct prime factors, and one of its prime factors would be greater than 10^6.

It is straightforward to see that certain numbers could never be perfect. For example, any prime number p is not perfect (why?); in fact, any prime power p^n is not perfect. To verify this, suppose p^n is perfect for some prime p and for some positive integer n. The proper divisors of p^n are $1, p, p^2, \ldots, p^{n-1}$ (apply the FTA), so by assumption,

$$1 + p + p^2 + \ldots + p^{n-1} = p^n.$$

However, this equality is absurd, since the equality's right-hand side is divisible by p, while the left side is not. Therefore, p^n is not perfect.

Classroom Discussion. The perfect number 6 is a product of two distinct primes. We show that it is the only perfect number with this property.

Analysis. To examine this problem, suppose that n is a perfect number such that $n = pq$, where p and q are distinct primes. To begin, let's see if this assumption forces one of the primes to be 2. Suppose this is not the case, and thus both p and q are odd.

The proper divisors of n are $1, p$, and q, so by assumption, $1 + p + q = pq$, which in turn gives $pq - p - q = 1$. Notice that by adding 1 to both sides of the equality, we are able to factor the left-hand expression in the following manner:

$$pq - p - q + 1 = (p-1)(q-1) = 2.$$

However, this contradicts the fact that both p and q are odd (why?), consequently, p or q must equal 2 (why?). There is no loss in generality assuming $p = 2$, and substituting this value in the equation $1 + p + q = pq$ shows that $q = 3$ as desired.

Note that we could have deduced the fact that $q = 3$ from the characterization of even perfect numbers (Theorem 3.5.8) (justify), but for this specific problem, it was simple to draw this conclusion directly. ◆

Summary. Combining the conclusions of the previous Classroom Discussion and the paragraph preceding it gives the statement: if $n = pq$, where p and q are odd primes (not necessarily distinct), then the odd number n is not perfect.

3.5.9 Classroom Discussion: Summing the Divisors of Odd Integers.

Throughout our studies we have often raised questions based upon limited data derived from specific cases. Valuable insights can be gleaned through such data inspection, however, we are well aware that this kind of information, even in voluminous quantity, does not constitute a proof for the general situation.

Complete and expand the following table, which consists of the sums of positive divisors of odd positive integers n, and try to formulate some questions based on this data.

Sum of positive divisors of odd integers

$n =$	1	3	5	7	9	11	13	15	17	19
$\sum_{d\mid n} d =$	1	4	6	8	13	12	14	24	18	20
$n =$	21	23	25	27	29	31	33	37	39	41
$\sum_{d\mid n} d =$	32	24	31	40	30	32	48	38	56	42
$n =$	43	45	47	49	51	53	55	57	59	61
$\sum_{d\mid n} d =$										

◆

The next question immediately comes to mind.

Question. Is $\sum_{d|n} d < 2n$ for each odd positive integer n?

The given data suggests that this question has an affirmative answer, and several additional rows of calculations also support this conclusion. Even though the evidence seems quite compelling, we have still only checked a finite number of the infinite possibilities. With a good deal of time and patience, you may have discovered that for $n = 945$,

$$\sum_{d|n} d = 1{,}920 > 2 \cdot 945 = 1{,}890.$$

So our initial speculation, while founded upon many cases, does not remain true in general.

Question. What method did you employ to find $\sum_{d|n} d$ for the given integers n?

To systematically find this sum, first express n in its prime factorization, then apply the FTA to identify all possible divisors of n, and, finally, use these divisors to compute $\sum_{d|n} d$.

In the exercises at the end of this section, we develop a nice formula for computing $\sum_{d|n} d$ in terms of the prime factors of n.

Classroom Connection. Let's apply our studies to a related problem from *Prime Time* of *Connected Mathematics* (see Figure 3.5.2 on page 142). As you work through this problem, clearly write down the definitions of the new concepts introduced, and give several more examples than those asked for in the problem.

Deficient and Abundant Numbers. In the problem on page 142 we learned that a positive integer is called **deficient** if $\sum_{d|n} d < 2n$ and **abundant** if $\sum_{d|n} d > 2n$.

As noted earlier, it is not yet known whether there are infinitely many perfect numbers or not; however, since each prime number is deficient (why?), it follows that there are infinitely many deficient numbers.

Question. Are there infinitely many abundant numbers?

In our work thus far, we have seen some examples of abundant numbers, such as the numbers 12, 24, and 36, but have not yet established that there are infinitely many abundant numbers. The form of the prime factorizations of these numbers $12 = 2^2 \cdot 3$, $24 = 2^3 \cdot 3$, and $36 = 2^4 \cdot 3$ suggest the next question.

Question. Are all integers of the form $n = 2^t \cdot 3$, where $t > 1$, abundant numbers? An affirmative answer to this question would show that there are infinitely many abundant numbers.

Analysis. Proceeding as in the proof of Euclid's theorem on even perfect numbers (Theorem 3.5.6), it follows that

$$\sum_{d|n} d = 1 + 2 + 2^2 + \ldots + 2^t + 3 + (2 \cdot 3) + (2^2 \cdot 3) + \ldots + (2^t \cdot 3)$$

20. The sum of the proper factors of a number may be greater than, less than, or equal to the number. Ancient mathematicians used this idea to classify numbers as **abundant, deficient,** and **perfect.** Each whole number greater than 1 falls into one of these three categories.

a. Draw and label three circles as shown below. The numbers 12, 15, and 6 have been placed in the appropriate circles. Use your factor list to figure out what each label means. Then, write each whole number from 2 to 30 in the correct circle.

b. Do the labels seem appropriate? Why or why not?
c. In which circle would 36 belong?
d. In which circle would 55 belong?

Reproduced from page 14, problem 20 of *Prime Time* in *Connected Mathematics*.

FIGURE 3.5.2

$$= (1 + 2 + 2^2 + \ldots + 2^t) + 3(1 + 2 + 2^2 + \ldots + 2^t)$$
$$= (1 + 2 + 2^2 + \ldots + 2^t)(1 + 3) = 4 \cdot (2^{t+1} - 1) = 4 \cdot 2^{t+1} - 4.$$

To see that n is abundant for $t > 1$, we must show that

$$\sum_{d|n} d = (4 \cdot 2^{t+1} - 4) > 2n = 2 \cdot (2^t \cdot 3) = 3 \cdot 2^{t+1} \text{ holds for } t > 1.$$

Note that this inequality does not hold for $t = 1$, since in this case, $n = 6$ is perfect. However, for $t > 1$, it follows that $2^{t+1} > 4 = 2^2$, so $2^{t+1} - 4 > 0$. Thus,

$$3 \cdot 2^{t+1} + (2^{t+1} - 4) > 3 \cdot 2^{t+1},$$

and this is what we wanted to prove. Therefore, $n = 2^t \cdot 3$ is abundant for each $t > 1$.

The study of prime numbers and related concepts have shaped the course of mathematics and continue to influence new directions. The literature abounds with important and deep results concerning these jewels, many of which were established by the preeminent mathematicians of their time. One of the most intriguing, and still unsolved problems of the last 140 years, the Riemann hypothesis, grew out of an effort to better understand the distribution of the primes among the positive integers. Although a description of this problem and its rich history is beyond the scope of these materials, it is important to understand that there is still much to be learned about the fascinating world of primes.

EXERCISES 3.5

1. For each of the following polynomials, find all positive integers x for which $f(x)$ is prime or show that the given polynomial has no prime value for any positive integer x. Justify your answers. Direction: In some of the cases, set $f(x) = p$ as a prime, and then proceed in a fashion similar to that used in Classroom problem 3.5.4.

   a. $f(x) = x^3 - 1$
   b. $f(x) = x^4 - 1$
   c. $f(x) = x^2 - 9$
   d. $f(x) = x^2 - 6x + 9$
   e. $f(x) = x^2 + x + 6$
   f. $f(x) = x^2 + 2x - 15$
   g. $f(x) = x^2 + 4x - 21$

2. For the polynomial $f(x) = x^2 - 1$ and a prime p, find all positive integers x so that $f(x) = 7p$. For each of those positive integers, determine p. Fill in the details of the sketched argument.

   Sketch: Since $(x - 1)(x + 1) = 7p$, then $x - 1$ either equals $1, 7, p,$ or $7p$. Why?
   Case 1. If $x - 1 = 1$, then $x = 2$, which is not possible. Why?
   Case 2. If $x - 1 = 7$, then $x = 8$, so $f(8) = 63 = 7p$, which is not possible under the given assumptions. Explain.
   Case 3. If $x - 1 = p$, then $x + 1 = 7$ (why?), so $x = 6$, and, in this case, $p = 5$. Why?
   Case 4. If $x - 1 = 7p$, then $x + 1 = 1$ (why?), which is not possible under the given assumptions. Explain.

   Therefore, $x = 6$ is the only positive integer solution and, in this case, $p = 5$.

3. For the polynomial $f(x) = x^2 - 4$ and a prime p, find all positive integers x so that $f(x) = 13p$. For each of those positive integers found, determine p. Direction: Argue as in Exercise 2 above.

4. This exercise outlines steps culminating in a formula for $\sum_{d|n} d$ in terms of the prime factors of n. It is standard to denote $\sum_{d|n} d = \sigma(n)$. This functional notation reflects that the correspondence $n \to \sigma(n)$ defines a function from $\mathbf{Z}^+$ into $\mathbf{Z}^+$. Verify each of the following statements:

   a. If $n = p^m$, then $\sigma(n) = 1 + p + \ldots + p^m = \frac{p^{m+1} - 1}{p - 1}$. Direction: Sum of a geometric series, see Chapter 1.
   b. If $n = p^3 q^4$, where $p \neq q$ are primes, then the positive divisors of n are precisely the terms in the expanded product of $(1 + p + p^2 + p^3)(1 + q + q^2 + q^3 + q^4)$.
   c. If $n = p^3 q^4$, where $p \neq q$ are primes, then $\sigma(n) = \left(\frac{p^4 - 1}{p - 1}\right)\left(\frac{q^5 - 1}{q - 1}\right)$.
   d. If $n = p_1^{m_1} p_2^{m_2} \cdots p_s^{m_s}$ is the canonical prime factorization of the positive integer n, then
   $$\sigma(n) = \frac{(p_1^{m_1+1} - 1)}{(p_1 - 1)} \frac{(p_2^{m_2+1} - 1)}{(p_2 - 1)} \cdots \frac{(p_s^{m_s+1} - 1)}{(p_s - 1)}.$$
   e. If m and n are relatively prime positive integers, then $\sigma(mn) = \sigma(m)\sigma(n)$.
   f. Use part d to check each of the computations in the table that appears in Classroom Discussion 3.5.9.

5. **a.** Prove: If $n = p^{2m}$, where p is a prime, then $\sigma(n)$ is odd. Direction: Write $\sigma(n) = 1 + p + p^2 + \ldots + p^{2m}$, and show that $(p + p^2 + \ldots + p^{2m})$ is even.
   b. Generalize part a. Show that if $n = a^2$, then $\sigma(n)$ is odd. Direction: Express $n = p_1^{2m_1} p_2^{2m_2} \cdots p_s^{2m_s}$, and show that each parenthetical term in the product $\sigma(n) = (1 + p_1 + \ldots + p_1^{2m_1}) \cdots (1 + p_s + \ldots + p_s^{2m_s})$ is odd.
   c. Is the converse of the statement in b true? If it is true, prove it; if it is false, provide a counterexample.
   d. Show that if $n = a^2$, then n is not perfect. Direction: Apply part b.
6. **a.** Prove: If $n = 6t$ for some integer $t > 1$, then n is abundant. Direction: Verify that $\sigma(n) = \sigma(6t) > (1t + 2t + 3t + 6t) = \sigma(6)t = 12t$.
   b. Prove: If $n = mt$, where m is perfect and $t > 1$, then n is abundant.

Arithmetic and Algebra of the Integers Modulo n

CHAPTER 4

4.1 CLASSROOM CONNECTIONS: DIVISIBILITY TESTS
4.2 REFLECTIONS ON CLASSROOM CONNECTIONS: JUSTIFYING THE DIVISIBILITY TESTS
4.3 CLOCK ADDITION
4.4 MODULAR ARITHMETIC
4.5 COMPARING ARITHMETIC PROPERTIES OF Z AND Z_n
4.6 MULTIPLICATIVE INVERSES IN Z_n
4.7 ELEMENTARY APPLICATIONS OF MODULAR ARITHMETIC
4.8 FERMAT'S LITTLE THEOREM AND WILSON'S THEOREM
4.9 LINEAR EQUATIONS DEFINED OVER Z_n
4.10 EXTENDED STUDIES: THE CHINESE REMAINDER THEOREM
4.11 EXTENDED STUDIES: QUADRATIC EQUATIONS DEFINED OVER Z_n

In Chapter 2, our work focused on the Fundamental Theorem of Arithmetic and its central importance in the arithmetic and algebra of the integers. We shall continue to broaden and deepen our knowledge of the integers by introducing and investigating an arithmetic system closely related to the integers. Questions concerning integers can often be settled by translating and analyzing them within the framework of this allied system and then interpreting the results in terms of the original question. The practice of building bridges between mathematical worlds is fundamental to the subject and, in a broad sense, is the essence of mathematics.

Dividing the Bill. My wife and I recently went to dinner with a group of friends, all of whom were nonmathematicians except for me. There were a total of nine of us at this meal, and we agreed in advance to split the bill plus tip evenly. I was really happy with this hassle-free agreement, which by the way, is in stark contrast to the usual mathematician's "infinitesimal approach" of bill dividing.

146 Chapter 4   Arithmetic and Algebra of the Integers Modulo *n*

It is a well-known theorem that given any banquet of mathematicians, there is at least one member of that dinner party who insists that each person compute their own individual charge and will not rest until this activity is accomplished. This ordeal is further complicated by the attempt to determine the cost/individual of shared appetizers and beverages, and often times an informal inquisition is needed to identify the sharing individuals. Finally, the culinary nightmare is usually concluded with a vigorous discussion on the percentage tip for each diner and whether the tip should be computed on the gross bill (bill before taxes) or on the net bill.

Food tastes so much better when you are relaxed and worry-free. Our party of nine ordered a variety of interesting dishes and completed the meal with wonderful desserts and coffee. The service was excellent, and after several refills on the coffee, the waiter brought us the bill totaling (including tax and gratuity) $377.37 and handed it to a stately looking gentleman sitting across from me. Almost instantaneously, the distinguished gent proclaimed, "We're in luck; it's exactly divisible by nine." Several at the table complimented the fine gentleman on his impressive mental arithmetic and asked him the equal amount that each person owed. He replied, "I don't know what each of us owes. I simply know that the total is evenly divisible by nine." Many seemed perplexed at his response, and the woman sitting directly to his right wondered aloud how he could know one answer without knowing the other. Before he had the opportunity to explain his methods, someone at the other end of the table had used good old-fashioned long division and proclaimed that each of us owed $41.93. This solution satisfied all and unintentionally ended the discussion concerning the gentleman's methods.

Question.   How did the gentleman know that the total was evenly divisible by 9 without knowing what the equal shares were?

Our work in the next section describes the gentleman's method and thoroughly explains why it always works.

4.1   CLASSROOM CONNECTIONS: DIVISIBILITY TESTS

Let's start our investigation by working through divisibility rules from the Numbers and Operations Toolbox in Book 3 of *MathThematics* (see Figure 4.1.1 on facing page). Perhaps you are familiar with some or all of the rules described on this page, but what about the mathematical reasoning that justifies each rule? The search for details that validate these rules takes us to a new fascinating arithmetic system, which in turn further enhances our understanding of the integers.

4.2   REFLECTIONS ON CLASSROOM CONNECTIONS: JUSTIFYING THE DIVISIBILITY TESTS

The examples from *MathThematics* clearly demonstrate the application of the divisibility rules, and the exercises provide additional practice problems. As with all the mathematics we learn, our interest is not only in the application of rules, but what makes them work. The foundations laid in Chapters 2 and 3 provide us with the necessary tools to immediately explain some of the given divisibility rules, whereas

Section 4.2   Reflections on Classroom Connections: Justifying Divisibility Tests   147

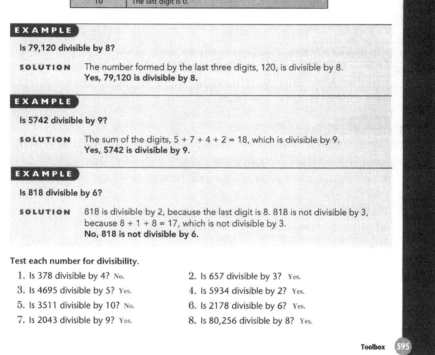

Reproduced from page 595 of *MathThematics*.

FIGURE 4.1.1

we handle others through the development of new concepts. Let's see which ones we can justify at this time.

Divisibility by 2.   If the last digit of an integer is 0, 2, 4, 6, or 8, then the integer is divisible by 2.

Analyzing a concrete example helps us gain a better understanding of the underlying ideas and leads to an argument supporting the general case.

148 Chapter 4 Arithmetic and Algebra of the Integers Modulo n

Since the integer 3,576 can be uniquely represented in expanded form as

$$3,576 = 3 \cdot 10^3 + 5 \cdot 10^2 + 7 \cdot 10^1 + 6 \cdot 10^0,$$

and each term is divisible by 2 (why?), then the sum of the terms is also divisible by 2 (why?).

Using the reasoning in this example, it is now easy to construct an argument justifying the rule for divisibility by 2.

Justification. An integer

$$a = a_m a_{m-1} \ldots a_2 a_1 a_0 = a_m 10^m + a_{m-1} 10^{m-1} + \ldots + a_2 10^2 + a_1 10^1 + a_0 10^0,$$

where a_0 is 0, 2, 4, 6, or 8 is divisible by 2 since each term is divisible by 2.

Classroom Problems. Show that the converse of the rule for divisibility by 2 is also true, i.e., if an integer n is divisible by 2, then its last digit must be 0, 2, 4, 6, or 8. Direction: Recall that if $d|b$ and $d|(b + c)$, then $d|c$. ◆

In view of the previous problem, we can restate the test for divisibility by 2.

Restating Divisibility by 2. For an integer $a = a_m a_{m-1} \ldots a_2 a_1 a_0$, $2|a \Leftrightarrow 2|a_0$.

Classroom Problems. Using reasoning similar to what was used in the rule for divisibility by 2, justify the following divisibility tests:

Divisibility by 5. For an integer $a = a_m a_{m-1} \ldots a_2 a_1 a_0$, $5|a \Leftrightarrow 5|a_0$.

Divisibility by 10. For an integer $a = a_m a_{m-1} \ldots a_2 a_1 a_0$, $10|a \Leftrightarrow a_0 = 0$.

Divisibility by 4. For an integer $a = a_m a_{m-1} \ldots a_2 a_1 a_0$, $4|a \Leftrightarrow 4|a_1 a_0$.

Divisibility by 8. For an integer $a = a_m a_{m-1} \ldots a_2 a_1 a_0$, $8|a \Leftrightarrow 8|a_2 a_1 a_0$.

Divisibility by 6. For an integer $a = a_m a_{m-1} \ldots a_2 a_1 a_0$, $6|a \Leftrightarrow 2|a$ and $3|a$. Direction: Recall that if $\gcd(d, b) = 1$ and $d|c$ and $b|c$, then $db|c$. ◆

The remaining tests needing justification are divisibility by 3 and by 9. Unfortunately, the methods used to establish the previous tests do not directly apply to these cases. However, the material introduced in the next section promptly resolves these two remaining cases.

4.3 CLOCK ADDITION

Our work in this brief section involves the introduction and study of an addition operation based on the 12-hour clock. The ideas we consider help set the stage for Section 4.4, where we develop a finite system called the *integers modulo n*. For each integer $n > 1$, we define a set of elements equipped with a newly defined addition and multiplication, and then use this structure to extend and deepen our knowledge of the integers. We undertake a systematic investigation of the arithmetic and algebraic properties of this system and utilize the resulting deductions in multiple ways and in various settings. In particular, the justifications for the divisibility rules involving 3 and 9 will follow easily from this work.

Clock Arithmetic and the Concerned Parent. It is a rare occasion when I receive a call or e-mail from a parent of one of my students. There is one instance, however, that I'll always remember. It involved a (mildly irate) father who was questioning some mathematics he observed in his daughter's notebook. It seems that he was browsing through his daughter's college notebooks one evening, when he came upon a page of notes from the mathematics course she was taking from me. He noticed some rather strange arithmetic statements, such as $8 + 6 = 2$ and $7 + 9 = 4$, and rather than questioning his daughter as to the meaning of these computations (maybe he was inspecting her work without her permission), he decided instead to directly question me. The phone conversation was quite brief, but he seemed very angry about (in his words), "the nonsensical mathematical computations" his daughter had written in her notebook. He insisted that $8 + 6 = 14$ and $7 + 9 = 16$ and that the computations she had written down were absolutely incorrect. My response to him was in the form of two questions:

1. Were you in the military? He answered no to this question, and with this information, I posed the next question.
2. If it was 8 o'clock now, what time would it be 6 hours from now? He quickly answered it would be 2 o'clock and asked what these questions had to do with his particular objections. Again I responded with another question.
3. How did you determine the answer to the second question?

He simply said that he added 6 to 8 and then subtracted 12 to get the answer of 2. A brief period of silence ensued, and then I heard him say, "So that's what it meant." He thanked me for my time and patience and apologized for his frosty tone. I indicated that no apology was needed and thanked him for taking such a serious interest in his daughter's education.

Question. Describe how the father made sense of the arithmetic statements, $8 + 6 = 2$ and $7 + 9 = 4$, and explain the relevance of Question 1 in terms of these computations.

Addition Table for a Single-Handed 12-Hour Clock (Ignoring AM and PM). Since our focus is on whole number hours, it is reasonable to simplify matters by using a 12-hour single-handed clock that only keeps track of hours. With such a clock in mind, it is easy to construct the "addition" table that displays all possible "clock additions" involving the integers 1 through 12.

+	1	2	3	4	5	6	7	8	9	10	11	12
1	2	3	4	5	6	7	8	9	10	11	12	1
2	3	4	5	6	7	8	9	10	11	12	1	2
3	4	5	6	7	8	9	10	11	12	1	2	3
4	5	6	7	8	9	10	11	12	1	2	3	4
5	6	7	8	9	10	11	12	1	2	3	4	5
6	7	8	9	10	11	12	1	2	3	4	5	6
7	8	9	10	11	12	1	2	3	4	5	6	7
8	9	10	11	12	1	2	3	4	5	6	7	8
9	10	11	12	1	2	3	4	5	6	7	8	9
10	11	12	1	2	3	4	5	6	7	8	9	10
11	12	1	2	3	4	5	6	7	8	9	10	11
12	1	2	3	4	5	6	7	8	9	10	11	12

Question. Can the table be used to determine all possible clock additions, e.g., what time will it be 43 hours from 6 o'clock?

Although this table does not directly list the "clock sum" 6 + 43, it does contain another sum that represents the same time on the clock. In particular, since 43 o'clock is the same time as 7 o'clock (why?), then the sum 6 + 43 yields the same time as 6 + 7.

To see that the single-handed 12-hour clock is in the same position at 43 o'clock and at 7 o'clock, start the hand at 12 o'clock and move it through 43 hours on the clock. This procedure involves three complete trips (a complete trip starts at 12 and ends at 12) around the clock and 7 additional hours, that is, the hand stops on 7. Our interest is where the hand ends up after starting on 12 and moving through 43 hours, not the number of times traveled around the clock.

In mathematical terms, our focus is on the unique remainder 7 (guaranteed by the division algorithm) that is obtained when 43 is divided by 12 (i.e., $43 = 3 \cdot 12 + 7$, where $0 \leq 7 < 12$). Using this language, we can precisely express when two positive integers represent the same time.

4.3.1 Positive Integers Representing the Same Time. Two positive integers represent the same time on the clock provided they have the same remainder when divided by 12.

For example, 75 o'clock represents the same time as 123 o'clock, since dividing each of the times by 12 yields the remainder of 3 (i.e., $75 = 6 \cdot 12 + 3$ and $123 = 10 \cdot 12 + 3$).

Another approach to computing the "clock sum" 6 + 43 is to start with the clock hand on 6, and then move it through 43 hours on the clock. This is equivalent to starting the hand at 12, and then shifting it through 49 hours on the clock. As before, 49 o'clock converts to 1 o'clock (why?), which agrees with the sum reached through the first method.

Section 4.4 Modular Arithmetic 151

Interpreting 0 and Negative Integers on the Clock. Thus far our discussion has focused on positive integer times, but 0 and negative integer times can also be interpreted on a single-handed 12-hour clock. In fact, the definition given in definition 4.3.1 can be applied to all integers.

0 o'clock. Since 0 and 12 have the remainder of 0 when divided by 12, then we say that 0 o'clock represents the same time as 12 o'clock.

$-n$ o'clock (where $n > 0$). A few examples help illustrate this case. What positive integer time t, where $1 \leq t \leq 12$, represents -3 o'clock (i.e., what is the remainder when -3 is divided by 12)? Recall from Chapter 3 that $-3 = -1 \cdot 12 + 9$, and so -3 o'clock represents the same time as 9 o'clock. Note that the same result can be achieved by starting with the hand at 12 o'clock and moving counterclockwise three hours.

For another example, let's determine what positive integer time t (for $1 \leq t \leq 12$) represents -95 o'clock? Employing the division algorithm gives $-95 = -8 \cdot 12 + 1$, and thus -95 o'clock and 1 o'clock represent the same time.

The notation and terminology we have employed to this point is somewhat imprecise and cumbersome. Without accurate definitions and unambiguous notation, confusion could occur as it did with the concerned parent who took issue with the statement $8 + 6 = 2$. To alleviate such problems, we carefully formalize clock arithmetic in the next section.

4.4 MODULAR ARITHMETIC

Our goal in this section is to extend and formalize the arithmetic of the clock that we introduced in the last section and then apply this work to complete the division tests stated in Section 4.1.

We used the 12-hour clock to introduce clock addition because of its familiarity, but we could have just as easily used an n-hour clock, where $n \geq 2$. Let's consider the 4-hour single-handed clock and use it as a guide to develop the n-hour single-handed clock.

The 4-Hour Single-Handed Clock. The integers 0, 1, 2, 3 indicate the four times on this clock. All integers can be represented on this clock in an analogous manner to the 12-hour clock. For example, 2 o'clock, 6 o'clock, 10 o'clock, −2 o'clock, etc., all represent the same time on the 4-hour clock, since they all have the same remainder when divided by 4.

To succinctly represent all times equivalent to a given time, we introduce the following sets of equivalent times:

$$[0]_4 = \{x \in \mathbf{Z} : x \text{ and } 0 \text{ represent the same time on the 4-hour clock}\}$$
$$= \{x \in \mathbf{Z} : x = 4q, q \text{ an integer}\};$$
$$[1]_4 = \{x \in \mathbf{Z} : x = 4q + 1, q \text{ an integer}\};$$
$$[2]_4 = \{x \in \mathbf{Z} : x = 4q + 2, q \text{ an integer}\};$$
$$[3]_4 = \{x \in \mathbf{Z} : x = 4q + 3, q \text{ an integer}\}.$$

For any integer a, $[a]_4 = \{x \in \mathbf{Z}: x \text{ and } a \text{ have the same remainder when divided by 4}\}$.

Classroom Problems. Show that each integer lies in exactly one of the sets $[0]_4, [1]_4, [2]_4, [3]_4$, i.e., the sets are pairwise disjoint and $[0]_4 \cup [1]_4 \cup [2]_4 \cup [3]_4 = \mathbf{Z}$. ◆

Congruent (Equivalent) Times. Integers a and b represent the same time on the 4-hour clock provided they have the same remainder when divided by 4. In this situation, it is said that ***a* and *b* are congruent modulo 4** and is denoted by ***a* ≡ *b*(mod 4)**.

It is important to note that

$$a \equiv b(\bmod 4) \Leftrightarrow [a]_4 = [b]_4.$$

Justification. ($\Rightarrow$) Assume that $a \equiv b(\bmod 4)$, i.e., $a = 4q_1 + r$ and $b = 4q_2 + r$, where $0 \leq r < 4$. To show that $[a]_4 = [b]_4$, we verify that $[a]_4 \subseteq [b]_4$ and $[b]_4 \subseteq [a]_4$.

If $x \in [a]_4$, then $x = 4z + r$, with $0 \leq r < 4$, so $x \in [b]_4$. (Why?) An analogous argument establishes reverse inclusion, and thus $[a]_4 = [b]_4$.

($\Leftarrow$) For this direction, suppose that $[a]_4 = [b]_4$. Since $a \in [a]_4$, then $a \in [b]_4$, which means that $a \equiv b \pmod{4}$.

Classroom Problems. Verify: $[91]_4 = [115]_4$, $74 \equiv -22 \pmod{4}$ and $[-103]_4 \neq [54]_4$. ◆

It would be incorrect (or an abuse of notation) to write $91 = 115$, since these numerals represent different integers, but it is perfectly legitimate to write the equality $[91]_4 = [115]_4$ or the congruence $91 \equiv 115 \pmod{4}$.

The definitions and notation used in connection to the 12-hour and 4-hour clocks extend easily to the n-hour clock, and in fact can be expressed in general terms without any reference to an n-hour clock.

4.4.1 Residues Modulo n. For a fixed integer $n \geq 2$ and any integer a, the remainder resulting from division of a by n is termed the **residue of a modulo n** (abbreviated $a \pmod{n}$), and determining $a \pmod{n}$ from a is called **reducing a modulo n**. If integers a and b have the same residue (mod n), then **a is said to be congruent to b modulo n**, and this is denoted by **$a \equiv b \pmod{n}$**. For example, $37 \equiv 226 \pmod 9$ and $-103 \equiv 51 \pmod 7$.

The set $[a]_n = \{x \in \mathbf{Z} : x \text{ and } a \text{ have the same residue modulo } n\}$ is called the **residue class of a (mod n)**, and the set $\mathbf{Z}_n = \{[0]_n, [1]_n, [2]_n, \ldots, [n-1]_n\}$, consisting of **$n$-distinct** residue classes (mod n), is referred to as the **integers (mod n)**. (Explain why the members of $\mathbf{Z}_n$ are distinct.)

Some useful properties of residue classes follow directly from the definitions just given.

Classroom Problems. Verify each of the following statements:

1. $a \equiv b \pmod{n} \Leftrightarrow [a]_n = [b]_n$.
2. For each integer a, $[a]_n \neq \emptyset$.
3. $[a]_n = \{x \in \mathbf{Z} : x = qn + a, q \text{ an integer}\}$.
4. Either $[a]_n \cap [b]_n = \emptyset$ or $[a]_n = [b]_n$.
5. $\mathbf{Z} = [0]_n \cup [1]_n \cup \cdots \cup [n-2]_n \cup [n-1]_n$.

We sketch the details for Problem 4 and leave the rest of the fun for you.

Analysis. If $[a]_n \cap [b]_n \neq \emptyset$, then there exists an integer x so that $x \in [a]_n \cap [b]_n$. Hence, $x \equiv a \pmod{n}$ and $x \equiv b \pmod{n}$, and this implies that $a \equiv b \pmod{n}$. (Why?) Therefore, by Problem 1, $[a]_n = [b]_n$. ◆

Observation. Since $a \equiv b \pmod{n} \Leftrightarrow a$ and b have the same remainder when divided by n, it follows that $n | (a - b)$. (Why?)

Question. Is the converse of the previous observation valid, i.e., if $n | (a - b)$, then is $a \equiv b \pmod{n}$?

Analysis. If $n|(a - b)$, then $a - b = nt$ for some integer t. To show that a and b have the same remainder when divided by n, we first employ the division algorithm to get:
$$a = q_1 n + r_1, 0 \le r_1 < n \text{ and } b = q_2 n + r_2, 0 \le r_2 < n.$$

Our goal is to demonstrate that $r_1 = r_2$, i.e., $(r_1 - r_2) = 0$. To reach this end, we may assume without loss of generality, that $r_1 \ge r_2$ (why?), so $0 \le (r_1 - r_2) < n$.

Using the previous divisions and the hypothesis gives,
$$a - b = (q_1 n + r_1) - (q_2 n + r_2) = (q_1 - q_2)n + (r_1 - r_2) = nt,$$

and, consequently,
$$(r_1 - r_2) = n[t - (q_1 - q_2)].$$

Since $0 \le (r_1 - r_2) < n$, then $[t - (q_1 - q_2)] = 0$ (why?), so $(r_1 - r_2) = 0$ as desired.

4.4.2 Equivalent Definition of $a \equiv b \pmod{n}$. For a fixed integer $n \ge 2$ and integers a and b,
$$a \equiv b \pmod{n} \Leftrightarrow n|(a - b).$$

The definition here requires a subtraction and one division, whereas definition 4.4.1 requires two divisions, thus, it is often simpler to use definition 4.4.2 to determine if two integers are congruent modulo n. For example, note that $4,529 \equiv -1,471 \pmod{4}$, since $4,529 - (-1,471) = 6,000$ and $4|6,000$. If the objective is to represent each of the given integers as an integer appearing on the face of the clock, then definition 4.4.1 is needed. In particular, $4,529 \equiv 1 \pmod{4}$, i.e., $[4,529]_4 = [1]_4$, likewise, $[-1,457]_4 = [1]_4$.

As we have just mentioned, it is valuable to have a cache of equivalent definitions that can be easily employed in a variety of contexts. Let's consider another statement that is "potentially" equivalent to the definition of $a \equiv b \pmod{n}$.

Note that if $a \equiv b \pmod{n}$, then $a = q_1 n + r$ and $b = q_2 n + r$, where $0 \le r < n$. Recall from our work on the Euclidean algorithm (Observation 3.2.1) that
$$\gcd(a, n) = \gcd(n, r) = \gcd(b, n).$$

Hence, $a \equiv b \pmod{n} \Rightarrow \gcd(a, n) = \gcd(b, n)$. For these statements to be equivalent, we must show that the reverse implication (converse) is also valid. Unfortunately, this is not the case.

Classroom Problems. Give an example showing that $\gcd(a, n) = \gcd(b, n) (n > 1)$, but where a is not congruent to $b \pmod{n}$.

Although we did not produce a new statement equivalent to the original definition, the implication that emerged contributes to our overall understanding and might prove useful in other deductions. ◆

Addition in Z_n. In Section 4.2 we considered an "addition" table for the 12-hour single-handed clock. The entries of this table were determined by the usual rules of telling time, and although the table only consisted of times 1 o'clock through 12 o'clock, we were able to resolve how any integer time was equivalent to a time on the face of the clock.

The notation we've just developed provides a convenient way to represent all possible integer times on the face of a clock. In particular, consider the new 4-hour single-handed clock. To simplify the notation, the notation $[a]$ is used rather than $[a]_4$, and we adopt this practice of excluding the modulus subscript when no confusion will arise.

A clock-addition table, similar to the one constructed for the 12-hour clock, can also be constructed using this notation.

+(mod 4)	[0]	[1]	[2]	[3]
[0]	[0]	[1]	[2]	[3]
[1]	[1]	[2]	[3]	[0]
[2]	[2]	[3]	[0]	[1]
[3]	[3]	[0]	[1]	[2]

The computations in this table and in the (mod 12) table were determined in a similar manner. In particular, the (mod 12) sum of [7] and [8] was calculated by first computing the integer sum $7 + 8$ and then reducing this (mod 12). Likewise, adding [3] and [3] (mod 4) was determined by finding $3 + 3$ and then reducing this (mod 4). These sums can be represented as follows:

$$[7]_{12} + [8]_{12} = [7 + 8]_{12} = [15]_{12} = [3]_{12} \text{ and}$$

$$[3]_4 + [3]_4 = [3 + 3]_4 = [6]_4 = [2]_4.$$

Note that we use the $+$ symbol to represent integer addition and modular addition. Sometimes, to avoid confusion, this addition is symbolized as $+(\text{mod } n)$ or $+_n$, however, to simplify notation, $+$ is most often used. Also, as we mentioned earlier, if the context is clear, the modulus subscripts will usually not be included.

4.4.3 General Definition of Addition in $\mathbf{Z}_n$. For elements $[a]$ and $[b]$ in $\mathbf{Z}_n$, **addition (mod n)** is defined as:

$$[a] + [b] = [a + b].$$

Question. Does the addition (mod n) definition depend on the choice of the representative of the set of equivalent times? More specifically, if $[a] = [a']$ and $[b] = [b']$, does it then follow that $[a] + [b] = [a'] + [b']$?

An invalid argument sometimes used to justify this equality is: "Equals added to equals gives equals." Actually, this is precisely what we are trying to establish, where "added" means addition (mod n). The next analysis demonstrates that addition (mod n) is **well-defined**.

Addition (mod n) Is Well-Defined. Given $[a] = [a']$ and $[b] = [b']$, we wish to show that $[a + b] = [a' + b']$. We know that $a \equiv a' \pmod{n}$ and $b \equiv b' \pmod{n}$, and so $n|(a - a')$ and $n|(b - b')$. Hence, $n|((a - a') + (b - b'))$, and thus $n|((a + b) - (a' + b'))$. Therefore, $[a + b] = [a' + b']$, and addition (mod n) is well-defined.

An "Addition" That Is Not Well-Defined. An arithmetic error that most mathematics teachers frequently encounter is the infamous "addition" of fractions, where the student simply adds the numerators and the denominators of the given fractions. It is easy to see that this procedure does not correspond to the number line meaning of the addition of fractions. For example, compare the notorious "addition" of $\frac{1}{2}$ and $\frac{1}{2}$ with the usual sum, and to avoid confusion, let's denote this disreputable "addition" by the symbol $+'$.

Notorious addition : $\frac{1}{2} +' \frac{1}{2} = \frac{2}{4} = \frac{1}{2}$,   **Usual addition :** $\frac{1}{2} + \frac{1}{2} = 1.$

Actually, $+'$ is different from $+$, and it is not even a well-defined operation. In fact, the following example demonstrates that its definition is dependent on the choice of equivalent fractions:

$$\frac{1}{2} +' \frac{1}{4} = \frac{2}{6} = \frac{1}{3}, \text{ and } \frac{2}{4} +' \frac{1}{4} = \frac{3}{8} \neq \frac{1}{3}.$$

Multiplication in $\mathbf{Z}_n$. Multiplication in the integers is defined as repeated addition. For example, the notation $6 \cdot 2$ represents two equal sums: $6 \cdot 2 = 2 + 2 + 2 + 2 + 2 + 2 = 6 + 6$. Multiplication in $\mathbf{Z}_n$ can also be defined as repeated addition (mod n). For example, in $\mathbf{Z}_{12}$, the sum $[3] + [3] + [3] + [3] + [3] = [3 + 3 + 3 + 3 + 3] = [5 \cdot 3] = [15] = [3]$. It is reasonable to represent this calculation with the notation $[5] \cdot [3]$, where $\cdot$ in this context stands for multiplication (mod n).

4.4.4 General Definition of Multiplication in $\mathbf{Z}_n$. For elements $[a]$ and $[b]$ in $\mathbf{Z}_n$, **multiplication (mod n)** is defined as:

$$[a] \cdot [b] = [a \cdot b].$$

As + was used to represent both integer addition and addition (mod n), the symbol • is also used to represent both integer multiplication and multiplication (mod n).

Convention. It is customary, when multiplying variables, to suppress the multiplication symbol and write, $[a][b] = [ab]$.

Multiplication (mod n) Is Well-Defined. Given $[a] = [a']$ and $[b] = [b']$, we wish to show that $[ab] = [a'b']$. Proceeding as in the addition (mod n) case, we have $n|(a - a')$ and $n|(b - b')$ and wish to show that $n|((ab) - (a'b'))$. To reach this end, note that

$$n|(a - a')b, \text{ i.e., } n|(ab - a'b) \text{ and } n|a'(b - b'), \text{ i.e., } n|(a'b - a'b').$$

Therefore, $n|((ab - a'b) + (a'b - a'b'))$, and finally $n|((ab) - (a'b'))$ as required.

Multiple and Exponent Notation. For $t \geq 1$, the notation $t[a]$ denotes the t-term sum $[a] + [a] + \ldots + [a] = [ta] = [t][a]$, and $[a]^t$ denotes the t-term product, $[a][a]\cdots[a] = [a^t]$. For $s \geq 1$ and $t \geq 1$, note that $[a]^s[a]^t = [a]^{s+t}$. (Justify.)

Classroom Problems. 1. Find the remainders in the following division problems:

a. $17 \overline{)2^{100}}$ b. $23 \overline{)(2^{11} - 1)}$ c. $17 \overline{)5^{33}}$

Analysis. a. Calculating in $\mathbf{Z}_{17}$, $[2^{100}] = [(2^4)^{25}] = [16^{25}] = [16]^{25} = [-1]^{25} = [-1] = [16]$, so the remainder of the division $17\overline{)2^{100}}$ is 16.

b. Calculating in $\mathbf{Z}_{23}$, $[2^{11}] = [2 \cdot (2^5)^2] = [2] \cdot [32]^2 = [2] \cdot [9]^2 = [2] \cdot [81] = [2] \cdot [12] = [24] = [1]$. Therefore, $23|(2^{11} - 1)$, and hence the remainder of this division is 0.

This calculation shows that it is possible for m to be prime while $2^m - 1$ is not prime. Recall (from Section 3.5) that in 1536, Hudalrichus Regius first demonstrated this fact through the factorization $2^{11} - 1 = 23 \cdot 89$.

c. Working in $\mathbf{Z}_{17}$, $[5^{33}] = [5 \cdot 5^{32}] = [5] \cdot [(5^2)^{16}] = [5] \cdot [8]^{16} = [5] \cdot [8^2]^8 = [5] \cdot [13]^8 = [5] \cdot [(-4)^2]^4 = [5] \cdot [16]^4 = [5] \cdot [-1]^4 = [5]$. ◆

Classroom Problems: Revisiting Even and Odd Numbers from a (mod 2) Point of View. In Classroom Problems 3.1.4, you were asked to determine whether certain computations produced even or odd integers, and then give arguments justifying your answers. Let's repeat the justification part of that exercise, but this time with arguments using $\mathbf{Z}_2$ arithmetic. Base your proofs on the observation that an integer x is even if and only if $[x] = [0]$ in $\mathbf{Z}_2$ (why?), and an integer y is odd if and only if $[y] = [1]$ in $\mathbf{Z}_2$ (why?).

Let a and b be even integers and c and d be odd integers. Show that

1. $a \pm b$ is even
2. $c \pm d$ is even
3. $a \pm c$ is odd
4. ab is even
5. cd is odd
6. ac is even

We provide the details for Problem 3 and leave the rest for you.

Analysis of Problem 3. Note that $[a \pm c] = [a] \pm [c] = [0] + [1] = [1]$, so $a \pm c$ is odd. ◆

Justifying the Remaining Division Tests. We are now prepared to justify the division tests for 3 and 9 that were first considered in Section 4.1. Before doing this, however, let's recast the proofs of the other division tests in terms of modular arithmetic.

Revisiting the Divisibility Tests for 2, 4, 5, 8, and 10. The arguments given to support the division tests for 2, 4, 5, 8, and 10 all hinged on a key point involving powers of 10, and can be restated in terms of modular arithmetic. Viewing them this way suggests a technique to handle the tests for 3 and 9. We consider a proof of the test for divisibility by 4 and leave the remaining proofs for the exercises.

Divisibility by 4. For an integer $a = a_m a_{m-1} \ldots a_2 a_1 a_0$, $4 \mid a \Leftrightarrow 4 \mid a_1 a_0$.

Working in $\mathbf{Z}_4$,

$$\begin{aligned}
[a] &= [a_m 10^m + a_{m-1} 10^{m-1} + \ldots + a_1 10 + a_0] \\
&= [a_m 10^m] + [a_{m-1} 10^{m-1}] + \ldots + [a_2 10^2] + [a_1 10] + [a_0] \\
&= [a_m][10^m] + [a_{m-1}][10^{m-1}] + \ldots + [a_2][10^2] + [a_1][10] + [a_0] \\
&= [a_m][10]^m + [a_{m-1}][10]^{m-1} + \ldots + [a_2][10]^2 + [a_1][10] + [a_0] \\
&= [a_m][0] + [a_{m-1}][0] + \ldots + [a_2][0] + [a_1][10] + [a_0] \\
&= [a_m 0 + a_{m-1} 0 + \ldots + a_2 0 + a_1 10 + a_0] \\
&= [a_1 10 + a_0].
\end{aligned}$$

Therefore, a and $a_1 a_0$ have the same remainder when divided by 4, and thus,

$$4 \mid a \Leftrightarrow 4 \mid a_1 a_0.$$

A similar approach can be used to justify the tests for divisibility by 3 and 9. The validation for both tests can be accomplished simultaneously.

Divisibility by 3. For an integer $a = a_m a_{m-1} \ldots a_2 a_1 a_0$,

$$3 \mid a \Leftrightarrow 3 \mid (a_m + a_{m-1} + \ldots + a_1 + a_0).$$

Divisibility by 9. For an integer $a = a_m a_{m-1} \ldots a_2 a_1 a_0$,

$$9 \mid a \Leftrightarrow 9 \mid (a_m + a_{m-1} + \ldots + a_1 + a_0).$$

Justification. Computing in $\mathbf{Z}_3$ and in $\mathbf{Z}_9$,

$$[a] = [a_m 10^m + a_{m-1} 10^{m-1} + \ldots + a_1 10 + a_0]$$
$$= [a_m 10^m] + [a_{m-1} 10^{m-1}] + \ldots + [a_2 10^2] + [a_1 10] + [a_0]$$
$$= [a_m][10^m] + [a_{m-1}][10^{m-1}] + \ldots + [a_2][10^2] + [a_1][10] + [a_0]$$
$$= [a_m][10]^m + [a_{m-1}][10]^{m-1} + \ldots + [a_2][10]^2 + [a_1][10] + [a_0]$$
$$= [a_m][1]^m + [a_{m-1}][1]^{m-1} + \ldots + [a_2][1]^2 + [a_1][1] + [a_0] \text{ (why?)}$$
$$= [a_m + a_{m-1} + \ldots + a_2 + a_1 + a_0].$$

In conclusion, a and $(a_m + a_{m-1} + \ldots + a_2 + a_1 + a_0)$ have the same remainder when divided by 3 and when divided by 9.
   Therefore,

$$3 \mid a \Leftrightarrow 3 \mid (a_m + a_{m-1} + \ldots + a_1 + a_0)$$

and $9 \mid a \Leftrightarrow 9 \mid (a_m + a_{m-1} + \ldots + a_1 + a_0)$.

Casting Out Nines. The divisibility test for 9 has been used for centuries as an aid in checking arithmetic calculations. This application is referred to as **casting out nines**, and although its origins are not exactly known, there is some evidence suggesting that it was discovered in India prior to AD 1000 and is sometimes called the "Hindu" or "Indian" check.

Example. Let's use arithmetic in $\mathbf{Z}_9$ to determine if the following multiplication is computed correctly:

$$(215,879) \cdot (504,836) = 108,983,470,844.$$

If this equality is true, then when we reduce it (mod 9), it should remain true. Reducing the left-hand side of the equality gives

$$[(215,879) \cdot (504,836)] = [215,879] \cdot [504,836] =$$
$$[2 + 1 + 5 + 8 + 7 + 9] \cdot [5 + 0 + 4 + 8 + 3 + 6] =$$
$$[32] \cdot [26] = [3 + 2] \cdot [2 + 6] = [5] \cdot [8] = [40] = [4].$$

Reducing the right-hand side of the equality gives,

$$[108,983,470,844] = [1 + 0 + 8 + 9 + 8 + 3 + 4 + 7 + 0 + 8 + 4 + 4]$$
$$= [56] = [11] = [2].$$

Since $[4] \neq [2]$ in $\mathbf{Z}_9$, it follows that the original multiplication problem is incorrect.

Limitations of the Casting-Out-Nines Method. Translating integer arithmetic statements into $\mathbf{Z}_9$ arithmetic statements is a simple way to detect some errors in computation, but it is not an infallible method. In particular, if the integer computation is correct, then so is the corresponding $\mathbf{Z}_9$ computation (i.e., if equality fails in $\mathbf{Z}_9$, then it also fails in $\mathbf{Z}$). (Why?) However, correctness of the $\mathbf{Z}_9$ computation does not necessarily imply correctness of the original integer computation.

For example, notice that $212 \cdot 546 \neq 117,552$, however, the respective quantities are equal (mod 9). To see this, reduce the left-hand side of this statement (mod 9) to get

$$[212 \cdot 546] = [212] \cdot [546] = [2 + 1 + 2] \cdot [5 + 4 + 6]$$
$$= [5] \cdot [15] = [5] \cdot [6] = [30] = [3],$$

and reduce the right-hand side of this statement to get

$$[117,552] = [1 + 1 + 7 + 5 + 5 + 2] = [21] = [3].$$

The reduced quantities are equal in $\mathbf{Z}_9$, but the original quantities are unequal in $\mathbf{Z}$. The problem here is that this method does not detect the interchange of digits (why?), and furthermore, it does not distinguish between the digits 0 and 9, e.g., $[2,081] = [2,981]$. (Check.)

Divisibility by 11 and Casting Out Elevens. Although casting out nines does not detect digit interchanges, there exists other tests (some that are more sophisticated) that do, to differing extents, detect transpositions of unequal adjacent digits. For example, casting out elevens detects single digit errors and single transpositions of unequal adjacent digits. To explain this in more detail, let's first investigate what happens to an integer when it is reduced (mod 11).

For an integer $n = a_m a_{m-1} \ldots a_2 a_1 a_0$, computing in $\mathbf{Z}_{11}$,

$$[n] = [a_m 10^m + a_{m-1} 10^{m-1} + \ldots + a_1 10 + a_0]$$
$$= [a_m 10^m] + [a_{m-1} 10^{m-1}] + \ldots + [a_2 10^2] + [a_1 10] + [a_0]$$
$$= [a_m][10^m] + [a_{m-1}][10^{m-1}] + \ldots + [a_2][10^2] + [a_1][10] + [a_0]$$
$$= [a_m][10]^m + [a_{m-1}][10]^{m-1} + \ldots + [a_2][10]^2 + [a_1][10] + [a_0]$$
$$= [a_m][-1]^m + [a_{m-1}][-1]^{m-1} + \ldots + [a_2][-1]^2$$
$$+ [a_1][-1] + [a_0] \text{ (why?)}$$
$$= [(-1)^m a_m + (-1)^{m-1} a_{m-1} + \ldots + (-1)^2 a_2 + (-1) a_1 + a_0].$$

In summary, n and $((-1)^m a_m + (-1)^{m-1} a_{m-1} + \ldots + (-1)^2 a_2 + (-1) a_1 + a_0)$ have the same remainder when divided by 11, therefore,

$$11 | n \Leftrightarrow 11 | ((-1)^m a_m + (-1)^{m-1} a_{m-1} + \ldots + (-1)^2 a_2 + (-1) a_1 + a_0).$$

Example. Using the previous test, let's determine whether 11 divides the following numbers:

a. 111,111
b. 4,578,209.

For part a, 11 divides 111,111, since 11 divides $(-1 + 1 - 1 + 1 - 1 + 1) = 0$.

For part b, 11 does not divide 4,578,209, since 11 does not divide $(4 - 5 + 7 - 8 + 2 - 0 + 9) = 9$.

Casting Out Elevens Detects Single Digit Errors. In $\mathbf{Z}_9$, we know that $[3,078] = [3,978]$, but $3,078 \neq 3,978$. However, this phenomenon does not occur in $\mathbf{Z}_{11}$. Namely, if

$$a = a_n 10^n + a_{n-1} 10^{n-1} + \ldots + a_j 10^j + \ldots + a_1 10^1 + a_0 10^0 \text{ and}$$

$$b = a_n 10^n + a_{n-1} 10^{n-1} + \ldots + b_j 10^j + \ldots + a_1 10^1 + a_0 10^0,$$

and $[a] = [b]$ in $\mathbf{Z}_{11}$, then $a = b$. In conclusion, if $a_j \neq b_j$ for some j, then $[a] \neq [b]$ in $\mathbf{Z}_{11}$.

Justification. Since $[a] = [b]$ in $\mathbf{Z}_{11}$, then

$$[(a_n - a_n)10^n + (a_{n-1} - a_{n-1})10^{n-1} + \ldots + (a_j - b_j)10^j$$
$$+ \ldots + (a_1 - a_1)10^1 + (a_0 - a_0)10^0] = [0],$$

$$[(a_j - b_j)10^j] = [(a_j - b_j)(-1)^j] = [0].$$

Thus, $11 | (a_j - b_j)$ (why?), and since $0 \leq a_j, b_j \leq 9$, it follows that $a_j = b_j$, so $a = b$.

Casting Out Elevens Detects a Single Transposition of Unequal Adjacent Digits. Notice that in $\mathbf{Z}_{11}$, $[29, 251] \neq [22, 951]$ (a single transposition of unequal adjacent digits), however $[2, 357] = [2, 753]$ (interchange of the nonadjacent digits 3 and 7) and $[29, 251] = [22, 915]$ (transposition of the digits 9 and 2, and 5 and 1). Thus, $\mathbf{Z}_{11}$ arithmetic does not detect a single transposition of unequal nonadjacent digits or multiple transpositions of unequal digits but does detect a single transposition of unequal adjacent digits. To verify this last assertion in complete generality, it is not difficult to develop an argument similar to the previous one.

Justification. Let

$$a = a_n 10^n + a_{n-1} 10^{n-1} + \ldots + a_j 10^j + a_{j-1} 10^{j-1} + \ldots + a_1 10^1 + a_0 10^0 \text{ and}$$

$$b = a_n 10^n + a_{n-1} 10^{n-1} + \ldots + a_{j-1} 10^j + a_j 10^{j-1} + \ldots + a_1 10^1 + a_0 10^0,$$

and suppose that $[a] = [b]$ in $\mathbf{Z}_{11}$. Our goal is to show that $a_j = a_{j-1}$. Note that

$$[a] - [b] = [(a_j - a_{j-1})10^j + (a_{j-1} - a_j)10^{j-1}] =$$
$$[(a_j - a_{j-1})(-1)^j + (a_{j-1} - a_j)(-1)^{j-1}] = [0],$$

so $[2(a_j - a_{j-1})] = [0]$. (Why?) It follows that $11|2(a_j - a_{j-1})$, and since 11 is prime, we know $11|(a_j - a_{j-1})$. (Why?) Therefore, as $0 \leq a_j, a_{j-1} \leq 9$, it must be that $a_j = a_{j-1}$, and thus $a = b$.

We complete this section with a few nice applications of modular arithmetic to perfect numbers (see Section 3.5).

Even Perfect Numbers Modulo 9. Notice that the perfect numbers 28, 496, and 8,128 all have the same remainder when divided by 9. In particular, calculating in $\mathbf{Z}_9$ gives,

$$[28] = [2 + 8] = [10] = [1],$$
$$[496] = [4 + 9 + 6] = [19] = [10] = [1], \text{ and}$$
$$[8,128] = [8 + 1 + 2 + 8] = [19] = [10] = [1].$$

The next observation employs the structure of even perfect numbers to show that the same conclusion can be reached for all even perfect numbers $n > 6$.

4.4.5 Observation. If $n > 6$ is an even perfect number, then $[n] = [1]$ in $\mathbf{Z}_9$.

Justification. An even perfect integer n can always be represented as $n = 2^{p-1}(2^p - 1)$ (see Theorem 3.5.6), where $2^p - 1$ is prime, and, consequently, p is also prime (see Summary 3.5.5). In view of the previous calculations, we can assume that $p > 7$, and thus by the division algorithm, it follows that $p = 6q + 1$ or $p = 6q + 5$, where $q \geq 1$. (Why?)

Case 1. $p = 6q + 1$. With this hypothesis,

$$2^{p-1} = 2^{6q}, \text{ so } 2^p - 1 = (2 \cdot 2^{p-1}) - 1 = (2 \cdot 2^{6q}) - 1.$$

Computing in $\mathbf{Z}_9$ and using the fact that $[2^6] = [1]$ gives

$$[n] = [2^{p-1}(2^p - 1)] = [2^{p-1}][2^p - 1] = [2^{6q}][2 \cdot 2^{6q} - 1]$$
$$= [2^6]^q([2] \cdot [2^6]^q - [1]) =$$
$$[1]^q([2] \cdot [1]^q - [1]) = ([2] - [1]) = [1].$$

Case 2. $p = 6q + 5$. Proceeding as in Case 1,

$$2^{p-1} = 2^{6q+4} = 2^{6q} \cdot 2^4, \text{ so } 2^p - 1 = (2 \cdot 2^{p-1}) - 1$$
$$= (2 \cdot 2^{6q+4}) - 1 = (2^5 \cdot 2^{6q}) - 1.$$

Computing in $\mathbf{Z}_9$ gives

$$[n] = [2^{p-1}(2^p - 1)] = [2^{p-1}][2^p - 1] = [2^{6q+4}][2 \cdot 2^{6q+4} - 1] =$$
$$[2^6]^q[2^4]([2^5] \cdot [2^6]^q - [1]) =$$
$$[2^4]([2^5] - [1]) = [2^3]([2^6] - [2]) = [8]([1] - [2]) = -[8] = [1].$$

The Digit in the Ones Place of Even Perfect Numbers. All the examples of perfect numbers that we encountered in Chapter 3 were even integers n ending in either 6 or 8, i.e., either $[n] = [6]$ or $[n] = [8]$ in $\mathbf{Z}_{10}$ (recall that it is not known if there are any odd perfect integers). Arguing as in Observation 4.4.5, we now demonstrate that this is true in general.

4.4.6 Observation. If n is an even perfect number, then the digit in the ones place of n is 6 or 8.

Justification. As in the previous observation, $n = 2^{p-1}(2^p - 1)$, where both $2^p - 1$ and p are prime. Since the statement under question is true for the even perfect number $n = 6$ (in this case $p = 2$) and the even perfect number $n = 28$ (in this case $p = 3$), we can assume that $p > 3$. Thus, by the division algorithm, p can be expressed as $p = 4q + 1$ or $p = 4q + 3$, where $q \geq 1$. (Why?)

Case 1. $p = 4q + 1$. Under this assumption,

$$2^{p-1} = 2^{4q} = 16^q, \text{ so } 2^p - 1 = (2 \cdot 2^{p-1}) - 1 = (2 \cdot 16^q) - 1.$$

Calculating in $\mathbf{Z}_{10}$ gives,

$$[n] = [2^{p-1}(2^p - 1)] = [2^{p-1}][2^p - 1] = [16^q][(2 \cdot 16^q) - 1]$$
$$= [16]^q([2][16]^q + [-1]).$$

Note that $[16] = [6]$, which implies that $[16]^2 = [6]^2 = [36] = [6]$. Similarly, $[16]^q = [6]$ for each positive integer q (see no. 4 of Exercises 4.4, which calls for an inductive proof of this fact).

Therefore, $[n] = [16]^q([2][16]^q + [-1]) = [6]([2][6] + [-1]) = [6][11] = [6][1] = [6]$.

Case 2. $p = 4q + 3$. Progressing as in Case 1,

$$2^{p-1} = 2^{4q+2} = 2^2 \cdot 16^q = 4 \cdot 16^q, \text{ so}$$

$$2^p - 1 = (2 \cdot 2^{p-1}) - 1 = (2 \cdot 4 \cdot 16^q) - 1 = (8 \cdot 16^q) - 1.$$

Working in $\mathbf{Z}_{10}$ produces the desired conclusion:

$$[n] = [2^{p-1}(2^p - 1)] = [2^{p-1}][2^p - 1] = [4 \cdot 16^q][(8 \cdot 16^q) - 1]$$
$$= [4][16]^q([8][16]^q + [-1]) =$$
$$[4][6]([8][6] + [-1]) = [24][47] = [4][7] = [28] = [8].$$

EXERCISES 4.4

1. Justify the following statements.
   a. $[29] = [501]$ in $\mathbf{Z}_8$
   b. $-53 \equiv -81 \pmod{4}$
   c. $309 \equiv -207 \pmod{12}$
   d. $[-136] = [136]$ in $\mathbf{Z}_{17}$

164 Chapter 4 Arithmetic and Algebra of the Integers Modulo n

2. Use modular arithmetic to find the remainders in the given divisions.
   a. $12\overline{)13^{113}}$
   b. $4\overline{)7^{543}}$ Direction: $[7] = [-1]$ in $\mathbf{Z}_4$
   c. $5\overline{)2^{250}}$ Direction: $250 = 2 \cdot 125$
   d. $47\overline{)2^{23} - 1}$

3. Use $\mathbf{Z}_9$ arithmetic (casting out nines) to show that the following computations are in error.
   a. $45{,}321 \cdot 794 = 35{,}974{,}874$
   b. $3{,}086 + 8{,}829 + 1{,}045 = 11{,}960$

4. Using mathematical induction, show that in $\mathbf{Z}_{10}$, $[16]^q = [6]$ for all $q \geq 1$.

5. a. For integers a, b, and c, justify the following equivalence in $\mathbf{Z}_n$: $[a] = [b] \Leftrightarrow [a] + [c] = [b] + [c]$.
   b. For integers a, b, and c, justify the following implication in $\mathbf{Z}_n$: if $[a] = [b]$, then $[a][c] = [b][c]$.
   c. If $[c] = [0]$, then the converse of part b is clearly not true (just as in $\mathbf{Z}$). Give an example in some $\mathbf{Z}_n$, where $[a][c] = [b][c] = [0]$, $[c] \neq [0]$, and $[a] \neq [b]$. Furthermore, find an example in some $\mathbf{Z}_n$ where $[a][c] = [b][c] \neq [0]$, and $[a] \neq [b]$.

6. a. For integers a and b, show that if $[a] = [b]$ in $\mathbf{Z}_n$, then $[a]^t = [b]^t$ for each integer $t \geq 1$.
   b. Show that the converse of 6a is not true in general. In particular, provide an example where $[a]^2 = [b]^2$, but $[a] \neq [b]$.

7. Let $f(x)$ be a polynomial with integer coefficients. Show that if $[a] = [b]$ in $\mathbf{Z}_n$, then $[f(a)] = [f(b)]$ in $\mathbf{Z}_n$.

8. a. Give an example where $[a] = [b]$ in $\mathbf{Z}_n$, but $[a] \neq [b]$ in $\mathbf{Z}_d$ for some integer $d > 1$.
   b. Show that if $[a] = [b]$ in $\mathbf{Z}_n$ and $d|n$ where $d > 0$, then $[a] = [b]$ in $\mathbf{Z}_d$.
   c. Observe that if $n > 6$ is an even perfect number, then $[n] = [1]$ in $\mathbf{Z}_3$. Direction: Use Observation 4.4.5 and 8b.

9. a. Give an example where $[a] = [b]$ in $\mathbf{Z}_m$ and $[a] = [b]$ in $\mathbf{Z}_n$, but $[a] \neq [b]$ in $\mathbf{Z}_{mn}$.
   b. Prove: $[a] = [b]$ in $\mathbf{Z}_m$ and $[a] = [b]$ in $\mathbf{Z}_n \Leftrightarrow [a] = [b]$ in $\mathbf{Z}_{lcm(m,n)}$.
   c. Prove: If $[a] = [b]$ in $\mathbf{Z}_m$ and $[a] = [b]$ in $\mathbf{Z}_n$ and $\gcd(m,n) = 1$, then $[a] = [b]$ in $\mathbf{Z}_{mn}$.

10. Let n be an integer greater than 1, a any integer, and $d = \gcd(a,n)$. Establish the following equivalence: $[a] = [b]$ in $\mathbf{Z}_n \Leftrightarrow [\frac{a}{d}] = [\frac{b}{d}]$ in $\mathbf{Z}_{\frac{n}{d}}$. Direction: ($\Rightarrow$) Show that $\frac{b}{d}$ is an integer. ($\Leftarrow$) For this part, it is assumed that $\frac{b}{d}$ is an integer, otherwise the notation $[\frac{b}{d}]$ is undefined.

4.5 COMPARING ARITHMETIC PROPERTIES OF Z AND $\mathbf{Z}_n$

In this section, we systematically investigate commonalities and differences of important arithmetic properties of $\mathbf{Z}$ and $\mathbf{Z}_n$. The analysis utilizes our knowledge of integer arithmetic properties and the definitions of addition and multiplication in $\mathbf{Z}_n$.

Addition Properties of Z and $\mathbf{Z}_n$

1a. Additive Identity in Z: $0 + a = a + 0 = a$, for each integer a.

Section 4.5 Comparing Arithmetic Properties of Z and Z_n

The integer 0 is the unique integer having this property and is called **the additive identity** of **Z**.

1b. Additive identity in Z_n: $[0] + [a] = [a] = [a] + [0]$, for each integer a.

These equalities follow from the additive identity property in **Z** (Explain).

Uniqueness of additive identity in Z_n: The element $[0]$ is the only member of Z_n with the property stated in 1b.

Justification: If $[b] + [a] = [a] + [b] = [a]$ for each integer a, then $[b] + [0] = [0]$ (here $[0]$ is playing the role of $[a]$). However, $[b] + [0] = [b]$, and thus $[b] = [0]$.
   The element $[0]$ is called **the additive identity** of Z_n.

2a. Additive inverses in Z: $a + (-a) = 0 = (-a) + a$, for each integer a.

Recall that $-a$ is the only element of **Z** having the property that $a + (-a) = 0$. The integer $-a$ is called **the additive inverse** of the integer a.

Subtraction notation in Z: For integers a and b, the sum $a + (-b)$ is denoted by $a - b$.

2b. Additive inverses in Z_n: $[a] + [-a] = [0] = [-a] + [a]$, for each integer a. (Justify.)

Uniqueness of additive inverses in Z_n: Note that $[-a]$ is the unique member of Z_n with the property described in 2b.

Justification: If $[a] + [b] = [0] = [b] + [a]$, then $[b] = [b] + [0] = [b] + ([a] + [-a]) = ([b] + [a]) + [-a]$ (use 4b below) $= [0] + [-a] = [-a]$.

The element $[-a]$ is called **the additive inverse** of $[a]$ in Z_n and is also denoted by $-[a]$.

Subtraction notation in Z_n: For members $[a]$ and $[b]$ of Z_n, the sum $[a] + [-b]$ is denoted by $[a] - [b]$.

Simplifying computations by using additive inverses: When simplifying arithmetic calculations in Z_n, it is sometimes advantageous to utilize the additive inverses of certain elements. For example, consider the following calculations (justify each equality):

In Z_{38}: $[29] + [3] + [32] + [25] = [-9] + [3] + [-6] + [-13]$
           $= [(-9) + 3 + (-6) + (-13)] = [-25] = [13]$.

In Z_{45}: $[39] \cdot [5] \cdot [41] = [-6] \cdot [5] \cdot [-4] = [(-6) \cdot 5 \cdot (-4)] = [120] = [30]$.

In Z_{26}: $[5]^{150} = ([5]^2)^{75} = [25]^{75} = [-1]^{75} = [-1] = [25]$.

3a. Commutative property of addition in Z: $a + b = b + a$, for each integer a and b.

3b. Commutative property of addition in Z_n: $[a] + [b] = [b] + [a]$, for each integer a and b.

This equality follows from the commutativity of addition in **Z**, i.e., $[a] + [b] = [a + b] = [b + a] = [b] + [a]$.

4a. Associative property of addition in Z: $a + (b + c) = (a + b) + c$, for each integer a, b, and c.

4b. Associative property of addition in Z_n: $[a] + ([b] + [c]) = ([a] + [b]) + [c]$, for each integer a, b, and c. (Justify.)

Multiplication Properties of Z and Z_n

1a. Multiplicative identity in Z: $1a = a1 = a$, for each integer a.

The integer 1 is the only integer with this property and is called **the multiplicative identity** of **Z**.

1b. Multiplicative identity in Z_n: $[1][a] = [a][1] = [a]$, for each integer a. (Justify.)

We have seen that the additive identity in Z_n is unique, and a similar argument shows the same is true for multiplication. (Justify.) The element [1] is called **the multiplicative identity** of Z_n.

2. **Multiplicative inverses in Z:** The **multiplicative inverse** of a nonzero integer a is an integer b so that $ab = 1$; thus, 1 and -1 are the only integers possessing a multiplicative inverse.

The discussion involving multiplicative inverses in Z_n is quite interesting and deserves an entire section for its complete development. Thus we first handle the rather straightforward commutative, associative, and distributive properties for multiplication and then pursue multiplicative inverses for Z_n in the next section.

3a. Commutative property of multiplication in Z: $ab = ba$, for each integer a and b.

3b. Commutative property of multiplication in Z_n: $[a][b] = [b][a]$, for each integer a and b. (Justify.)

Classroom Problems. For $t \geq 1$, show that $([a][b])^t = [a]^t[b]^t$. ◆

4a. Associative property of multiplication in Z: $a(bc) = (ab)c$, for each integer a, b, and c.

4b. Associative property of multiplication in Z_n: $[a]([b][c]) = ([a][b])[c]$, for each integer a, b, and c. (Justify.)

The next property relates multiplication and addition and is the backbone of multidigit multiplication.

Section 4.5 Comparing Arithmetic Properties of Z and Z_n

5a. Distributive property of multiplication over addition in Z: $a(b + c) = ab + ac$, for each integer a, b, c.

Since multiplication in the integers is commutative, it also follows that $(b + c)a = ba + ca$. (Explain.)

The multiplication procedure used for multidigit multiplication is a direct application of the distributive property. For example, to compute the product $67 \cdot 45$, the standard multiplication process is:

$$\begin{array}{r} 67 \\ 45 \\ \hline 335 \\ 268 \\ \hline 3{,}015 \end{array}$$

This multiplication method is nothing more than an abbreviated application of the distributive property, i.e.,

$$67 \cdot 45 = 67 \cdot (40 + 5) = (67 \cdot 40) + (67 \cdot 5) = 2{,}680 + 335 = 3{,}015.$$

The distributive property, in combination with the commutative property of multiplication, is especially useful for multiplying general expressions involving integers. For example, the following expansions hold for all integers a and b:

$$(a + b)^2 = (a + b)(a + b) = (a + b)a + (a + b)b = a^2 + ba + ab + b^2$$
$$= a^2 + 2ab + b^2,$$
$$(a + b)^3 = (a + b)^2(a + b) = (a^2 + 2ab + b^2)a + (a^2 + 2ab + b^2)b$$
$$= a^3 + 3a^2b + 3ab^2 + b^3.$$

Recall that in Chapter 1 we considered the Binomial Theorem, which is the general statement of the previous expansions. In particular, we showed that

$$(a + b)^m = \sum_{k=0}^{m} \binom{m}{k} a^{m-k} b^k$$

holds for all integers a and b.

5b. Distributive property of multiplication over addition in Z_n: $[a]([b] + [c]) = [a][b] + [a][c]$, for each integer a, b, c. (Justify.)

Note that computing the expansions $([a] + [b])^2$ and $([a] + [b])^3$ in Z_n amounts to reducing the previous integer expansions (mod n). In particular,

$$([a] + [b])^2 = [a + b]^2 = [(a + b)^2] = [a^2 + 2ab + b^2]$$
$$= [a^2] + [2ab] + [b^2] = [a]^2 + [2][a][b] + [b]^2,$$

168 Chapter 4 Arithmetic and Algebra of the Integers Modulo n

$$([a] + [b])^3 = [a + b]^3 = [(a+b)^3] = [(a+b)^2][a+b]$$
$$= [(a+b)^2]([a] + [b]) =$$
$$([a]^2 + [2][a][b] + [b^2])[a] + ([a]^2 + [2][a][b] + [b^2])[b]$$
$$= \ldots = [a]^3 + [3][a]^2[b] + [3][a][b]^2 + [b]^3.$$

4.5.1 The Binomial Theorem in $\mathbf{Z}_n$. As in the previous expansions, a modulo n reduction of the Binomial Theorem in $\mathbf{Z}$ (see Section 1.7) produces the analogous result in $\mathbf{Z}_n$, i.e.,

$$([a] + [b])^m = \sum_{r=0}^{m} [\binom{m}{r}][a]^{m-r}[b]^r.$$

4.5.2 Modular Redemption. A common error committed by middle-grade through college-level mathematics students is the infamous incorrect expansion $(a + b)^2 = a^2 + b^2$ or more generally, $(a+b)^m = a^m + b^m$, where a and b represent either integers, rational numbers, or real numbers. For all those students who have been "marked down" for perpetrating this egregious error, there is some salvation to be found in the modular world.

In particular, notice that in $\mathbf{Z}_2$, the expansion

$$([a] + [b])^2 = [a]^2 + [b]^2$$

is true for all integers a and b (why?); similarly, in $\mathbf{Z}_3$ the expansion

$$([a] + [b])^3 = [a]^3 + [b]^3$$

holds for all integers a and b. (Why?) However, the redemption fails in $\mathbf{Z}_4$. To see this, use the Binomial Theorem in $\mathbf{Z}_4$ to obtain the next equality:

$$([a] + [b])^4 = [a]^4 + [4][a]^3[b] + [6][a]^2[b]^2 + [4][a][b]^3 + [b]^4$$
$$= [a]^4 + [2][a]^2[b]^2 + [b]^4,$$

and then choose integers a and b so that the middle term is nonzero in $\mathbf{Z}_4$, e.g., $a = b = 1$ produces a middle term of $[2] \neq [0]$.

Question. In which $\mathbf{Z}_m$ ($m \geq 2$) is it true that $([a] + [b])^m = [a]^m + [b]^m$ for all integers a and b?

Let's look at some examples in $\mathbf{Z}_5$ and $\mathbf{Z}_6$. Applying the Binomial Theorem, we see that

$$([a] + [b])^5 = [a]^5 + [b]^5 \text{ (verify) and that}$$
$$([a] + [b])^6 = [a]^6 + [3][a]^4[b]^2 + [2][a]^3[b]^3 + [3][a]^2[b]^4 + [b]^6 \text{ (verify)}.$$

From the evidence thus far, it is reasonable to investigate the binomial expansions in $\mathbf{Z}_p$ for p a prime number.

What occurred in $\mathbf{Z}_2$, $\mathbf{Z}_3$, and $\mathbf{Z}_5$ is that the binomial coefficients $\binom{p}{r}$ for $1 \leq r \leq p - 1$ were divisible by p, so the (mod p) binomial coefficients of the "in between" terms equaled [0] in $\mathbf{Z}_2$, [0] in $\mathbf{Z}_3$, and [0] in $\mathbf{Z}_5$. More generally, this property of binomial coefficients is true for all primes p.

4.5.3 Observation. If p is a prime number, then $p | \binom{p}{r}$ for each $1 \leq r \leq p - 1$.

Justification. Recall that $\binom{p}{r} = \dfrac{p!}{r!(p-r)!} = \dfrac{p(p-1)\cdots s[p-(r+1)]}{r!}$ (see Section 1.6), so by multiplying both sides of the equality by $r!$, we see that $p | r! \binom{p}{r}$. Since p is a prime, it follows that $p | r!$ or $p | \binom{p}{r}$ (see Corollary 3.3.2). However, p does not divide $r!$, as $1 \leq r \leq p - 1$ (explain), so $p | \binom{p}{r}$ as desired.

4.5.4 Summary of Modular Redemption. If p is a prime number, then $([a] + [b])^p = [a]^p + [b]^p$ in $\mathbf{Z}_p$ for all integers a and b.

EXERCISES 4.5

1. Perform the following calculations by using the method illustrated in the examples following the discussion of additive inverses.
   a. In $\mathbf{Z}_{99}$: $[84] + [16] + [93] + [5] =$
   b. In $\mathbf{Z}_{47}$: $[45]^5 =$
   c. In $\mathbf{Z}_{64}$: $[60] \cdot ([59] + [4]) =$
2. a. Show that if $n > 1$ is odd, then $[1] + [2] + \cdots + [n-1] = [0]$ in $\mathbf{Z}_n$.
      Direction: Recall from Chapter 1 the sum of a general arithmetic progression.
   b. Show that if $n > 1$ is even, then $[\frac{n}{2}] = [-\frac{n}{2}]$ in $\mathbf{Z}_n$.
   c. Show that if $n > 1$ is even, then $[1] + [2] + \cdots + [n-1] = [\frac{n}{2}]$ in $\mathbf{Z}_n$.
      Direction: Use part b and the sum of an arithmetic progression.

4.6 MULTIPLICATIVE INVERSES IN $\mathbf{Z}_n$

We are now ready to tackle inverses relative to multiplication and thus complete the discussion of arithmetic properties initiated in the previous section. As mentioned earlier, the integer story is uninteresting, whereas the situation in $\mathbf{Z}_n$ is quite fascinating and is related to our work in Chapters 2 and 3.

2b. Multiplicative inverses in $\mathbf{Z}_n$: Clearly $[1]$ and $[-1]$ have multiplicative inverses in $\mathbf{Z}_n$ (why?), but unlike the integers, it is possible, given a nonzero element $[a] \neq \pm[1]$ in $\mathbf{Z}_n$ ($[a] \neq [0]$ in $\mathbf{Z}_n$), for there to exist a nonzero $[b]$ in

$\mathbf{Z}_n$ so that $[a][b] = [1]$, i.e., $[a]$ possesses a multiplicative inverse in $\mathbf{Z}_n$ (and likewise $[b]$ has a multiplicative inverse in $\mathbf{Z}_n$).

4.6.1 Example. In $\mathbf{Z}_9$, $[2]$ has a multiplicative inverse, i.e., $[2] \cdot [5] = [1]$, and consequently $[5]$ has a multiplicative inverse, since $[5] \cdot [2] = [1]$. Similarly, $[4]$ and $[7]$ are inverses of each other, as $[4] \cdot [7] = [1] = [7] \cdot [4]$. In addition, $[8] \cdot [8] = [1]$, and thus $[8]$ is its own multiplicative inverse.

Note that the remaining nonzero members $[3]$ and $[6]$ do not have multiplicative inverses in $\mathbf{Z}_9$. One way to verify this assertion, say for $[3]$, is to show that the equation

$$[3]X = [1],$$

has no solutions in $\mathbf{Z}_9$. This can be accomplished by the calculations

$$[3] \cdot [0] = [0] \neq [1], \quad [3] \cdot [1] = [3] \neq [1], \quad [3] \cdot [2] = [6] \neq [1],$$
$$[3] \cdot [3] = [0] \neq [1], \quad [3] \cdot [4] = [3] \neq [1], \quad [3] \cdot [5] = [6] \neq [1],$$
$$[3] \cdot [6] = [0] \neq [1], \quad [3] \cdot [7] = [3] \neq [1], \quad [3] \cdot [8] = [6] \neq [1],$$

and therefore there is no integer b so that $[3][b] = [1]$. (Why do the previous nine calculations show that there is no arbitrary integer b so that $[3][b] = [1]$?)

We can also see that $[3]$ does not have a multiplicative inverse in $\mathbf{Z}_9$ through an indirect approach.

Justification. Suppose there is some integer b so that $[3][b] = [1]$. Multiplying both sides of the equation by $[3]$ yields, $[3]([3][b]) = [3]$, which in turn leads to the contradiction $[0] = [3]$. Therefore, $[3]$ does not have a multiplicative inverse in $\mathbf{Z}_9$.

The previous example touches upon some important points, the first of which is the issue of uniqueness of multiplicative inverses.

4.6.2 The Uniqueness of Multiplicative Inverses in $\mathbf{Z}_n$. If a nonzero element $[a]$ in $\mathbf{Z}_n$ has a multiplicative inverse $[b]$ in $\mathbf{Z}_n$, then it is the only element of $\mathbf{Z}_n$ with the property that $[a][b] = [1]$.

Justification. Suppose that $[a][b] = [1]$ and $[a][c] = [1]$ for some elements $[b]$ and $[c]$ in $\mathbf{Z}_n$. We will show that $[b] = [c]$. Multiply each side of the equality $[a][b] = [a][c]$ by $[b]$ to obtain the equality $[b]([a][b]) = [b]([a][c])$, and then use associativity and commutativity to conclude that $[b] = [c]$. (Explain.)

Notation. If a nonzero element $[a]$ in $\mathbf{Z}_n$ has a multiplicative inverse $[b]$ in $\mathbf{Z}_n$, then the unique element $[b]$ is usually denoted by $[a]^{-1}$, and analogously, $[b]^{-1} = [a]$.

A peculiarity that occurred in Example 4.6.1 is the existence of nonzero elements in $\mathbf{Z}_9$ whose products are zero, and this prompts the following definition.

Zero Divisors. The nonzero elements $[3]$ and $[6]$ in $\mathbf{Z}_9$ each "divide" $[0]$ with *nonzero quotients*, e.g., $[3] \cdot [3] = [0]$ and $[6] \cdot [3] = [0]$. This strange phenomenon does not occur in $\mathbf{Z}$ and is a striking arithmetic distinction between $\mathbf{Z}$ and $\mathbf{Z}_9$.

Section 4.6 Multiplicative Inverses in Z_n 171

In general, a nonzero element $[a]$ in $\mathbf{Z}_n$ is called a **zero-divisor** or a **divisor of zero** if there is a nonzero $[b]$ in $\mathbf{Z}_n$ so that $[a][b] = [0]$ (in this case, $[b]$ is also a zero-divisor).

For a given zero-divisor $[a]$ there can exist several distinct $[b]$s such that $[a][b] = [0]$, e.g., in $\mathbf{Z}_{15}$, $[5] \cdot [3] = [15] = [0]$, $[5] \cdot [6] = [0]$, and $[5] \cdot [9] = [0]$.

If there does not exist such a nonzero $[b]$ in $\mathbf{Z}_n$, then $[a]$ is called a **nonzero-divisor** (this terminology is restricted to nonzero elements of $\mathbf{Z}_n$).

The cancellation process used in the justification of result 4.6.2 can be expressed in general terms.

Cancellation in Z and $\mathbf{Z}_n$. If $a \neq 0$ and $ab = ac$ in $\mathbf{Z}$, then it follows that $b = c$. One way to see this, by using only integer arithmetic, is by noting that $ab - ac = a(b - c) = 0$ implies that $a = 0$ or $(b - c) = 0$, since $\mathbf{Z}$ has no divisors of zero. By assumption, $a \neq 0$, so $b = c$.

Another way to make the same deduction is to work inside the rational numbers, where both sides of the equality could be multiplied by $1/a$ (the multiplicative inverse of a in the rationals) to deduce that $b = c$.

The cancellation picture is much different in $\mathbf{Z}_n$, since zero divisors might be present. For example, in $\mathbf{Z}_6$, $[2]$ cannot be cancelled from the equality $[2][1] = [2][4]$, whereas $[5]$ can be cancelled from the equality $[5][b] = [5][c]$.

Analysis. Note that $[5]$ is a nonzero-divisor in $\mathbf{Z}_6$, and since the assumption gives $[5]([b] - [c]) = [0]$, then $[b] - [c] = [0]$, i.e., $[b] = [c]$. (Why?) Another way to draw the same conclusion is to multiply both sides of the equality by $[5]^{-1}$ (note that $[5]^{-1} = [5]$ in $\mathbf{Z}_6$) and then conclude that $[b] = [c]$.

We soon see that in the setting of modular arithmetic, these methods are actually equivalent.

4.6.3 Relating Multiplicative Inverses and Zero Divisors. The indirect argument considered in Example 4.6.1 shows that if a nonzero element $[a]$ in $\mathbf{Z}_n$ has a multiplicative inverse in $\mathbf{Z}_n$, then $[a]$ is a nonzero-divisor in $\mathbf{Z}_n$ (explain), or, equivalently, a zero-divisor $[a]$ in $\mathbf{Z}_n$ does not have a multiplicative inverse in $\mathbf{Z}_n$.

Classroom Discussion. Let's determine the zero-divisors in $\mathbf{Z}_{12}$ and also the elements that have multiplicative inverses in $\mathbf{Z}_{12}$, and then do the same for $\mathbf{Z}_{15}$ and $\mathbf{Z}_7$.

Analysis. 1. Notice that since $2|12$, then $12 = 2 \cdot 6$, so $[2] \cdot [6] = [12] = [0]$, i.e., $[2]$ and $[6]$ are zero-divisors in $\mathbf{Z}_{12}$. Thus, each divisor d of 12 with $d \neq \pm 1$ and $d \neq \pm 12$ gives rise to a zero-divisor $[d]$ in $\mathbf{Z}_{12}$ (explain the stated restrictions). Hence, the elements $[2]$, $[-2] = [10]$, $[3]$, $[-3] = [9]$, $[4]$, $[-4] = [8]$, and $[6]$ (note that $[6] = [-6]$) are zero-divisors in $\mathbf{Z}_{12}$, in fact, these account for all the zero-divisors in $\mathbf{Z}_{12}$. (Verify.)

Since we have just observed that zero-divisors do not have multiplicative inverses, the only remaining candidates to be checked for multiplicative inverses are the nonzero-divisors [5], [7], and [11] (clearly, [1] is its own inverse). A quick verification shows that $[5]^{-1} = [5]$, $[7]^{-1} = [7]$, and $[11]^{-1} = [11]$.

2. Utilizing the same strategy as in 1, note that [3], [−3] = [12], [5], [−5] = [10] are zero-divisors in $\mathbf{Z}_{15}$, but unlike the previous example, these are not the only zero-divisors in $\mathbf{Z}_{15}$. In particular, [6] and [9] are also zero-divisors in $\mathbf{Z}_{15}$, since [6] • [5] = [30] and [9] • [5] = [45] = [0]. This completes the list of all zero-divisors in $\mathbf{Z}_{15}$. (Verify.)

As in 1, it turns out that the nonzero-divisors all have inverses. By checking different possibilities (trial and error), it follows that $[2]^{-1} = [8]$, $[4]^{-1} = [4]$, $[7]^{-1} = [13]$, $[11]^{-1} = [11]$, and $[14]^{-1} = [14]$. (Verify.)

3. The initial strategy used in the last two examples to find zero-divisors does not apply to $\mathbf{Z}_7$, since 7 has no nontrivial divisors. Other strategies to discover zero-divisors in $\mathbf{Z}_7$ also fail, since it turns out that each nonzero element in $\mathbf{Z}_7$ has a multiplicative inverse. In particular, $[2]^{-1} = [4]$, $[3]^{-1} = [5]$, and $[6]^{-1} = [6]$.

Some Observations Concerning the Previous Examples. We have seen that the zero-divisors in $\mathbf{Z}_9$ (respectively, in $\mathbf{Z}_{12}$, in $\mathbf{Z}_{15}$, and in $\mathbf{Z}_7$) are precisely the nonzero elements [a] in $\mathbf{Z}_9$ (respectively, in $\mathbf{Z}_{12}$, in $\mathbf{Z}_{15}$, and in $\mathbf{Z}_7$) such that $\gcd(a, 9) > 1$ (respectively, $\gcd(a, 12) > 1$, $\gcd(a, 15) > 1$, and $\gcd(a, 7) > 1$).

We also observed that the elements having multiplicative inverses in $\mathbf{Z}_9$ (respectively, in $\mathbf{Z}_{12}$, in $\mathbf{Z}_{15}$, and in $\mathbf{Z}_7$) are the nonzero elements [a] in $\mathbf{Z}_9$ (respectively, [a] in $\mathbf{Z}_{12}$, [a] in $\mathbf{Z}_{15}$, and [a] in $\mathbf{Z}_7$) with the property that $\gcd(a, 9) = 1$ [respectively, $\gcd(a, 12) = 1$, $\gcd(a, 15) = 1$, and $\gcd(a, 7) = 1$].

These conclusions can be drawn in general, and our next consideration characterizes those elements of $\mathbf{Z}_n$ that have multiplicative inverses and also provides a method for finding them. ◆

4.6.4 Characterizing Multiplicative Inverses in $\mathbf{Z}_n$. A nonzero element [a] in $\mathbf{Z}_n$ has a multiplicative inverse [b] in $\mathbf{Z}_n$ if and only if $\gcd(a, n) = 1$.

Justification. First assume a nonzero element [a] in $\mathbf{Z}_n$ has a multiplicative inverse [b] in $\mathbf{Z}_n$. Since [a][b] = [1], then $n|(ab - 1)$, so $ab - 1 = nt$, for some integer t or equivalently, $ab - nt = 1$. From our work in Chapter 3, it is now easy to conclude that $\gcd(a, n) = 1$. (Explain.)

Conversely, assume $\gcd(a, n) = 1$. We know, by the Euclidean algorithm (see Summary 3.2.2), that there exists integers b and c so that $1 = ab + nc$. Reducing this equality (mod n) produces a multiplicative inverse for [a]:

$$[1] = [ab + nc] = [ab] + [nc] = [ab] + [0] = [ab] = [a][b].$$

Notation. If [a] has a multiplicative inverse in $\mathbf{Z}_n$, then $[a]^t$ has a multiplicative inverse in $\mathbf{Z}_n$ for each $t \geq 1$ (why?), and we write $([a]^t)^{-1} = [a]^{-t}$.

The next two corollaries follow directly from the characterization given in result 4.6.4 (Justify.)

4.6.5 Corollary: Cancellation in Z_n. If $[a][b] = [a][c]$ and $\gcd(a,n) = 1$, then $[b] = [c]$.

An extension of this corollary is considered in Exercises 4.6, no. 6.

4.6.6 Corollary. If p is a prime, then each $[a] \neq [0]$ in Z_p has a multiplicative inverse. (The converse of this statement is addressed in Exercise 4.6, no. 2.)

Classroom Problems. Using result 4.6.4, determine those members of Z_{42} that have multiplicative inverses and then find their inverses by using the Euclidean algorithm.

For example, $[25]$ has a multiplicative inverse in Z_{42}, since $\gcd(25, 42) = 1$. To find its inverse, use the Euclidean algorithm to determine integers b and c so that $1 = 25b + 42c$, then reduce this equality (mod 42) to show that $[b]$ is the inverse of $[a]$.

$$\begin{aligned}
42 &= 1 \cdot 25 + 17 & 1 &= 17 - 2 \cdot 8 \\
25 &= 1 \cdot 17 + 8 & &= 17 - 2(25 - 1 \cdot 17) \\
17 &= 2 \cdot 8 + 1 & &= 3 \cdot 17 - 2 \cdot 25 \\
8 &= 8 \cdot 1 + 0 & &= 3(42 - 1 \cdot 25) - 2 \cdot 25 \\
& & &= 3 \cdot 42 - 5 \cdot 25 \\
& & &= 25 \cdot (-5) + 42 \cdot 3.
\end{aligned}$$

Therefore, $[1] = [25 \cdot (-5) + (42 \cdot 3)] = [25][-5] = [25][37]$, and so $[25]^{-1} = [37]$. ◆

The previous examples led to a multiplicative inverse characterization in Z_n (see Section 4.6.4) and inspire the next characterization of zero-divisors in Z_n.

4.6.7 Characterizing Zero-Divisors in Z_n. A nonzero element $[a]$ in Z_n is a zero-divisor in Z_n if and only if $\gcd(a, n) > 1$.

Justification. If $[a]$ in Z_n is a zero-divisor in Z_n, then it does not have a multiplicative inverse (see result 4.6.3), so $\gcd(a, n) > 1$. (Why?)

Conversely, suppose that $\gcd(a, n) = d > 1$ for some $[a]$ in Z_n. Thus, $a = ds$ and $n = dt$ for some integers s and t. Notice that

$$[a][t] = [ds][t] = ([d][s])[t] = ([s][d])[t] = [s]([d][t]) = [s][dt]$$
$$= [s][n] = [s][0] = [0],$$

so $[a]$ is a zero-divisor in Z_n.

Connecting Multiplicative Inverses and Nonzero-Divisors. By combining results 4.6.4 and 4.6.7, it follows that a nonzero element $[a]$ in Z_n has a multiplicative inverse $\Leftrightarrow [a]$ is a nonzero-divisor. (Explain.)

EXERCISES 4.6

1. For each of the listed Z_n's, find all the elements that have multiplicative inverses and determine those inverses.
    a. Z_{25}

b. $\mathbf{Z}_{19}$

c. $\mathbf{Z}_{30}$

2. Show that if each nonzero element in $\mathbf{Z}_p$ has a multiplicative inverse, then p is prime. Direction: Suppose that $p = ab$, for integers a and b satisfying $1 < a < p$ and $1 < b < p$, and derive a contradiction.

3. Prove: If $[a]$ and $[b]$ have multiplicative inverses in $\mathbf{Z}_n$, then $[a][b]$ has a multiplicative inverse in $\mathbf{Z}_n$.

4. a. Give an example of an element $[a]$ in some $\mathbf{Z}_n$, where $[a] \neq [0]$, $[a] \neq [1]$, and $[a]^2 = [a]$. An element $[a]$ in $\mathbf{Z}_n$ with the property that $[a]^2 = [a]$ is called **idempotent**.

   b. Show that if $[a]^2 = [a]$ in $\mathbf{Z}_n$ and $\gcd(a, n) = 1$, then $[a] = [1]$.

5. a. Find an example of an element $[a]$ in some $\mathbf{Z}_n$, where $[a]^3 = [0]$, but $[a]^2 \neq [0]$. An element $[a]$ of $\mathbf{Z}_n$ is called **nilpotent** if there is some positive integer t so that $[a]^t = [0]$.

   b. Show that if $[a]$ and $[b]$ are nilpotent elements in $\mathbf{Z}_n$, then $[a] + [b]$ is nilpotent in $\mathbf{Z}_n$.

6. **Generalized Cancellation.** If $[a][b] = [a][c]$ in $\mathbf{Z}_n$, then $[b] = [c]$ in $\mathbf{Z}_{\frac{n}{d}}$, where $d = \gcd(a, n)$. Direction: Since $ab - ac = nt$ for some integer t, then $[\frac{a}{d}][b] = [\frac{a}{d}][c]$ in $\mathbf{Z}_{\frac{n}{d}}$. Now apply Corollary 4.6.5.

4.7 ELEMENTARY APPLICATIONS OF MODULAR ARITHMETIC

We have seen, and will continue to see throughout this book, how modular arithmetic can be used to expand our understanding of integer arithmetic and provide us with a potent computational and theoretical tool. In particular, modular arithmetic is a powerful tool in the study of several important mathematical and "real world" problems. In this section, we consider some applications that can be explored in a middle-school mathematics classroom.

Application 1: A Children's Puzzle. My uncle Sam (who was actually born on the 4th of July) liked to give his nieces and nephews puzzles to think about. One of his favorites was to have us choose any whole number a, with at least two distinct digits, and then use this number to make a new number b by rearranging the digits of a in any order.

Once we accomplished this, he told us to subtract the smaller number from the larger number and then add the digits of the subtraction answer c. From this point on, he instructed us to continue to add the digits of the resulting differences until we ended up with a single-digit number.

He claimed that we would always end up with the number 9 and challenged us to prove him wrong. Of course, we spent a great deal of time checking various examples, but we never managed to find one that didn't work or to explain why his assertion was always true.

To illustrate this puzzle, suppose we choose the integer $a = 5,361$ and form the integer resulting from any rearrangement of the digits of $a = 5,361$, e.g., $b = 1,536$. Then $a - b = 3,825 = c$. Adding the digits of c gives $3 + 8 + 2 + 5 = 18$, and then adding the digits of 18 results in the integer 9, as predicted by my uncle Sam.

If I only knew then what I know now, it would have been easy to show that my uncle's claim was correct.

Analysis. First of all, the condition that a has at least two distinct digits guarantees that it is possible to rearrange the digits of a to get a new number b that is distinct from a. Thus, the subtraction answer is $c \neq 0$, so the sum of the digits of c, and the sum of the digits of the resulting sums, must be greater than 0.

Without loss of generality, we may assume that $a > b$, i.e., $c = (a - b) > 0$, and we claim that $9|c$. To see this, express

$$a = a_m 10^m + a_{m-1} 10^{m-1} + \ldots + a_1 10 + a_0,$$

$$\text{and } b = b_m 10^m + b_{m-1} 10^{m-1} + \ldots + b_1 10 + b_0,$$

where the digits $b_m, b_{m-1}, \ldots, b_1, b_0$ are a (nontrivial) permutation of the digits of a, so $a \neq b$.

Calculating in $\mathbf{Z}_9$ gives

$$[c] = [a - b] = [a] + [-b] = [a_m + a_{m-1} + \ldots + a_1 + a_0]$$
$$+ [-(b_m + b_{m-1} + \ldots + b_1 + b_0)] = [0],$$

and thus $9|c$. Therefore, 9 divides the sum of the digits of $c = a - b$, i.e., if

$$c = c_t 10^t + c_{t-1} 10^{t-1} + \ldots + c_1 10^1 + c_0 10^0$$

$$\text{and } d = c_t + c_{t-1} + \ldots + c_1 + c_0,$$

then $[c] = [d] = [0]$. Using the same reasoning, 9 divides the sum of the digits of d and the sum of the digits of the resulting sums. Note that $c > d$, and in each subsequent stage, the sums of digits continue decreasing and remain divisible by 9. Eventually, the process ends with the integer 9.

The condition that a have at least two distinct digits was only necessary for the given puzzle, and the previous proof shows that if a is any integer and b is an integer formed by permuting the digits of a, then $9|(a - b)$ (order of subtraction is unimportant).

Instead of 9 dividing the difference of the integers a and b defined above ($a \neq b$ have the same digits but not in the same order), what can be said about 9 dividing their sum? In the previous example, $a + b = 5,361 + 1,536 = 6,897$, which is not divisible by 9. Again, using arithmetic in $\mathbf{Z}_9$, we see that $9|(a + b)$ if and only if $9|a$.

Justification. One direction is clear, since if $9|a$, then by the definition of b, $9|b$ (why?), and so $9|(a + b)$.

For the other direction, assume $9|(a + b)$, i.e., $[a + b] = [0]$ in $\mathbf{Z}_9$. Note that

$$[a + b] = [2(a_m + a_{m-1} + \ldots + a_2 + a_1 + a_0)]$$
$$= [2][a_m + a_{m-1} + \ldots + a_2 + a_1 + a_0] = [0],$$

and since [2] is a nonzero-divisor in $\mathbf{Z}_9$ (why?), then $[a_m + a_{m-1} + \ldots + a_2 + a_1 + a_0] = [0]$. Therefore, $9|a$ as asserted.

Application 2: Kaprekar's Constant: An Amazing 4-Digit Number. We have just made some interesting observations concerning integers a and b, where b is formed as some permutation of the digits of a. Let's consider the situation in which a is a 4-digit number having at least 2 distinct digits, e.g., $a = 3,851$.

Stage 1. Construct the number $b_1 = 8,531$ by arranging the digits of a in decreasing order; construct the number $c_1 = 1,358$ by arranging the digits of a in increasing order; then compute $b_1 - c_1 = 8,531 - 1,358 = 7,173$.

Stage 2. Repeat Stage 1 with the number 7,173, i.e., $b_2 = 7,731$, $c_2 = 1,377$, and $b_2 - c_2 = 6,354$.

Stage 3. Working with the number 6,354, $b_3 = 6,543$, $c_3 = 3,456$, and $b_3 - c_3 = 3,087$.

Stage 4. Using 3,087 as the seed, $b_4 = 8,730$, $c_4 = 0,378$, and $b_4 - c_4 = 8,352$.

Stage 5. The number 8,352 gives $b_5 = 8,532$, $c_5 = 2,358$, and $b_5 - c_5 = 6,174$.

Stage 6. The number 6,174 gives $b_6 = 7,641$, $c_6 = 1,467$, and $b_6 - c_6 = 6,174$.

Notice that all stages after Stage 4 result in the number **6,174**. Actually, it turns out that this iterative process always ends in 7 or fewer stages with the number 6,174 (once 6,174 is reached, then all successive iterations result in 6,174). This mysterious number was discovered in 1949 by the Indian mathematician D. R. Kaprekar, and is called **Kaprekar's constant**.

Question. Are there analogous Kaprekar constants for 2-digit or 3-digit numbers (where not all the digits are the same)? We'll settle the 2-digit case now and consider the 3-digit case in the exercises (see Exercises 4.7, no. 1c).

Observation. There are no 2-digit Kaprekar constants.

Justification (Indirect). Suppose there is a 2-digit (distinct digits) Kaprekar constant, i.e., a number a, so that when b and c are formed as above, it follows that $b - c = a$ (b is formed by arranging the digits of a in decreasing order, and c is formed by arranging the digits of a in increasing order). Note that under the given conditions, c must equal a (why?), and $b = 10a_1 + a_0$, where $a_1 > a_0$ and $0 < a_1 \leq 9$, $0 \leq a_0 \leq 9$. (Why?) Thus, by definition, $c = 10a_0 + a_1$, and since $b = 2c$, we have $10a_1 + a_0 = 2(10a_0 + a_1)$.

Collecting like terms gives the equalities

$$10a_1 - 2a_1 = a_1(10 - 2) = 8a_1 = 20a_0 - a_0 = a_0(20 - 1) = 19a_0.$$

The Fundamental Theorem of Arithmetic tells us that $19|a_1$ (why?), and this contradicts the fact that $0 < a_1 \leq 9$. Therefore, there is no 2-digit Kaprekar constant.

Section 4.7   Elementary Applications of Modular Arithmetic   177

Application 3: Coding.   One person I remember most from junior high is my friend Bob (whose nickname was Bobcat Bobby, but I'm not sure why), who loved to send secret messages to his other friends. He would invent elaborate codes to disguise his messages from unintended readers, such as the teachers of our classes, and would give each of his friends the necessary rules to decode his messages. Bob absolutely loved algebra class, since it provided him with many new tools to encode messages. Actually, without formally knowing it, Bob employed modular arithmetic in some of his coding schemes.

Bob's Algebraic Coding Method.   The first thing Bob did was to assign an integer to each letter of the alphabet. For example, one assignment might be as in the following table, where each letter in the alphabet is associated to the integer below it.

TABLE 4.7.1

A	B	C	D	E	F	G	H	I	J	K	L	M
0	1	2	3	4	5	6	7	8	9	10	11	12
N	O	P	Q	R	S	T	U	V	W	X	Y	Z
13	14	15	16	17	18	19	20	21	22	23	24	25

The next stage of his process involved shifting the numbers by a certain fixed amount and then using the resulting numbers to shift the letters (the new letter assignment is determined by using the shifted number and the original letter assignments).

Encoding Expressed Algebraically.   If t is an integer with $0 \leq t \leq 25$, then the shifted integer s, where $0 \leq s \leq 25$, is determined by the function $F: \mathbf{Z}_{26} \to \mathbf{Z}_{26}$ defined by $[s] = F([t]) = [t] + [12] = [t + 12]$ in $\mathbf{Z}_{26}$. For example,
$B \to 1 \to F([1]) = [13] \to 13 \to N$ and $R \to 17 \to F([17]) = [29] = [3] \to 3 \to D$.
The next table collects all the assignments determined by the function F.

Shift by 12: Encoding and decoding table

Original letter	A	B	C	D	E	F	G	H	I	J	K	L	M
$t =$	0	1	2	3	4	5	6	7	8	9	10	11	12
$s = t + 12$ mod 26	12	13	14	15	16	17	18	19	20	21	22	23	24
Shifted letter	M	N	O	P	Q	R	S	T	U	V	W	X	Y

Original letter	N	O	P	Q	R	S	T	U	V	W	X	Y	Z
$t =$	13	14	15	16	17	18	19	20	21	22	23	24	25
$s = t + 12$ mod 26	25	0	1	2	3	4	5	6	7	8	9	10	11
Shifted letter	Z	A	B	C	D	E	F	G	H	I	J	K	L

Encoding a Message. Before encoding a message using the shifted correspondence, it is helpful to further disguise the original message by breaking it into smaller segments of some fixed length, say four letters.

Let's encode the message **MATHEMATICS IS BEAUTIFUL**. First we subdivide it into segments of length 4 separated by spaces:

MATH EMAT ICSI SBEA UTIF UL.

Next we employ the correspondence between original letters and shifted letters to get the encoded message:

YMFT QYMF UOEU ENQM GFUR GX.

Decoding a Message. Decoding a message is not difficult if you know the original assignment of letters and numbers and the shifting constant. Of course, after decoding the message, the segments must be merged appropriately to form meaningful words.

Decoding Expressed Algebraically. If s is a shifted number, where $0 \leq s \leq 25$, then the original number t, where $0 \leq t \leq 25$ is determined by the function $G : \mathbf{Z}_{26} \to \mathbf{Z}_{26}$, defined by $[t] = G([s]) = [s] - [12] = [s - 12]$ in $\mathbf{Z}_{26}$. Recall that the function G is called the **inverse of the function** F, i.e., $F[G([s])] = [s]$ for each integer $[s] \in \mathbf{Z}_{26}$, and $G[F([t])] = [t]$ for each $[t] \in \mathbf{Z}_{26}$.

Note that by using Table 4.7.1 and the inverse function G, the decoding letter assignments can be completely determined (this is equivalent to reading the "shift by 12" coding tables from shifted letters back to original letters).

For example,

$$E \to 4 \to G([4]) = [4 - 12] = [-8] = [18] \to 18 \to S, \text{ and}$$
$$U \to 20 \to G([20]) = [20 - 12] = [8] \to 8 \to I.$$

Classroom Problems. Using Bob's method as previously described (reading the coding tables from shifted letters back to original letters), decode the following message:

MXSQ NDMU EIMK OAAX ◆

Bob's coding methods worked really well to protect the secrecy of his junior high school messages but are much too simple (easy to break) to be used to protect government and other sensitive information from unintended recipients. To handle such confidential information, very secure encryption methods, some using large prime numbers and tools of modular arithmetic, are currently being used (the field of study pertaining to encryption is called **cryptology** or **cryptography**). Although we do not describe these ideas in detail, we note that they are directly related to our studies in modular arithmetic.

Before moving to the next application, let's consider a natural generalization of Bob's coding method. In Chapter 6, we will push Bob's methods even further by using matrices.

Section 4.7   Elementary Applications of Modular Arithmetic   179

Coding with General Linear Functions. Can Bob's procedure be extended to all linear functions $F : \mathbf{Z}_{26} \to \mathbf{Z}_{26}$, where $[s] = F([t]) = [a][t] + [b] = [at + b]$ in $\mathbf{Z}_{26}$, $[a] \neq [0]$?

For instance, can it be extended to the function, $[s] = F([t]) = [4][t] + [7] = [4t + 7]$ in $\mathbf{Z}_{26}$?

Constructing the correspondence tables as before leads to the following new tables:

Shift by 12: Encoding and decoding table

Original letter	A	B	C	D	E	F	G	H	I	J	K	L	M
$t =$	0	1	2	3	4	5	6	7	8	9	10	11	12
$s = 4t + 7 \bmod 26$	7	11	15	19	23	1	5	9	13	17	21	25	3
Shifted letter	H	L	P	T	X	B	F	J	N	R	V	Z	D

Original letter	N	O	P	Q	R	S	T	U	V	W	X	Y	Z
$t =$	13	14	15	16	17	18	19	20	21	22	23	24	25
$s = 4t + 7 \bmod 26$	7	11	15	19	23	1	5	9	13	17	21	25	3
Shifted letter	H	L	P	T	X	B	F	J	N	R	V	Z	D

Notice that multiple letters correspond to a single letter. For example, both A and N correspond to H. This predicament is a serious problem, since decoding a message would not lead to a unique answer. Thus, it is not the case that Bob's method can be extended to all linear functions $F: \mathbf{Z}_{26} \to \mathbf{Z}_{26}$, but only to those that avoid this "several to one" problem.

In mathematical terms, this circumstance is described by saying that F is not a one-to-one function. Recall that a function f defined from a set X into a set Y is called **one-to-one** or **injective** if $f(x_1) \neq f(x_2)$ whenever $x_1 \neq x_2$ (x_1, x_2 are elements of X).

In our example, $F: \mathbf{Z}_{26} \to \mathbf{Z}_{26}$ is not one-to-one, since $F([0]) = [7] = F([13])$, but $[0] \neq [13]$ in $\mathbf{Z}_{26}$, hence, F does not have an inverse function (decoding rule). (Why?)

In view of the previous discussion, our goal is to characterize one-to-one linear functions $F: \mathbf{Z}_n \to \mathbf{Z}_n$, and thus determine which linear functions can be used to extend Bob's method. In comparison, it should be noted that all linear functions $f(t) = at + b, a \neq 0$, defined on the integers are one-to-one. (Justify.)

Characterizing One-to-One Linear Functions $[s] = F([t]) = [a][t] + [b] = [at+b]$ in $\mathbf{Z}_{26}$, where $[a] \neq [0]$. In the previous example, F failed to be one-to-one, and in particular,

$$F([13]) = [4] \cdot [13] + [7] = [4 \cdot 13] + [7] = [7] = [4] \cdot [0] + [7] = F([0]).$$

This equality happened because [4] was a zero-divisor in $\mathbf{Z}_{26}$. More generally, if $[a]$ is a zero-divisor in $\mathbf{Z}_n$, then there is a $[c] \neq [0]$ in $\mathbf{Z}_n$ so that $[a][c] = [0]$, consequently,

$$F([c]) = [a][c] + [b] = [b] = [a][0] + [b] = F([0]),$$ i.e., F is not one-to-one.

By applying the same reasoning, we see more generally that if F is one-to-one, then $[a]$ is a nonzero-divisor in $\mathbf{Z}_n$, i.e., $\gcd(a, n) = 1$.

Let's investigate whether the converse of this statement is valid. To this end, suppose that $[a] \neq [0]$ is a nonzero-divisor in $\mathbf{Z}_n$. We wish to show that F is one-to-one, i.e., if $F([c_1]) = F([c_2])$ for some $[c_1], [c_2]$ in $\mathbf{Z}_n$, then $[c_1] = [c_2]$. Consider the following implications:

$$F([c_1]) = F([c_2]) \Rightarrow [a][c_1] + [b] = [a][c_2] + [b]$$
$$\Rightarrow [a]([c_1] - [c_2]) = [0]$$
$$\Rightarrow [c_1] - [c_2] = [0] \text{ (since } [a] \text{ is a nonzero-divisor)}$$
$$\Rightarrow [c_1] = [c_2].$$

Therefore, F is one-to-one. Hence, a linear function $F: \mathbf{Z}_n \to \mathbf{Z}_n$ is one-to-one if and only if $\gcd(a, n) = 1$.

Inverse Function. In addition to this equivalence, it also follows that $F: \mathbf{Z}_n \to \mathbf{Z}_n$, defined by $[s] = F([t]) = [a][t] + [b] = [at + b]$ in $\mathbf{Z}_n$, where $[a] \neq [0]$, is one-to-one if and only if it has an inverse function $G: \mathbf{Z}_n \to \mathbf{Z}_n$.

Justification. First suppose that F is one-to-one. We know by the previous equivalence that $\gcd(a, n) = 1$, so $[a]^{-1}$ exists. Hence,

$$[a]^{-1}[s] = [a]^{-1}([a][t] + [b]) = [t] + [a]^{-1}[b], \text{ so, } [t] = [a]^{-1}[s] - [a]^{-1}[b].$$

The function $G: \mathbf{Z}_n \to \mathbf{Z}_n$ defined by $G([s]) = [a]^{-1}[s] - [a]^{-1}[b]$ is the inverse of F, since

$$F[G([s])] = F([a]^{-1}[s] - [a]^{-1}[b]) = [a]([a]^{-1}[s] - [a]^{-1}[b]) + [b] = [s], \text{ and}$$
$$G[F([t])] = G([a][t] + [b]) = [a]^{-1}([a][t] + [b]) - [a]^{-1}[b] = [t].$$

Conversely, assume that $F: \mathbf{Z}_n \to \mathbf{Z}_n$ has an inverse function $G: \mathbf{Z}_n \to \mathbf{Z}_n$. This implies that F is one-to-one, since

$$F([c_1]) = F([c_2]) \Rightarrow G[F([c_1])] = G[F([c_2])] \Rightarrow [c_1] = [c_2].$$

The equivalences just determined are summarized in the next theorem.

4.7.1 Theorem. Let $F: \mathbf{Z}_n \to \mathbf{Z}_n$ be defined by $F([t]) = [a][t] + [b] = [at + b]$ in $\mathbf{Z}_n$, where $[a] \neq [0]$. The following statements are equivalent:

   a. F is one-to-one.
   b. $\gcd(a, n) = 1$.
   c. F has an inverse function $G: \mathbf{Z}_n \to \mathbf{Z}_n$.

Comparison Comment: Linear Functions Defined on the Integers. Even though linear functions $f: \mathbf{Z} \to \mathbf{Z}$ defined by $f(t) = at + b$, with a, b integers and $a \neq 0$

Section 4.7   Elementary Applications of Modular Arithmetic   181

have linear inverse functions g: Range$(f) \rightarrow \mathbf{Z}$, these inverses might have rational (noninteger) coefficients. For example, $s = f(t) = 2t + 1$ has the inverse function g: Odd integers $\rightarrow \mathbf{Z}$ defined by $g(s) = \frac{s}{2} - \frac{1}{2}$.

Classroom Problems.   **1.** Using the alphabet assignments in Table 4.7.1 and the function $[s] = F([t]) = [5][t] + [4] = [5t + 4]$ in $\mathbf{Z}_{26}$, encode in 4-letter blocks a quotation of the mathematician Siméon Poisson (1781–1840): "Life is good for only two things, discovering mathematics and teaching mathematics."
2. Find the inverse function G of the function $[s] = F([t]) = [9][t] + [2] = [9t + 2]$ in $\mathbf{Z}_{26}$, and use it in combination with Table 4.7.1 to decode the following encoded quotation from the mathematician Blaise Pascal (1623–1662):

RNMG YZMW IMMY VGMP RNML MRRM ZWXW OMGK DYE.   ◆

Application 4: Identification Codes.   For the past 30 years, most books published in the United States and in several other countries have had an **International Standard Book Number (ISBN)** printed on their back covers. These numbers encode specific information that identifies each published book, and they come in a variety of different digit configurations (see Exercise 4.7, no. 5).

For example, the ISBN 0-13-044941-5 appears on the book *Mathematics for High School Teachers* by Usiskin, Peressini, Marchisotto, and Stanley and was published by Prentice Hall in 2003. The first digit indicates that the book is in English; the next two digits identify the publisher, Prentice Hall; the subsequent six digits specifically identify the book; the last digit is called the *check digit* and is defined in terms of the previous nine digits. Its purpose is to help detect data errors that might occur during data entering or data transmission.

The ISBN check digit is the last digit in the ISBN code, e.g., the check digit in the book previously mentioned is 5. The rule for determining the check digit involves $\mathbf{Z}_{11}$ arithmetic and was designed to detect single digit errors and single transposition errors.

ISBN Check Digit Rule.   Given the ISBN 0-13-044941-d, the check digit $0 \leq d \leq 10$ is defined to be the number satisfying the following equality in $\mathbf{Z}_{11}$:

$$[d] = -[(10 \cdot 0) + (9 \cdot 1) + (8 \cdot 3) + (7 \cdot 0) + (6 \cdot 4) + (5 \cdot 4)$$
$$+ (4 \cdot 9) + (3 \cdot 4) + (2 \cdot 1)].$$

Calculating in $\mathbf{Z}_{11}$ gives

$$[d] = -[0 + 9 + 24 + 0 + 24 + 20 + 36 + 12 + 2] = -[127]$$
$$= -[1 + (-2) + 7] = -[6] = [5],$$

since we are assuming that $0 \leq d \leq 10$, it follows that $d = 5$.

Convention.   If $d = 10$, the accepted practice is to write $d = X$. For example, consider the ISBN 0-13-085199-X of the book *Elementary Linear Algebra* by Kolman and Hill and published by Prentice Hall in 2000.

To see how the check digit of X was determined, let's use the ISBN rule to find d:

$$[d] = -[(10 \cdot 0) + (9 \cdot 1) + (8 \cdot 3) + (7 \cdot 0) + (6 \cdot 8)$$
$$+ (5 \cdot 5) + (4 \cdot 1) + (3 \cdot 9) + (2 \cdot 9)]$$
$$= -[155] = -[1 + (-5) + 5] = -[1] = [10].$$

Therefore, since our assumption requires d to be in the range $0 \leq d \leq 10$, we see that $d = 10 = X$.

Before moving on to some other identification codes, let's see why the ISBN check digit detects single transposition errors.

Detecting Single Transposition Errors in the ISBN Code. Suppose $C_1 = a_1 a_2 a_3 a_4 a_5 a_6 a_7 a_8 a_9$ is an ISBN code without a check digit a_{10} appended ($0 \leq a_i \leq 9$ for each i and $0 \leq a_{10} \leq 10$). We assert that if any two distinct digits of this code are transposed, then the resulting code C_2 will have a check digit d different from a_{10}. In particular, assume a_2 and a_7 have been transposed (where $a_2 \neq a_7$), and thus $C_2 = a_1 a_7 a_3 a_4 a_5 a_6 a_2 a_8 a_9$.

Justification. We claim that $d \neq a_{10}$. Approaching this indirectly, suppose that $d = a_{10}$. Calculating in $\mathbf{Z}_{11}$,

$$[d] = -[10a_1 + 9a_2 + 8a_3 + 7a_4 + 6a_5 + 5a_6 + 4a_7 + 3a_8 + 2a_9]$$
$$= -[10a_1 + 9a_7 + 8a_3 + 7a_4 + 6a_5 + 5a_6 + 4a_2 + 3a_8 + 2a_9] = [a_{10}].$$

Multiplying both sides of the equality by $-[1]$ and simplifying, results in the equality

$$[9(a_2 - a_7) + 4(a_7 - a_2)] = [9(a_2 - a_7) + (-4)(a_2 - a_7)]$$
$$= [5(a_2 - a_7)] = [0].$$

Since $\mathbf{Z}_{11}$ has no divisors of zero, we know that $[5][(a_2 - a_7)] = [0]$ implies that $[a_2 - a_7] = [0]$. Hence, $11 | (a_2 - a_7)$, so $a_2 = a_7$ as $0 \leq a_2, a_7 \leq 9$. This contradiction completes the argument.

Notice that the previous proof can easily be adapted to handle the general case involving the transposition of distinct digits a_i and a_j.

Universal Product Code (UPC). Another important identification code that is affixed to merchandise is the Universal Product Code. Most purchases at the grocery store or the mall involve scanning this code.

There are different versions of the UPC, but the most frequently used ones involve an 8-digit or 12-digit code. For example, consider the UPC 049000032628 on a carton of diet soda. The first digit 0 indicates this is a grocery item; the next five digits identify the manufacturer; the second list of five digits characterizes the product; the last digit 8 is the check digit, and it is defined using $\mathbf{Z}_{10}$ arithmetic.

UPC Check Digit Rule. Given the UPC 04900003262d, the check digit $0 \le d \le 9$ is defined to be the number satisfying the following equality in $\mathbf{Z}_{10}$:

$$[d] = -[(3 \cdot 0) + (1 \cdot 4) + (3 \cdot 9) + (1 \cdot 0) + (3 \cdot 0) + (1 \cdot 0) + (3 \cdot 0)$$
$$+ (1 \cdot 3) + (3 \cdot 2) + (1 \cdot 6) + (3 \cdot 2)] = -[52] = -[2] = [8].$$

Since we are assuming that $0 \le d \le 9$, then $d = 8$.

The UPC identification scheme does not detect as many types of errors as does the ISBN identification system. In particular, the UPC check digit does not detect all single (adjacent or nonadjacent) transposition errors. For example, the distinct numbers 04900003262 and 94000003262 (interchanging the first and third digits) have the same check digit 8.

For an example demonstrating the failure of check digit detection in a single adjacent transposition, consider the UPC 04900003262 and the number 09400003262 that results from a transposition of the second and third digits of the given UPC. We know that the check digit of the specified UPC is 8, and the following calculation in $\mathbf{Z}_{10}$ shows that the transposed number also has a check digit of 8:

$$[d] = -[(3 \cdot 0) + (1 \cdot 9) + (3 \cdot 4) + (1 \cdot 0) + (3 \cdot 0) + (1 \cdot 0) + (3 \cdot 0)$$
$$+ (1 \cdot 3) + (3 \cdot 2) + (1 \cdot 6) + (3 \cdot 2)] = -[42] = -[2] = [8].$$

In addition to the ISBN and UPC identification codes, there are codes for money orders, airline tickets, credit cards, mail, etc. Some involve check digits that are sensitive to a variety of error types, while others are not as effective in detecting errors. In Exercise 4.7, no. 7, we consider an identification code associated with a money order.

Application 5: Modular Art. The cyclical structure of modular arithmetic can be utilized to create fascinating pictures. We now consider a few methods that result in this "modular art."

Circle Art. Our first example involves a row of the multiplication table in $\mathbf{Z}_{13}$. We focus on the nonzero elements of $\mathbf{Z}_{13}$ multiplied by a fixed nonzero member of $\mathbf{Z}_{13}$, say [6].

184 Chapter 4 Arithmetic and Algebra of the Integers Modulo *n*

$\cdot_{13}$	[1]	[2]	[3]	[4]	[5]	[6]	[7]	[8]	[9]	[10]	[11]	[12]
[6]	[6]	[12]	[5]	[11]	[4]	[10]	[3]	[9]	[2]	[8]	[1]	[7]

Note that each nonzero member of $\mathbf{Z}_{13}$ appears in the second row of the table (elements of $\mathbf{Z}_{13}$ multiplied by [6]), although not in the original order. This is the case, since $[6][a] \neq [0]$ for $[a] \neq [0]$ (why?), and if $[a] \neq [b]$, then $[6][a] \neq [6][b]$. (Explain.)

To create our first work of modular art, we start by spacing the nonzero elements of $\mathbf{Z}_{13}$ on a circle, like the hours of a clock, and connect the nonzero elements $[a]$ to $[6a]$ with straight lines. The intersection of these lines produces a collection of regions in the circle's interior, and coloring alternate regions adds the final touch (see the following figure).

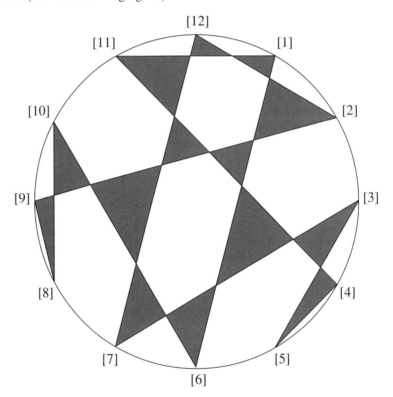

Repeating this process with the row obtained through multiplication by [5], i.e.,

$\cdot_{13}$	[1]	[2]	[3]	[4]	[5]	[6]	[7]	[8]	[9]	[10]	[11]	[12]
[5]	[5]	[10]	[2]	[7]	[12]	[4]	[9]	[1]	[6]	[11]	[3]	[8]

produces a different picture from the previous one (see the following figure).

Section 4.7   Elementary Applications of Modular Arithmetic   185

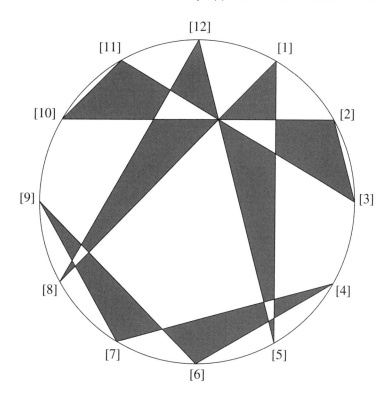

Square Art. The "circle art" examples utilized the cyclic structure of the multiplication table of $\mathbf{Z}_{13}$ to partition the interior of a circle into symmetrical regions. Our "square art" example exploits the cyclic structure of addition in $\mathbf{Z}_7$ to create interesting patterns in the interior of a square.

The first step is to construct the modulo 7 addition table.

$+_7$	[0]	[1]	[2]	[3]	[4]	[5]	[6]
[0]	[0]	[1]	[2]	[3]	[4]	[5]	[6]
[1]	[1]	[2]	[3]	[4]	[5]	[6]	[0]
[2]	[2]	[3]	[4]	[5]	[6]	[0]	[1]
[3]	[3]	[4]	[5]	[6]	[0]	[1]	[2]
[4]	[4]	[5]	[6]	[0]	[1]	[2]	[3]
[5]	[5]	[6]	[0]	[1]	[2]	[3]	[4]
[6]	[6]	[0]	[1]	[2]	[3]	[4]	[5]

Addition in $\mathbf{Z}_7$

Next, by deleting the first row and first column, we obtain the 7 × 7 part of the table consisting of the modulo 7 sums.

186 Chapter 4 Arithmetic and Algebra of the Integers Modulo *n*

[0]	[1]	[2]	[3]	[4]	[5]	[6]
[1]	[2]	[3]	[4]	[5]	[6]	[0]
[2]	[3]	[4]	[5]	[6]	[0]	[1]
[3]	[4]	[5]	[6]	[0]	[1]	[2]
[4]	[5]	[6]	[0]	[1]	[2]	[3]
[5]	[6]	[0]	[1]	[2]	[3]	[4]
[6]	[0]	[1]	[2]	[3]	[4]	[5]

At this stage, we replace each element of $\mathbf{Z}_7$ with the following shaded blocks.

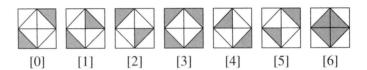

This substitution results in the following square art pattern.

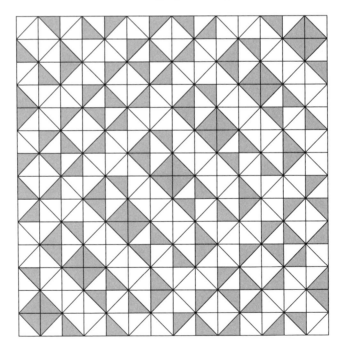

Finally, by placing this square at the origin in the first quadrant and then reflecting it about the *x*-axis and *y*-axis, we obtain a quilt pattern associated to addition in $\mathbf{Z}_7$.

EXERCISES 4.7

1. **a.** Apply Kaprekar's iterative process to the integers $a = 0001$; $a = 2{,}398$.
   b. Find a 4-digit number that requires 6 stages to reach 6,174.
   c. Find a Kaprekar constant for 3-digit numbers (where all the digits are not the same), and illustrate your answer with four distinct 3-digit numbers.
2. Decode the following message using Bob's shift-by-12 coding table:

   MXSQ NDMU EYKX URQ.

3. Find the inverse function G of the function $[s] = F([t]) = [7][t] + [8]$, and use it in combination with Table 4.7.1 to decode the following encoded quotation attributed to René Descartes (1596–1650):

   JKXR KWLV SOPK XEHM AKJK XRKW LOKV IXKZ KXUX IXK.

4. Using the assignments in Table 4.7.1 and the function $[s] = F([t]) = [15][t] + [11]$ in $\mathbf{Z}_{26}$, encode this quotation of Albert Einstein (1879–1955): "Imagination is more important than knowledge."

5. Which of the following ISBN codes are valid? For those codes that are not valid, indicate the correct check digit.
   a. 0-19-510519-3
   b. 0-345-30918-9
   c. 0-8247-9816-4
   d. 3-540-90333-X

6. Which of the following UPC identification numbers are valid? For those numbers that are not valid, indicate the correct check digit.
   a. 033317198207
   b. 071641800746
   c. 095205320466

7. MoneyGram money orders contain 10-digit identification codes with an eleventh digit as a check digit. The check digit x_{11}, where $0 \leq x_{11} \leq 10$ is defined in $\mathbf{Z}_{11}$ as: $[x_{11}] = [x_1 x_2 x_3 x_4 x_5 x_6 x_7 x_8 x_9 x_{10}]$ (digit representation), where $0 \leq x_i \leq 9$ for $0 \leq i \leq 10$.

   Verify that the check digits on the following two money orders are correct.

8. Create a circle art drawing utilizing the following multiplication rows (excluding zero):
   a. Multiplication by [8] in $\mathbf{Z}_{23}$.
   b. Multiplication by [7] in $\mathbf{Z}_{30}$.

9. Create a square art pattern by using the addition table of $\mathbf{Z}_9$.

4.8 FERMAT'S LITTLE THEOREM AND WILSON'S THEOREM

In the previous section we determined when a nonzero element in $\mathbf{Z}_n$ had a multiplicative inverse, and we showed how to use the Euclidean algorithm to find that inverse. Our initial focus in this section centers on a useful computational and theoretical result that provides us with another method for computing multiplicative inverses of the nonzero elements in $\mathbf{Z}_p$, where p is a prime number.

In $\mathbf{Z}_5$ notice, with the help of result 4.5.4, the following calculations:

$$[2]^5 = ([1] + [1])^5 = [1]^5 + [1]^5 = [2];$$
$$[3]^5 = ([2] + [1])^5 = [2]^5 + [1]^5 = [2] + [1] = [3];$$
$$[4]^5 = ([3] + [1])^5 = [3]^5 + [1]^5 = [3] + [1] = [4].$$

These calculations also suggest a general method, other than the Euclidean algorithm, for finding multiplicative inverses of nonzero elements in $\mathbf{Z}_p$. More specifically, consider the equality $[2]^5 = [2]$ in $\mathbf{Z}_5$. Since $\gcd(2,5) = 1$, $[2]^{-1}$ exists in $\mathbf{Z}_5$, thus

$$[2]^{-1}[2]^5 = [2]^{-1}[2] \text{ implies that } [2]^4 = [1].$$

Therefore, $[2][2]^3 = [1]$, and since multiplicative inverses are unique, $[2]^{-1} = [2]^3 = [8] = [3]$, i.e., $[2][3] = [1]$.

In a similar fashion, it follows that $[3]^4 = [1]$ and $[4]^4 = [1]$, which in turn gives,

$$[3]^{-1} = [3]^3 = [27] = [2] \text{ and } [4]^{-1} = [4]^3 = [64] = [4].$$

The stage has now been set to state and prove an important result generalizing the previous example.

4.8.1 Theorem. If p is a prime number, then $[a]^p = [a]$ in $\mathbf{Z}_p$ for each integer a.

Proof. Considering the calculations of the $\mathbf{Z}_5$ example, an inductive argument is ideally suited to our purpose. Note that it is good enough to prove the statement for $a \geq 0$, since for any integer a there is some integer b, with $0 \leq b \leq p - 1$, so that $[a] = [b]$.

As the statement is clear for $a = 0$ and $a = 1$, assume that it is true for $a = k > 1$. Note that

$$[k + 1]^p = [k]^p + [1]^p = [k] + [1] = [k + 1] \text{ (justify each equality)},$$

and so by induction, the conclusion holds for all integers $a \geq 0$. In view of our earlier comments, the result is established for all integers a.

In the $\mathbf{Z}_5$ example, once we knew that $[a]^5 = [a]$, it followed by cancellation that if $[a] \neq [0]$, then $[a]^4 = [1]$. The next corollary, which is called Fermat's Little Theorem, puts this in a general context.

4.8.2 Fermat's Little Theorem. If p is a prime and $[a] \neq [0]$ in $\mathbf{Z}_p$ (i.e., p does not divide a), then $[a]^{p-1} = [1]$.

Proof. Suppose that $[a] \neq [0]$ in $\mathbf{Z}_p$, and thus $[a]^{-1}$ exists in $\mathbf{Z}_p$. Multiplying both sides of the equality $[a]^p = [a]$ by $[a]^{-1}$ yields the desired equality $[a]^{p-1} = [1]$.

Other treatments of Fermat's Little Theorem first establish its validity using a different argument than the one previously given, and then deduce

Theorem 4.8.1 as a corollary by multiplying both sides of the equality $[a]^{p-1} = [1]$ by $[a]$.

Using Fermat's Little Theorem to Find Multiplicative Inverses in Z_p (p Is a Prime). We have seen that the Euclidean algorithm is an effective tool for finding multiplicative inverses in Z_n. If the situation that $n = p$ is prime, Fermat's Little Theorem provides an alternate method for determining multiplicative inverses in Z_p.

For example, given the element $[2]$ in Z_{17}, we know by Fermat's Little Theorem that $[2]^{16} = [1]$. Hence, $[2][2]^{15} = [1]$, and therefore, by the uniqueness of multiplicative inverses,

$$[2]^{-1} = [2]^{15} = [2]^{12} \cdot [2]^3 = ([2]^4)^3 \cdot [2]^3 = [16]^3 \cdot [8] = [-1]^3 \cdot [8] = -[8] = [9].$$

Classroom Problems. Using Fermat's Little Theorem, find the multiplicative inverse of each nonzero element of Z_{17}. ◆

Calculating Powers with Fermat's Little Theorem. Fermat's Little Theorem is an effective computational and theoretical tool (see result 5.9.5) in modular arithmetic. To help demonstrate its arithmetic clout, we consider a few calculations now and reserve others for the exercises.

1. Determine the remainder in the division $13 \overline{)4^{125}}$.

Analysis. We know that $[4]^{12} = [1]$ in Z_{13}, so $[4]^{120} = ([4]^{12})^{10} = [1]$. Thus,

$$[4]^{125} = [4]^{120}[4]^5 = [4]^5 = [4]^2[4]^2[4] = [3][3][4] = [3][-1] = [-3] = [10].$$

Therefore, 10 is the remainder resulting from the division $13 \overline{)4^{125}}$.

2. Show that $7|(a^{25} - a)$ for each integer a, i.e., $[a]^{25} = [a]$ in Z_7.

Analysis. If $7|a$ then we are done (why?), so we may assume that 7 does not divide a. Under this assumption, we know that $[a]^6 = [1]$ in Z_7, so $[a]^{25} = ([a]^6)^4[a] = [a]$. Therefore, $[a^{25} - a] = [0]$ in Z_7, i.e., $7|(a^{25} - a)$ for each integer a.

We complete this section by utilizing our work on multiplicative inverses to arrive at a modular arithmetic characterization of prime numbers.

Self-Inverses. Note that $[1]$ and $[n-1]$ always have multiplicative inverses in Z_n, since $\gcd(1, n) = 1 = \gcd(n-1, n)$ (why?), and in particular, they are their own inverses (**self-inverses**). To see this, consider the following calculation:

$$[n-1]^2 = [(n-1)^2] = [n^2 - 2n + 1] = [n^2] + [-2n] + [1] = [1],$$

or more simply:

$$[n-1] = [-1], \text{ so } [n-1]^2 = [-1]^2 = [1].$$

In general, these may not be the only members of Z_n that are self-inverses. For example, recall that in Z_{15}, $[1]^{-1} = [1]$, $[4]^{-1} = [4]$, $[11]^{-1} = [11]$, and $[14]^{-1} = [14]$.

Section 4.8   Fermat's Little Theorem and Wilson's Theorem

Question. In which $\mathbf{Z}_n$'s are $[1]$ and $[n-1]$ the only self-inverse elements?

Of the previous examples considered, only $\mathbf{Z}_7$ exhibits this property, however, it is not difficult to see that $\mathbf{Z}_2$, $\mathbf{Z}_3$, $\mathbf{Z}_5$, and $\mathbf{Z}_{13}$ also possess the property. Perhaps a good starting point is to determine if the question has an affirmative answer in $\mathbf{Z}_p$, where p is a prime.

Let's consider an element $[a]$ in $\mathbf{Z}_p$ that is its own inverse, i.e., $[a]^2 = [1]$. This equality gives $[a]^2 - [1] = [0]$ and hence factors into $([a] - [1])([a] + [1]) = [0]$. Since $\mathbf{Z}_p$ has no zero-divisors (why?), then $[a] - [1] = [0]$ or $[a] + [1] = [0]$. Therefore, $[a] = [1]$ or $[a] = -[1] = [-1] = [p-1]$.

Conclusions.   **1.** Each nonzero element of $\mathbf{Z}_p$, where p is a prime, has a multiplicative inverse, and $[1]$ and $[p-1]$ are the only self-inverse elements.

2. The product of the nonzero elements of $\mathbf{Z}_p$ equals $[-1]$, i.e.,

$$[1] \cdot [2] \cdot [3] \cdots [p-1] = [p-1] = [-1].$$

This conclusion is immediate, since each nonzero element of $\mathbf{Z}_p$ has a unique multiplicative inverse, and $[1]$ and $[p-1]$ are the only self-inverse elements. (Explain.)

We have just established the following theorem, which is credited to a mathematician who stated the result but did not supply a proof.

4.8.3   Wilson's Theorem. If p is a prime number, then $[(p-1)!] = [-1]$ in $\mathbf{Z}_p$.

Question. What conclusions can be drawn about $[(n-1)!]$ in $\mathbf{Z}_n$ if n is a composite number?

Let's explore this question using some of the examples considered previously.

$\mathbf{Z}_6$: $[(6-1)!] = [5!] = [5 \cdot 4 \cdot 3 \cdot 2 \cdot 1] = [0]$ (Why?)

$\mathbf{Z}_{12}$: $[(12-1)!] = [11!] = [11 \cdot 10 \cdot 9 \cdot 8 \cdot 7 \cdot 6 \cdot 5 \cdot 4 \cdot 3 \cdot 2 \cdot 1] = [0]$ (Why?)

$\mathbf{Z}_{15}$: $[(15-1)!] = [14!] = [0]$ (Why?)

In each of these examples, we used the fact that $n = ab$, where $1 < a < n$, $1 < b < n$, and $a \neq b$ to conclude that $n | (n-1)!$, and this is, in fact, a general argument for all such composites. However, this argument does not directly work for all positive composite numbers greater than 1. (Why?) For instance, 4 does not divide 3!, and in particular, $[(4-1)!] = [3!] = [2]$ in $\mathbf{Z}_4$. This observation directs our focus to composites that are squares. Perhaps a few such examples are warranted.

$\mathbf{Z}_9$: $[(9-1)!] = [8!] = [8 \cdot 7 \cdot 6 \cdot 5 \cdot 4 \cdot 3 \cdot 2 \cdot 1] = [0]$ (Why?)

$\mathbf{Z}_{16}$: $[(16-1)!] = [15!] = [0]$ (Why?)

The final equalities in the last two examples were arrived at by first using the fact that $n = a^2$, so a is a factor of $(n-1)!$. Additionally, $(n-1)!$ also has $2a$ as a factor, since $a > 2$, and so $n = a^2 > 2a$. Reiterating, if $n = a^2$ where $a > 2$, then

$$[(n-1)!] = [(n-1)(n-2) \cdots 2a \cdots a \cdots 2 \cdot 1] = [0].$$

Therefore, except for the extraneous case of $n = 4$, we have seen that if $n > 1$ is composite, then $[(n - 1)!] = [0]$ in $\mathbf{Z}_n$. It is interesting to note that this conclusion actually establishes the converse of Wilson's theorem.

4.8.4 Converse of Wilson's Theorem. If n is a positive integer greater than 1 and $[(n - 1)!] = [-1]$ in $\mathbf{Z}_n$, then n is a prime number.

Proof. If $n > 1$ is not prime, then we have just shown that $[(4 - 1)!] = [2]$ in $\mathbf{Z}_4$ and $[(n - 1)!] = [0]$ in $\mathbf{Z}_n$, where $n > 4$, so these contradictions imply that n must be prime.

Wilson's theorem completely characterizes when an integer $n > 1$ is prime, but it is not a practical test because it is difficult (even with a computer) to compute $[(p - 1)!]$ for large p.

EXERCISES 4.8

1. Using Fermat's Little Theorem, find the multiplicative inverse of each nonzero element of:

   a. $\mathbf{Z}_{13}$.   **b.** $\mathbf{Z}_{19}$.   **c.** $\mathbf{Z}_{23}$.

2. Find the remainder of the following divisions by employing Fermat's Little Theorem:

   a. $19 \overline{)6^{150}}$. Direction: $[6]^{150} = ([6]^{18})^8 \cdot [6]^6$; to compute $[6]^6$, use the fact that $[18] = [-1]$.

   b. $31 \overline{)9^{375}}$. Direction: $[9]^{375} = ([9]^{30})^{12} \cdot [9]^{15}$; to compute $[9]^{15}$, note that $[9]^{15} = [3]^{30}$.

3. Show that in $\mathbf{Z}_{10}$, $[a]^5 = [a]$ for any integer a, i.e., the ones place digit of a^5 and a are equal. Direction: Show that $5|(a^5 - a)$ and $2|(a^5 - a)$; to prove that $2|(a^5 - a)$, consider if a is even and if a is odd.

4. **a.** Explain why there is no positive integer t so that $[3]^t = [1]$ in $\mathbf{Z}_{12}$. Direction: Use result 4.6.4.

   b. Justify: If $[a]^t = [1]$ in $\mathbf{Z}_n$, then $\gcd(a, n) = 1$.

   c. Find an element $[a]$ in $\mathbf{Z}_{12}$, where $2 \leq a \leq 11$, and a positive integer t so that $[a]^t = [1]$.

   d. Prove: If $\gcd(a, n) = 1$, then there exists a positive integer t so that $[a]^t = [1]$. Direction: Since $\mathbf{Z}_n$ is finite, there exists positive integers r and s (say $r > s$) so that $[a]^r = [a]^s$.

5. Use the converse of Wilson's theorem to show that 13 is a prime. Direction: In $\mathbf{Z}_{13}$, $[12!] = [-1] \cdot [-2] \cdot [-3] \cdot [-4] \cdot [-5] \cdot [-6] \cdot [6] \cdot [5] \cdot [4] \cdot [3] \cdot [2] \cdot [1] = [6!]^2$.

6. **a.** Use Wilson's theorem to verify that $[35!] = [1]$ in $\mathbf{Z}_{37}$.

   b. Generalization of a: Show that if p is an odd prime, then $[(p - 2)!] = [1]$ in $\mathbf{Z}_p$. Direction: By Wilson's theorem, $[(p - 1)!] = [(p - 1) \cdot [(p - 2)!] = [-1] = [p - 1]$ in $\mathbf{Z}_p$.

7. Find an integer $0 \leq a < 23$ so that $[20!] = [a]$ in $\mathbf{Z}_{23}$. Direction: By Wilson's theorem, $[-1] = [22!] = [22 \cdot 21] \cdot [20!]$ in $\mathbf{Z}_{23}$.

4.9 LINEAR EQUATIONS DEFINED OVER Z_n

In Section 4.6, we concluded that a nonzero element $[a]$ of $\mathbf{Z}_n$ had a unique multiplicative inverse in $\mathbf{Z}_n$ if and only if $\gcd(a, n) = 1$. In the language of algebra, this conclusion can be restated in the following way:

4.9.1 The equation $[a]X = [1]$ defined over $\mathbf{Z}_n$ has a unique solution in $\mathbf{Z}_n$ if and only if $\gcd(a, n) = 1$.

Question. What are the solutions in $\mathbf{Z}_n$ of the linear equation $[a]X = [b]$ defined over $\mathbf{Z}_n$, where $[a] \neq [0]$ and $[b]$ is arbitrary?

Terminology. When describing an equation in one or more variables, it is important to specify where the coefficients come from (if it is not clear) and what permissible values the variables may take on. For example, the coefficients of the equation $2x = 3$ are integers, but the equation has no integer solutions and exactly one rational solution $x = 3/2$. Thus, the set of solutions may vary, depending upon the domain of x, and so to avoid confusion, it is critical to identify the set of allowable values for x. Recall that we applied this practice in our study of linear Diophantine equations in Chapter 3.

For some other examples illustrating the importance of specifying the permissible values for x, consider the equations $x^2 = 5$ and $x^2 = -1$, which are both defined over the integers, i.e., they have integer coefficients. The first equation has no solutions in the integers and the rationals but does have the solutions $x = \pm\sqrt{5}$ in the irrationals. The second equation does not have solutions in the real numbers (and hence has no solutions in the integers and the rationals) but does have the solutions $x = \pm i$ in the complex numbers.

The integer analogue of the question raised here has a simple answer.

Linear Equations with Integer Coefficients. Note that the linear equation $ax = b$, where $a \neq 0$ and b are in $\mathbf{Z}$ has solutions in $\mathbf{Z}$ if and only if $a|b$. (Verify.)

To better understand the question under consideration, let's look at some examples (as we always do) to help us formulate a general approach.

Classroom Problems. Find all solutions in $\mathbf{Z}_9$ (provided there are any) to the following equations:

    a. $[4]X = [3]$ **b.** $[6]X = [3]$ **c.** $[3]X = [4]$

Analysis. a. Since $\gcd(4, 9) = 1$, then $[4]$ has a multiplicative inverse, i.e., $[4]^{-1} = [7]$, so multiplying both sides of the equation by $[7]$ results in the unique solution $X = [21] = [3]$. (Check.)

General Conclusion. The method used in a applies to the general equation $[a]X = [b]$, where $\gcd(a, n) = 1$. Under this condition, such an equation always has the unique solution $X = [a]^{-1}[b]$.

    b. By inspection, the only solutions of the equation $[6]X = [3]$ in $\mathbf{Z}_9$ are $X = [2]$, $X = [5]$, and $X = [8]$. (Check.)

c. By replacing X with each element of $\mathbf{Z}_9$, we see that there are no solutions of the equation $[3]X = [4]$. Although this "plug and chug" method worked nicely in this example, it would not be very effective if we were looking for solutions, say, in $\mathbf{Z}_{217}$.

Let's see if we can use reasoning to reach the same conclusion that we arrived at through computation.

If there was a solution $X = [u]$ in $\mathbf{Z}_9$, i.e., $[3][u] = [4]$, then this means that $9|(3u - 4)$, and so there is an integer v so that $3u - 4 = 9v$. However, this is impossible, since $3(u - 3v)$ cannot equal 4. (Why?) Therefore, the equation $[3]X = [4]$ has no solutions in $\mathbf{Z}_9$. ◆

The previous examples illustrate that it is possible for a linear equation $[a]X = [b]$ defined over $\mathbf{Z}_n$ to have either a. no solutions; b. exactly one solution; or c. more than one solution in $\mathbf{Z}_n$. Determining general conditions guaranteeing one of these possibilities and discovering efficient methods to find explicit solutions is next on the docket.

Linear Equations Defined over $\mathbf{Z}_n$. Note that $[a]X = [b]$ has a solution $X = [u]$ in $\mathbf{Z}_n$ if and only if $n|(au - b)$, i.e., $au - b = nv$, for some integer v. (Why?) Equivalently, $[a]X = [b]$ has a solution in $\mathbf{Z}_n$ if and only if the equation $au - nv = b$ defined over the integers has an integer solution, i.e., if there exist integers u and v so that $au - nv = b$.

We have dealt with equations of this type in Chapter 3 (Diophantine equations) and established necessary and sufficient conditions for them to have a solution. Furthermore, we explicitly described the structure of all such solutions. In particular, recall that

$$ax + (-n)y = b \text{ has a solution if and only if } d|b, \text{ where } d = \gcd(a, n),$$

and all the solutions are of the form

$$x = x_0 - (-n/d)t \text{ and } y = y_0 + (a/d)t,$$

where $x = x_0$ and $y = y_0$ are particular solutions and t is an arbitrary integer (see result 3.3.4).

Existence of Solutions to Linear Equations. The statement characterizing when a Diophantine equation has a solution in the integers can be directly translated into a description of when an arbitrary linear equation defined over $\mathbf{Z}_n$ has a solution in $\mathbf{Z}_n$.

4.9.2 Theorem. *The linear equation $[a]X = [b]$ has a solution in $\mathbf{Z}_n$ if and only if $d|b$, where $d = \gcd(a, n)$.*

Applying this result to the previous examples confirms our conclusions and helps us develop methods for solving such equations.

For the equation $[4]X = [3]$, it was already noted that since $\gcd(4, 9) = 1$, then the equation has a unique solution in $\mathbf{Z}_9$.

With regards to the equation $[6]X = [3]$, since $\gcd(6, 9) = 3$, the theorem indicates that this equation has a solution in $\mathbf{Z}_9$. We will specify how to determine all of its solutions later.

Finally, the equation $[3]X = [4]$ does not have a solution in $\mathbf{Z}_9$, since $\gcd(3, 9) = 3$ does not divide 4.

Finding the Solutions of Linear Equations. We have observed that $[a]X = [b]$ has a solution in $\mathbf{Z}_n$ if and only if the equation $ax - ny = b$ defined over the integers has an integer solution, and if $x = x_0$ and $y = y_0$ are particular solutions, then all the solutions are of the form

4.9.3
$$x = x_0 - (-n/d)t \text{ and } y = y_0 + (a/d)t,$$

where t is an arbitrary integer. Note that

$$X = [x_0] \text{ is a solution of } [a]X = [b] \text{ in } \mathbf{Z}_n,$$

since $[b] = [ax_0 - ny_0] = [a][x_0] - [n][y_0] = [a][x_0]$.

Question. How do the solutions given in 4.9.3 relate to the distinct solutions of $[a]X = [b]$ in $\mathbf{Z}_n$?

Calculating in $\mathbf{Z}_n$, notice that

$$[a][x_0 - (-n/d)t] = [ax_0 + a(n/d)t] = [(b + ny_0) + a(n/d)t] =$$

$$[b] + [ny_0] + [(at)/d][n] = [b] \text{ (note that } (at)/d \text{ is an integer, since } d|a),$$

and so for each integer t, $[x_0 - (-n/d)t]$ is a solution to $[a]X = [b]$ in $\mathbf{Z}_n$. Of course, not all of these are distinct, since $\mathbf{Z}_n$ only has n-distinct members. It remains to determine the distinct solutions.

Classroom Discussion. Before we uncover the distinct solutions of a linear equation in the general case, let's again find the distinct solutions of $[6]X = [3]$ in $\mathbf{Z}_9$ by the method previously described.

Analysis. The corresponding Diophantine equation $6x - 9y = 3$ has the solutions $x = x_0 - (-9/3)t = x_0 - (-3)t$ and $y = y_0 + (6/3)t = y_0 + 2t$, where $x = x_0$ and $y = y_0$ are particular solutions and t is an arbitrary integer. For example, using the particular solutions $x_0 = 2$ and $y_0 = 1$, it follows that all the solutions are of the form $x = 2 + 3t$ and $y = 1 + 2t$, where t is an arbitrary integer. (Verify.)

We have seen that for each integer t, $[2 + 3t]$ is a solution of $[6]X = [3]$ in $\mathbf{Z}_9$, the question is, which of these solutions are distinct in $\mathbf{Z}_9$? Note that for $t = 0, 1, 2$, the solutions $x = [2 + (3 \cdot 0)] = [2]$, $x = [2 + (3 \cdot 1)] = [5]$, and $x = [2 + (3 \cdot 2)] = [8]$ are all distinct and agree with the solutions previously determined.

In fact, these are the only distinct solutions to the equation $[6]X = [3]$ in $\mathbf{Z}_9$. To help illustrate this point, let's consider the solutions generated when $t = 16$ and $t = 29$.

For $t = 16$, the solution

$$[2 + (3 \cdot 16)] = [2 + 3 \cdot ((3 \cdot 5) + 1)] = [2 + (9 \cdot 5) + (3 \cdot 1)]$$
$$= [2 + (3 \cdot 1)] = [5],$$

is the same as for $t = 1$.

196 Chapter 4 Arithmetic and Algebra of the Integers Modulo n

The strategy used in this computation was to divide 16 by $3 = \gcd(3, 9)$ and then use the quotient and remainder to reduce the computation to one of the solutions arising from $t = 0, 1, 2$.

Using the same strategy for $t = 29$, i.e., dividing 29 by $3 = \gcd(3, 9)$ and employing the quotient and remainder to simplify the computations results in the calculation

$$[2 + (3 \cdot 29)] = [2 + 3 \cdot ((3 \cdot 9) + 2)]$$
$$= [2 + (9 \cdot 9) + (3 \cdot 2)] = [2 + (3 \cdot 2)] = [8],$$

which agrees with the solution for $t = 2$. ◆

The previous computations make it clear how to proceed in the general case.

4.9.4 Theorem. The linear equation $[a]X = [b]$ defined over $\mathbf{Z}_n$, with $[a] \neq [0]$ and $\gcd(a, n) | b$, has exactly $d = \gcd(a, n)$ distinct solutions in $\mathbf{Z}_n$ that are given by $X = [x_0 + (n/d)t]$, where $X = [x_0]$ is a solution of $[a]X = [b]$ and $t = 0, 1, 2, \ldots, d - 1$.

Note that if $d = 1$, then there is a unique solution to the equation $[a]X = [b]$ in $\mathbf{Z}_n$.

Proof. Since we have already verified that $X = [x_0 + (n/d)t]$ is a solution of the equation $[a]X = [b]$ under the given hypotheses (see hypotheses of result 4.9.3), it suffices to show that for any integer t, the solution $X = [x_0 + (n/d)t]$ equals $[x_0 + (n/d)t]$ for some $t = 0, 1, 2, \ldots, d - 1$. To accomplish this goal, we proceed as in the previous examples.

Consider an arbitrary solution $X = [x_0 + (n/d)t]$ in $\mathbf{Z}_n$ of $[a]X = [b]$ for some integer t. As in the specific examples, we divide t by d to get $t = qd + r$, where $0 \leq r < d$. Replacing t in the solution X with this equivalent expression gives

$$X = [x_0 + (n/d)t] = [x_0 + (n/d)(qd + r)]$$
$$= [x_0 + nq + (n/d)(r)] = [x_0 + (n/d)(r)].$$

Since $0 \leq r < d$, we have shown that any solution $X = [x_0 + (n/d)t]$ in $\mathbf{Z}_n$ of $[a]X = [b]$ for some integer t is equal to one where t is restricted to the set of values $\{0, 1, 2, \ldots, d - 1\}$, and the proof is now complete.

Let's solve some specific equations using the ideas just developed.

4.9.5 Classroom Problems. Find all distinct solutions (or indicate why there are none) to the following equations using the methods that emerged from the previous discussions.

1. $[207]X = [198]$ in $\mathbf{Z}_{245}$.

Analysis. First, we determine if the given equation has a solution in $\mathbf{Z}_{245}$. To accomplish this, we must calculate $d = \gcd(207, 245)$ and see if it divides 198 or not (Theorem 4.9.2). We employ the Euclidean algorithm to find d.

$$245 = 1 \cdot 207 + 38$$
$$207 = 5 \cdot 38 + 17$$
$$38 = 2 \cdot 17 + 4$$
$$17 = 4 \cdot 4 + 1$$
$$4 = 1 \cdot 4 + 0$$

Hence, $d = 1$, and since $1 | 198$, we know that there is exactly one solution to the equation $[207]X = [198]$ in $\mathbf{Z}_{245}$. To explicitly find this unique solution, we use the equations that appeared in the application of the previous Euclidean algorithm.

$$\begin{aligned} 1 &= 17 - 4 \cdot 4 \\ &= 17 - 4 \cdot (38 - 2 \cdot 17) \\ &= 9 \cdot 17 - 4 \cdot 38 \\ &= 9 \cdot (207 - 5 \cdot 38) - 4 \cdot 38 \\ &= 9 \cdot 207 - 49 \cdot 38 \\ &= 9 \cdot 207 - 49 \cdot (245 - 207) \\ &= 58 \cdot 207 + (-49) \cdot 245 \end{aligned}$$

Our goal is to find $X = [x_0]$ so that $[207][x_0] = [198]$ in $\mathbf{Z}_{245}$, i.e., $207x_0 - 245t = 198$ for some integer t.

To accomplish this goal, multiply both sides of the equality $1 = 58 \cdot 207 + (-49) \cdot 245$ by 198 and obtain the equality $198 = 198 \cdot (58 \cdot 207) + 198 \cdot [(-49) \cdot 245]$. From this point we reduce (mod 245) to get $[198] = [198]([58][207]) = [207]([198][58])$, and so $X = [198][58] = [11,484] = [214]$ in $\mathbf{Z}_{245}$. A substitution check verifies that this is indeed the unique solution to $[207]X = [198]$ in $\mathbf{Z}_{245}$.

Equivalent Explanation: An equivalent way of describing the previous process, in the language of multiplicative inverses, is to note that in $\mathbf{Z}_{245}$, $[1] = [58 \cdot 207 + (-49) \cdot 245] = [58][207] + [-49][245] = [58][207]$, and thus, $[207]^{-1} = [58]$. Multiplying both sides of the original equation by $[58]$ gives $[58][207]X = [58][198]$, so $X = [11,484] = [214]$.

2. $[38]X = [45]$ in $\mathbf{Z}_{95}$.

Analysis. As in Problem 1, to guarantee the existence of at least one solution, we must first determine $d = \gcd(38, 95)$ and then check to see if it is a divisor of 95. In this situation, prime factorization can easily be used to find d. Namely, $38 = 2 \cdot 19$ and $95 = 5 \cdot 19$, so $d = 19$. Finally, since 19 does not divide 45, it follows that there are no solutions to the equation $[38]X = [45]$ in $\mathbf{Z}_{95}$.

3. $[16]X = [24]$ in $\mathbf{Z}_{28}$.

Analysis. It is easy to see that $\gcd(16, 28) = 4$, and since $4 | 24$, we know that the given equation has the 4 distinct solutions given by $X = [x_0 + (28/4)t]$, where $X = [x_0]$ is a solution of $[16]X = [24]$ and $t = 0, 1, 2, 3$.

Notice, by inspection, that $X = [5]$ is a solution to $[16]X = [24]$ in $\mathbf{Z}_{28}$, and thus the other distinct solutions are $X = [5 + 7 \cdot 1] = [12]$, $X = [5 + 7 \cdot 2] = [19]$, and $X = [5 + 7 \cdot 3] = [26]$. ◆

Finding an Initial Solution $X = [x_0]$ of $[a]X = [b]$ in $\mathbf{Z}_n$. In Problem 3, it was not difficult to find an initial solution $X = [5]$ by inspection, but in cases involving large numbers, employing the Euclidean algorithm is a much more efficient approach.

Some Alternate Methods for Solving Linear Equations Defined over $\mathbf{Z}_n$

Method 1. To illustrate this method, consider the task of finding an initial solution for the equation $[369]X = [205]$ in $\mathbf{Z}_{451}$ (once an initial solution is determined, then it is easy to find the other distinct solutions). By the way, this equation has 41 distinct solutions, since $\gcd(369, 451) = \gcd(3^2 \cdot 41, 11 \cdot 41) = 41$ and $41 | 205$.

Notice that if $X = [x_0]$ is a solution of $[369]X = [205]$ in $\mathbf{Z}_{451}$, then $X = [x_0]$ is also a solution of the equation $[9]X = [5]$ in $\mathbf{Z}_{11}$, and conversely (see Exercises 4.4, no. 8b).

To see this, suppose $X = [x_0]$ is a solution of $[369]X = [205]$ in $\mathbf{Z}_{451}$, i.e., $[369][x_0] = [205]$. Hence, $451 | (369x_0 - 205)$, which means that $369x_0 - 205 = 451t$, for some integer t. Thus, $41(9x_0 - 5) = 41 \cdot 11t$, and so by cancellation in the integers, $9x_0 - 5 = 11t$, i.e., $[9][x_0] = [5]$ in $\mathbf{Z}_{11}$.

Conversely, assume that $[9][x_0] = [5]$ in $\mathbf{Z}_{11}$. Translating this to an integer statement results in $9x_0 - 5 = 11s$ for some integer s, so $41 \cdot (9x_0 - 5) = 41 \cdot 11s$. Therefore, $[369][x_0] = [205]$ in $\mathbf{Z}_{451}$.

In summary, to find a solution to $[369]X = [205]$ in $\mathbf{Z}_{451}$, it suffices to find a solution to $[9]X = [5]$ in $\mathbf{Z}_{11}$. Since $\gcd(9, 11) = 1$, the equation $[9]X = [5]$ has a unique solution in $\mathbf{Z}_{11}$, namely $X = [3]$ (determined by inspection). Hence, $X = [3]$ is a solution to $[369]X = [205]$ in $\mathbf{Z}_{451}$. (Check by substitution.) Thus, the distinct solutions of $[369]X = [205]$ in $\mathbf{Z}_{451}$ are given by $X = [3 + 11t]$, where $t = 0, 1, 2, \ldots, 40$.

The simplification process just described can be stated in general terms and is justified by arguments analogous to those given in the specific example.

Simplifying Linear Equations Defined over $\mathbf{Z}_n$. Let $\gcd(a, n) = d$ and consider the equation $[a]X = [b]$ defined over $\mathbf{Z}_n$, where $d | b$. Then,

4.9.6 $X = [x_0]$ is a solution to $[a]X = [b]$ in $\mathbf{Z}_n$ if and only if $X = [x_0]$ is a solution to the equation $[\frac{a}{d}]X = [\frac{b}{d}]$ in $\mathbf{Z}_{\frac{n}{d}}$.

Notice that the equation $[a]X = [b]$ has d distinct solutions in $\mathbf{Z}_n$, while the equation $\left[\frac{a}{d}\right]X = \left[\frac{b}{d}\right]$ has a unique solution in $\mathbf{Z}_{\frac{n}{d}}$. (Why?) In particular, the equation $[369]X = [205]$ has 41 distinct solutions in $\mathbf{Z}_{451}$, whereas $\left[\frac{369}{41}\right]X = \left[\frac{205}{41}\right]$ has a unique solution in $\mathbf{Z}_{\frac{451}{41}}$.

Method 2. To illustrate this method, we focus on the equation $[15]X = [16]$ in $\mathbf{Z}_{143}$, which has a unique solution, as $\gcd(15, 143) = \gcd(15, 11 \cdot 13) = 1$.

Note that $X = [x_0]$ is a common solution to $[15]X = [16]$ in $\mathbf{Z}_{11}$, and $[15]X = [16]$ in $\mathbf{Z}_{13}$ if and only if $X = [x_0]$ is a solution to $[15]X = [16]$ in $\mathbf{Z}_{143}$ (see Exercise 4.4, no. 9c).

To verify the claim in the previous paragraph, suppose that $[15][x_0] = [16]$ in $\mathbf{Z}_{11}$ and $[15][x_0] = [16]$ in $\mathbf{Z}_{13}$, so $11|(15x_0 - 16)$ and $13|(15x_0 - 16)$. Therefore, $143|(15x_0 - 16)$ (why?), which means that $[15][x_0] = [16]$ in $\mathbf{Z}_{143}$.

Conversely, assume that $[15][x_0] = [16]$ in $\mathbf{Z}_{143}$. Thus, $143|(15x_0 - 16)$, so $11|(15x_0 - 16)$ and $13|(15x_0 - 16)$, i.e., $[15][x_0] = [16]$ in $\mathbf{Z}_{11}$ and $[15][x_0] = [16]$ in $\mathbf{Z}_{13}$.

In particular, a common solution to the system of equations

$$[15]X = [16] \text{ in } \mathbf{Z}_{11}$$
$$[15]X = [16] \text{ in } \mathbf{Z}_{13}$$

provides a solution to the original equation $[15]X = [16]$ in $\mathbf{Z}_{143}$. Equivalently, a common solution to the system

$$[4]X = [5] \text{ in } \mathbf{Z}_{11}$$
$$[2]X = [3] \text{ in } \mathbf{Z}_{13}$$

gives a solution to the original equation $[15]X = [16]$ in $\mathbf{Z}_{143}$.

Through inspection, we see that $X = [4]$ is the unique solution to $[4]X = [5]$ in $\mathbf{Z}_{11}$, and $X = [8]$ is the unique solution to $[2]X = [3]$ in $\mathbf{Z}_{13}$. We are looking for an element of the set $[4] = \{4 + 11t : t \text{ is an integer}\}$ that satisfies $[2]X = [3]$ in $\mathbf{Z}_{13}$. In particular, we are searching for an integer t so that $[2][4 + 11t] = [3]$ in $\mathbf{Z}_{13}$, i.e.,

$$[8] + [22][t] = [3] \text{ in } \mathbf{Z}_{13} \text{ or equivalently, } [9][t] = [-5] = [8] \text{ in } \mathbf{Z}_{13}.$$

Observe that the last equality holds if $t = 11$, and thus $[4 + 11 \cdot 11] = [125]$ satisfies both $[2]X = [3]$ in $\mathbf{Z}_{13}$ and $[4]X = [5]$ in $\mathbf{Z}_{11}$. (Check.) Therefore, $X = [125]$ is the unique solution of the equation $[15]X = [16]$ in $\mathbf{Z}_{143}$. (Check.)

EXERCISES 4.9

1. Using Theorem 4.9.4, find all distinct solutions to the following equations in the given modular systems or indicate why there are no solutions.
   a. Z_{11}: $[4]X = [9]$
   b. Z_{12}: $[8]X = [7]$
   c. Z_{15}: $[9]X = [6]$
   d. Z_{30}: $[12]X = [14]$
   e. Z_{49}: $[21]X = [35]$

4.10 EXTENDED STUDIES: THE CHINESE REMAINDER THEOREM

Method 2 of the previous section involved finding a common solution to the system of linear equations:

$$[4]X = [5] \text{ defined over } Z_{11}$$

$$[2]X = [3] \text{ defined over } Z_{13}$$

For such a system to have a common solution, it is of course necessary that each equation in the system have at least one solution, which is certainly the case in the previous example.

The study of systems involving two or more linear equations defined over modular structures dates back to ancient Chinese writings (circa first century). A typical problem was to find the smallest positive integer x that has remainders of 2, 3, and 2 when divided by 3, 4, and 5, respectively. In modular language, this problem amounts to finding the smallest positive integer x satisfying the system:

$$[x] = [2] \text{ in } Z_3$$

$$[x] = [3] \text{ in } Z_4$$

$$[x] = [2] \text{ in } Z_5$$

Analysis. We employ reasoning similar to that used in the previous Method 2 discussion. In particular, if $[x] = [2]$ in Z_3, then $x = 3t + 2$, where t is an integer. Hence we are searching for an integer t so that $[3t + 2] = [3]$ in Z_4, i.e., $[3t] = [1]$ in Z_4, and $[3t + 2] = [2]$ in Z_5, i.e., $[3t] = [0]$ in Z_5.

We have reduced the original problem of solving a system of three linear equations to the problem of solving a system of two linear equations [each of which has a unique solution, since $\gcd(3,4) = 1$ and $\gcd(3,5) = 1$]:

$$[3][t] = [1] \text{ in } Z_4$$

$$[3][t] = [0] \text{ in } Z_5$$

Again, we can employ the same reasoning as above to find a common solution to this system. Note that $[0] = \{5r: r \text{ is an integer}\}$ is a solution to the second equation, and if it is to satisfy the first equation, then $[3][5r] = [1]$ in $\mathbf{Z}_4$ for some integer r, i.e., $[3][r] = [1]$ $\mathbf{Z}_4$. By inspection, $r = 3$ produces a solution, so with $t = 5 \cdot 3$, we see that

$$[3] \cdot [15] = [3] \cdot [3] = [9] = [1] \text{ in } \mathbf{Z}_4 \text{ and } [3] \cdot [15]$$
$$= [3] \cdot [0] = [0] \text{ in } \mathbf{Z}_5.$$

Finally, $x = 3 \cdot 15 + 2 = 47$ generates a common solution to the original equations, i.e.,

$$[47] = [2] \text{ in } \mathbf{Z}_3$$
$$[47] = [3] \text{ in } \mathbf{Z}_4$$
$$[47] = [2] \text{ in } \mathbf{Z}_5.$$

It remains to show that $x = 47$ is the smallest positive integer producing a common solution. One way of handling this task is to manually show that no integer a, with $0 < a < 47$, produces a common solution. Another less painful approach involves the observation that any other solution to the original system equals $[47]$ in $\mathbf{Z}_{3 \cdot 4 \cdot 5} = \mathbf{Z}_{60}$ (thus there could not exist an integer a, with $0 < a < 47$, such that $[47] = [a]$ in $\mathbf{Z}_{60}$). In particular, suppose z is an integer so that

$$[z] = [2] \text{ in } \mathbf{Z}_3$$
$$[z] = [3] \text{ in } \mathbf{Z}_4$$
$$[z] = [2] \text{ in } \mathbf{Z}_5.$$

Hence, $[z] = [47]$ in $\mathbf{Z}_3$, $[z] = [47]$ in $\mathbf{Z}_4$, $[z] = [47]$ in $\mathbf{Z}_5$, and since $\gcd(3, 4, 5) = 1$, then $[z] = [47]$ in $\mathbf{Z}_{3 \cdot 4 \cdot 5} = \mathbf{Z}_{60}$ (see Exercises 4.4, no. 9c).

These ideas can be stated in general terms. For our purposes, we consider systems involving only two equations.

4.10.1 Chinese Remainder Theorem. If $\gcd(m, n) = 1$, then the system of equations

$$X = [b] \text{ in } \mathbf{Z}_m$$
$$X = [c] \text{ in } \mathbf{Z}_n$$

has a common solution $X = [a]$, and any other common solution equals $[a]$ in $\mathbf{Z}_{mn}$.

Proof. Our proof is modeled after the approach employed in the previous example. To start with, note that any solution to the first equation is of the form $X = [mt + b]$ in $\mathbf{Z}_m$, where t is an integer. Our task is to show that an integer t exists with $[mt + b] = [c]$ in $\mathbf{Z}_n$, or, equivalently, $[m][t] = [c - b]$ in $\mathbf{Z}_n$. This mission is easily accomplished by utilizing the hypothesis $\gcd(m, n) = 1$ to apply Theorem 4.9.4 to the equation $[m][t] = [c - b]$ in $\mathbf{Z}_n$.

To establish the uniqueness statement, suppose $X = [a]$ and $X = [d]$ are solutions to the system of equations. Hence, $[a] = [d]$ in $\mathbf{Z}_m$ and $[a] = [d]$ in $\mathbf{Z}_n$. (Why?) Therefore, since $\gcd(m, n) = 1$, it follows that $[a] = [d]$ in $\mathbf{Z}_{mn}$ (see Exercises 4.4, no. 9c).

Question. What can be said about the solutions of linear systems

$$X = [b] \text{ in } \mathbf{Z}_m$$
$$X = [c] \text{ in } \mathbf{Z}_n$$

where $\gcd(m, n) > 1$? Each equation is solvable (why?), but does this imply that the system is always solvable?

Example. Consider the system

$$X = [5] \text{ in } \mathbf{Z}_8$$
$$X = [7] \text{ in } \mathbf{Z}_{12},$$

where $\gcd(8, 12) = 4 > 1$. If $X = [a]$ were a common solution, i.e.,

$$[a] = [5] \text{ in } \mathbf{Z}_8$$
$$[a] = [7] \text{ in } \mathbf{Z}_{12},$$

then $8 | (a - 5)$ and $12 | (a - 7)$. Thus, $4 | ((a - 5) - (a - 7))$, implying that $4 | 2$, which is an absurd conclusion. Therefore, the given system of equations does not have a common solution.

The reasoning used in the last example, and the proof of the Chinese remainder theorem, establish the general case.

4.10.2 A system of linear equations

$$X = [b] \text{ in } \mathbf{Z}_m$$
$$X = [c] \text{ in } \mathbf{Z}_n$$

has a common solution $X = [a]$ if and only if $\gcd(m, n) | (c - b)$, and any other common solution equals $[a]$ in $\mathbf{Z}_{lcm(m,n)}$.

Justification. If the system of linear equations

$$X = [b] \text{ in } \mathbf{Z}_m$$
$$X = [c] \text{ in } \mathbf{Z}_n$$

has a common solution $X = [a]$, then an argument similar to the one used in the last example shows that $\gcd(m,n)|(c - b)$. (Verify.)

For the converse, suppose we are given the system

$$X = [b] \text{ in } \mathbf{Z}_m$$
$$X = [c] \text{ in } \mathbf{Z}_n$$

and that $\gcd(m,n)|(c - b)$.

Proceeding as in the proof of the Chinese remainder theorem, any solution to the first equation is of the form $X = [mt + b]$ in $\mathbf{Z}_m$, where t is an integer, so we wish to establish the existence of an integer t, with $[m][t] = [c - b]$ in $\mathbf{Z}_n$. Achieve this by noting that the equation $[m][t] = [c - b]$ has a solution in $\mathbf{Z}_n$, since $\gcd(m,n)|(c - b)$ (Theorem 4.9.4).

The uniqueness part of the proof follows from Exercises 4.4, no. 9b, and we have the details of this for the exercises.

EXERCISES 4.10

1. Find the unique solution to each of the following systems of linear equations:

   a. $X = [2]$ in $\mathbf{Z}_5$
      $X = [3]$ in $\mathbf{Z}_7$

   b. $X = [5]$ in $\mathbf{Z}_6$
      $X = [9]$ in $\mathbf{Z}_{11}$

2. Establish the uniqueness statement in result 4.10.2. Direction: Use no. 9b of Exercises 4.4.

4.11 EXTENDED STUDIES: QUADRATIC EQUATIONS DEFINED OVER $\mathbf{Z}_n$

In this section, we initiate an investigation of quadratic equations defined over $\mathbf{Z}_n$. Our approach is in the spirit of mathematical discovery. We start by analyzing a variety of basic examples, and then use what we have observed to help formulate conjectures and proofs. The results we arrive at may not be the most general possible, but nevertheless, they give us new insights into algebra over $\mathbf{Z}_n$ and afford us the opportunity to use our knowledge and creativity to reach new levels of understanding.

The study of linear equations $[a]X = [b]$ defined over $\mathbf{Z}_n$ dramatically illustrated the complexity involved in solving such equations, especially in comparison to linear equations $ax = b$ defined over $\mathbf{Z}$. As one might expect, determining methods to find the distinct solutions of general quadratic and higher-degree polynomial equations defined over $\mathbf{Z}_n$ is mathematically challenging and often involves sophisticated mathematics beyond the scope of these materials. However, in some special contexts, it is possible to employ elementary methods to arrive at the distinct solutions.

Let's start by first considering some examples that illustrate various significant differences between polynomial equations defined over $\mathbf{Z}$ with solutions in $\mathbf{Z}$ and polynomial equations defined over $\mathbf{Z}_n$ with solutions in $\mathbf{Z}_n$.

Examples. The polynomial equation $x^2 - 7x + 10 = 0$ defined over $\mathbf{Z}$ factors into the equivalent equation $(x - 2)(x - 5) = 0$, thus it follows that $x = 2$ and $x = 5$ are the *only* solutions to this equation. It is easy to observe that 2 and 5 are solutions (substitute them into the factored equation), but why are these the only (integer) solutions?

Important Point. To see that 2 and 5 are the only solutions, note that if $x = u$ is any integer solution to the equation $x^2 - 7x + 10 = 0$, then $(u - 2)(u - 5) = 0$. Since $\mathbf{Z}$ has no zero divisors, it must be that $u - 2 = 0$ or $u - 5 = 0$, i.e., $u = 2$ or $u = 5$ as desired. This point is extremely critical, and its importance can best be exemplified and appreciated in a situation where there are divisors of zero.

For such an example, consider the equation $X^2 - [7]X + [10] = [0]$ defined over $\mathbf{Z}_{20}$. Notice that this equation can be replaced with the equivalent equation $(X - [2])(X - [5]) = [0]$, so $X = [2]$ and $X = [5]$ are solutions in $\mathbf{Z}_{20}$. However, since $\mathbf{Z}_{20}$ possesses zero divisors, it does not automatically follow that these are the only solutions in $\mathbf{Z}_{20}$.

In particular, by inspection of the elements of $\mathbf{Z}_{20}$, notice that

$$([10] - [2]) \cdot ([10] - [5]) = [8] \cdot [5] = [40] = [0],$$

and

$$([17] - [2]) \cdot ([17] - [5]) = [15] \cdot [12] = [180] = [0],$$

so $X = [2]$, $X = [5]$, $X = [10]$, and $X = [17]$ are precisely the solutions of the equation $(X - [2])(X - [5]) = [0]$ in $\mathbf{Z}_{20}$. (Verify.)

Summary. We know from high school algebra that a quadratic equation defined over $\mathbf{Z}$ has at most two solutions in $\mathbf{Z}$, whereas in contexts with zero divisors present, it is possible for quadratic equations to have more than two distinct roots.

Of course, more can be said about the solutions of a quadratic equation $ax^2 + bx + c = 0$ with coefficients in $\mathbf{Z}$ (or other related number systems) if the set in which the solutions reside is expanded to the complex numbers. In particular, the **quadratic formula** provides an explicit description of the roots. Namely,

$$x = \frac{-b \pm \sqrt{b^2 - 4ac}}{2a}.$$

Unfortunately, there is no corresponding quadratic formula for *all* quadratic equations defined over an arbitrary $\mathbf{Z}_n$, but under special assumptions, it is possible to derive an analogous formula.

To further illustrate the algebraic disparity of polynomial equations defined over $\mathbf{Z}$ and $\mathbf{Z}_n$, we now focus on the simplest type of quadratic equations, namely $x^2 - c = 0$ defined over $\mathbf{Z}$, and similarly, $X^2 - [c] = [0]$ defined over $\mathbf{Z}_n$.

Squares in $\mathbf{Z}_n$. Recall than an integer c is called a **square** if there is an integer solution $x = a$ to the quadratic equation $x^2 - c = 0$, i.e., $a^2 = c$. Note that when

Section 4.11 Extended Studies: Quadratic Equations Defined over Z_n

$x = a$ is a solution, then $x = -a$ is also a solution, so the equation $x^2 - c = 0$ either has no integer solutions, exactly one integer solution (in the case $c = 0$), or exactly two integer solutions.

Analogously, an element $[c]$ in $\mathbf{Z}_n$ is called a **square in $\mathbf{Z}_n$** if the equation $X^2 - [c] = [0]$ has a solution $X = [a]$ in $\mathbf{Z}_n$. For example, $[2]$ is a square in $\mathbf{Z}_7$, since $[3]^2 = [2]$ in $\mathbf{Z}_7$, but $[2]$ is not a square in $\mathbf{Z}_5$, since the equation $X^2 - [2] = [0]$ has no solutions in $\mathbf{Z}_5$. (Check.)

As in the previous integer context, observe that if $X = [a]$ in $\mathbf{Z}_n$ is a solution to $X^2 - [c] = [0]$, then $X = -[a]$ in $\mathbf{Z}_n$ is also a solution, in particular, $-[3] = [4]$ is also a solution to $X^2 - [2] = [0]$ in $\mathbf{Z}_7$. Furthermore, $X = [3]$ and $X = [4]$ are the only distinct solutions of $X^2 - [2] = [0]$ in $\mathbf{Z}_7$. (Check.)

From the comments just made, we can see at this point that there are at least two solution scenarios for the equation $X^2 - [c] = [0]$: no solutions in $\mathbf{Z}_n$ or exactly two distinct solutions in $\mathbf{Z}_n$. However, if $[c] = [0]$ in $\mathbf{Z}_n$, then the equation $X^2 - [0] = [0]$, unlike its integer counterpart, can have exactly one distinct solution or more than one distinct solution in $\mathbf{Z}_n$. For example, the equation $X^2 = [0]$ has the unique solution $X = [0]$ in $\mathbf{Z}_5$ (check) but has the three distinct solutions $X = [0]$, $X = [3]$, and $X = [6]$ in $\mathbf{Z}_9$ (check).

In addition to the equation $X^2 = [0]$ having a unique solution or more than one distinct solution in some $\mathbf{Z}_n$, it is also possible for an equation $X^2 - [c] = [0]$, where $[c] \neq [0]$, to have these same solution outcomes. In particular, $X^2 - [3] = [0]$ has the unique solution $X = [3]$ in $\mathbf{Z}_6$ (check), whereas, $X^2 - [1] = [0]$ has the four distinct solutions $X = [1]$, $X = -[1] = [7]$, $X = [3]$, $X = -[3] = [5]$ in $\mathbf{Z}_8$. (Check.)

4.11.1 Summary of Solution Possibilities for $X^2 - [c] = [0]$. The equation $X^2 - [c] = [0]$ defined over $\mathbf{Z}_n$ can either have no solutions in $\mathbf{Z}_n$, exactly one distinct solution in $\mathbf{Z}_n$, exactly two distinct solutions in $\mathbf{Z}_n$, or more than two distinct solutions in $\mathbf{Z}_n$.

Table 4.11.1 collects some of the examples just considered and some additional examples illustrating a variety of solution possibilities in $\mathbf{Z}_n$ of the equation $X^2 - [c] = [0]$. This consolidation assists us in formulating some general statements concerning such equations. (Check the results given in each row.)

A comprehensive analysis characterizing the complete solution spectrum of quadratic equations of the form $X^2 - [c] = [0]$ or more general quadratic equations defined over $\mathbf{Z}_n$ involves mathematics outside the extent of these materials, thus we focus on some particular cases through a progression of elementary observations.

The solution picture for the equation $X^2 = [0]$ is a good place to start. Some easy deductions can be made when the modulus is prime or a square.

Observation 1. If p is a prime, then the equation $X^2 = [0]$ has the unique solution $X = [0]$ in $\mathbf{Z}_p$ (but not conversely, see row 6 of Table 4.11.1).

This observation is clear, since $\mathbf{Z}_p$ has no zero-divisors (see result 4.6.7).

TABLE 4.11.1: Solution possibilities in Z_n for the Equation $X^2 - [c] = [0]$

	Modular structure	Equation	Distinct solutions
1.	Z_6	$X^2 - [2] = [0]$	No solutions
2.	Z_7	$X^2 - [3] = [0]$	No solutions
3.	Z_2	$X^2 - [1] = [0]$	$X = [1]$ (1 distinct solution)
4.	Z_6	$X^2 - [3] = [0]$	$X = [3]$ (1 distinct solution)
5.	Z_5	$X^2 = [0]$	$X = [0]$ (1 distinct solution)
6.	Z_6	$X^2 = [0]$	$X = [0]$ (1 distinct solution)
7.	Z_{10}	$X^2 - [5] = [0]$	$X = [5] = -[5]$ (1 distinct solution)
8.	Z_4	$X^2 = [0]$	$X = [0], X = [2] = -[2]$ (2 distinct solutions)
9.	Z_8	$X^2 = [0]$	$X = [0], X = [4] = -[4]$ (2 distinct solutions)
10.	Z_7	$X^2 - [2] = [0]$	$X = [3], X = -[3] = [4]$ (2 distinct solutions)
11.	Z_{15}	$X^2 - [9] = [0]$	$X = [3], X = -[3] = [12]$ (2 distinct solutions)
12.	Z_9	$X^2 = [0]$	$X = [0], X = [3], X = -[3] = [6]$ (3 distinct solutions)
13.	Z_{18}	$X^2 = [0]$	$X = [0], X = [6], X = -[6] = [12]$ (3 distinct solutions)
14.	Z_{18}	$X^2 - [9] = [0]$	$X = [3], X = -[3] = [15]$, $X = [9]$ (3 distinct solutions)
15.	Z_8	$X^2 - [1] = [0]$	$X = [1], X = -[1] = [7]$, $X = [3], X = -[3] = [5]$ (4 distinct solutions)
16.	Z_{15}	$X^2 - [4] = [0]$	$X = [2], X = -[2] = [13]$, $X = [7], X = -[7] = [8]$ (4 distinct solutions)
17.	Z_{16}	$X^2 = [0]$	$X = [0], X = [4]$, $X = -[4] = [12], X = [8]$ (4 distinct solutions)

Section 4.11   Extended Studies: Quadratic Equations Defined over Z_n

Observation 2. If $n > 1$ is a square, then $X^2 = [0]$ has more than one distinct solution in Z_n (but not conversely, see row 9 of the table).

Justification. If $n > 1$ is a square, then $n = a^2$ for some integer $1 < a < n$. Therefore, $X = [0]$ and $X = [a]$ are two distinct solutions of $X^2 = [0]$ in Z_n.

To help motivate the next two general observations, which extend both Observations 1 and 2, we examine some specific examples. Once the basic ideas of these examples are exposed, giving the general results is straightforward.

Example.   **a.** Notice that in row 6 of Table 4.11.1, the equation $X^2 = [0]$ has the unique solution $X = [0]$ in Z_6. In this case, the two prime factors of $n = 2 \cdot 3$ each appear exactly once in the factorization. Let's consider another example having this property and see if the same unique solution conclusion follows.

For example, does the equation $X^2 = [0]$ have the unique solution $X = [0]$ in Z_{42} (here $n = 2 \cdot 3 \cdot 7 = 42$)? We could answer this question by laboriously substituting the elements of Z_{42} into the given equation, but perhaps an indirect approach is more efficient.

Specifically, if we try to show that the equation $X^2 = [0]$ has the unique solution $X = [0]$ in Z_{42}, we can suppose otherwise (i.e., that the given equation has a solution distinct from $[0]$ in Z_{42}) and show that this leads to a contradiction. Proceeding with this reasoning gives an integer a, with $1 < a < 42$ such that $[a]^2 = [0]$, i.e., $a^2 = 42t = (2 \cdot 3 \cdot 7)t$, for some integer t. Applying the Fundamental Theorem of Arithmetic implies that $(2 \cdot 3 \cdot 7)|a$, which is impossible. Therefore, $X = [0]$ is the unique solution of $X^2 = [0]$ in Z_{42}.

Observation 3: Extending Observation 1.   If the canonical prime factorization of $n = p_1 p_2 \cdots p_s$, $(p_1 < p_2 < \ldots < p_s)$, then the argument just given in Example a shows that the equation $X^2 = [0]$ has the unique solution $X = [0]$ in Z_n.

Example.   **b.** The equation $X^2 = [0]$ in row 13 of Table 4.11.1 has more than one distinct solution in Z_{18} (three distinct solutions). In contrast to Observation 3, one of the prime factors of $n = 2 \cdot 3^2$ appears more than once in the factorization. Let's examine another example with this factorization property and determine if the given equation has more than one distinct solution in this case.

For instance, does the equation $X^2 = [0]$ have more than one distinct solution in Z_{200} (here $n = 2^3 \cdot 5^2 = 200$)?

Note that if $a = 2^2 \cdot 5^2 = 100$, then $200 = 2^3 \cdot 5^2$ does not divide a, but

$$(2^3 \cdot 5^2)|(2^4 \cdot 5^4) = a^2.$$

Hence, $[a] \neq [0]$ in Z_{200} and $[a]^2 = [0]$ in Z_{200}. Thus, $X = [0]$ and $X = [100] = -[100]$ are two distinct solutions of the equation $X^2 = [0]$ in Z_{200}.

Similarly, if $b = 2^2 \cdot 5$, then $[b]^2 = [0]$, where $[b] \neq [0]$ in Z_{200}. Hence, $X = [20]$ and $X = -[20] = [180]$ are two more distinct solutions to $X^2 = [0]$ in Z_{200}.

Employing this process one more time produces two more distinct solutions, $X = [2^3 \cdot 5] = [40]$ and $X = -[40] = [160]$.

The next observation puts this example in a general context.

Observation 4: Extending Observation 2. If $n > 1$ has a prime factorization $n = p_1^{m_1} p_2^{m_2} \cdots p_s^{m_s}$, where at least one of the $m_i > 1$, then $X^2 = [0]$ has more than one distinct solution in $\mathbf{Z}_n$.

Using the ideas in the previous example, it is straightforward to develop an argument showing that $X^2 = [0]$ has more than one distinct solution in $\mathbf{Z}_n$.

Observations 3 and 4 provide conditions for the equation $X^2 = [0]$ to have the unique solution $X = [0]$ in $\mathbf{Z}_n$ or to have more than one distinct solution in $\mathbf{Z}_n$. Before conducting a similar investigation for the equation $X^2 - [c] = [0]$, where $[c] \neq [0]$ in $\mathbf{Z}_n$, it is interesting to note that with a more detailed analysis it is possible to determine all the distinct solutions in $\mathbf{Z}_n$ of the equation $X^2 = [0]$. Although we will not provide the specifics of this result in these pages, we will examine a crucial special case $n = p^m$, where p is a prime and $m > 1$, that is central to the general solution (Observation 1 resolves the case $m = 1$). It is worth noting that this special case explains in general terms the conclusions reached in rows 8, 9, 12, and 17 of Table 4.11.1.

Determining the Distinct Solutions in $\mathbf{Z}_n$ of $X^2 = [0]$, Where $n = p^m$, $m > 1$, and p Is a Prime. Let's consider two cases.

Case 1. Assume m is even, i.e., $m = 2t$ for some integer t. Under this assumption, note that

$$([p^t])^2 = [p^{2t}] = [p^m] = [0] \text{ in } \mathbf{Z}_{p^m},$$

so $[p^t]$ is a solution to the equation $X^2 = [0]$ in $\mathbf{Z}_{p^m}$. Moreover, $[p^t r]$ is also a solution for each integer r. (Why?) Note that for r in the range $0 \leq r < p^t$, these solutions are distinct (why?), and so we have shown that the equation $X^2 = [0]$ has at least p^t distinct solutions given by $[p^t r]$, where $0 \leq r < p^t$.

We now show that these are the only solutions of $X^2 = [0]$ in $\mathbf{Z}_{p^m}$. To this end, suppose that $[a]^2 = [0]$, where $[a] \in \mathbf{Z}_{p^m}$ and $0 \leq a < p^m$. Our goal is to demonstrate that $[a] = [p^t r]$ for some integer r with $0 \leq r < p^t$.

Since $[a]^2 = [0]$ in $\mathbf{Z}_{p^m}$ and $m = 2t$, we know that $p^{2t} | a^2$, i.e., $(p^t)^2 | a^2$. Therefore, $p^t | a$ (see Observation 2.6.9), so $a = p^t r$ for some integer r. More specifically, since $0 \leq a < p^m = p^{2t}$ then $0 \leq r < p^t$ (why?), thus we have established our goal.

We analyze the next case in an analogous fashion.

Case 2. Assume m is odd, i.e., $m = 2t + 1$ for some integer t. With this hypothesis, notice that

$$([p^{t+1}])^2 = [p^{2t+2}] = [p^{2t+1}][p] = [p^m][p] = [0][p] = [0] \text{ in } \mathbf{Z}_{p^m},$$

and thus $[p^{t+1}]$ is a solution to the equation $X^2 = [0]$ in $\mathbf{Z}_{p^m}$. As in Case 1, $[p^{t+1} r]$ is also a solution for each integer r, and for r in the range $0 \leq r < p^t$, these solutions are distinct (why?). Therefore, the equation $X^2 = [0]$ has at least p^t distinct solutions given by $[p^{t+1} r]$, where $0 \leq r < p^t$.

It remains to show that these are the only solutions of $X^2 = [0]$ in $\mathbf{Z}_{p^m}$. As in Case 1, assume that $[a]^2 = [0]$, where $[a] \in \mathbf{Z}_{p^m}$ and $0 \le a < p^m$. We want to prove that $[a] = [p^{t+1}r]$ for some integer r with $0 \le r < p^t$.

As $[a]^2 = [0]$ in $\mathbf{Z}_{p^m}$ and $m = 2t + 1$, then $p^{2t+1}|a^2$. Recall that the prime factors of a square must occur an even number of times, so it follows that $p^{2t+2}|a^2$. Hence, $p^{t+1}|a$, and so $a = p^{t+1}r$ for some integer r. Finally, since $0 \le a < p^m = p^{2t+1}$, then $0 \le r < p^t$.

4.11.2 Solution Summary of the Equation $X^2 = [0]$ in $\mathbf{Z}_{p^m}$, p Is a Prime.

1. If $m = 2t$, then there are p^t distinct solutions of the equation $X^2 = [0]$ in $\mathbf{Z}_{p^m}$, and they are given by $X = [p^t r]$, where $0 \le r < p^t$.
2. If $m = 2t + 1$, then there are p^t distinct solutions of the equation $X^2 = [0]$ in $\mathbf{Z}_{p^m}$, and they are given by $X = [p^{t+1} r]$, where $0 \le r < p^t$.

The Equation $X^2 - [c] = [0]$, Where $[c] \ne [0]$ in $\mathbf{Z}_n$. We are now ready to proceed with the analysis of the solution portrait of the equation $X^2 - [c] = [0]$, where $[c] \ne [0]$ in $\mathbf{Z}_n$. Unlike the equation $X^2 = [0]$, the equation $X^2 - [c] = [0]$, where $[c] \ne [0]$ can have no solutions in $\mathbf{Z}_n$ (e.g., rows 1 and 2 of Table 4.11.1). However, in likeness to the equation $X^2 = [0]$, it is possible for the equation $X^2 - [c] = [0]$, where $[c] \ne [0]$ to have exactly one solution in $\mathbf{Z}_n$ (rows 3 and 4 of Table 4.11.1) and more than one distinct solution in $\mathbf{Z}_n$ (rows 10, 11, 14, 15, and 16).

First, notice that if the equation $X^2 - [c] = [0]$ has a unique solution $X = [a]$ in $\mathbf{Z}_n$ for some $[c] \ne [0]$, then $[a] = -[a]$. Thus to analyze the unique solution case, it is necessary to know when it is possible for a nonzero element $[a]$ to have the property that $[a] = -[a]$, i.e., $[2][a] = [0]$. Thus we are interested in determining when the linear equation $[2]X = [0]$ has a nonzero solution $X = [a]$ in $\mathbf{Z}_n$. This task can be accomplished easily by applying our previous work on linear equations.

In particular, Theorem 4.9.4 tells us that if $\gcd(2, n) = d > 1$, then the equation has d-distinct solutions and the solutions are given in terms of a particular solution $X = [x_0]$. More specifically, since

$$\gcd(2, n) = 1 \text{ if and only if } n \text{ is odd}$$

$$\gcd(2, n) = 2 \text{ if and only if } n \text{ is even,}$$

then the following observations are immediate. (Justify each observation.)

Observation 5. The equation $[2]X = [0]$ has the unique solution $X = [0]$ in $\mathbf{Z}_n$ if and only if n is odd and has the two distinct solutions $X = [0]$ and $X = [\frac{n}{2}]$ in $\mathbf{Z}_n$ if and only if n is even.

Observation 6. If $X^2 - [c] = [0]$ has a unique solution in $\mathbf{Z}_n$ for some $[c] \ne [0]$, then n is even, and the unique solution is $X = [\frac{n}{2}]$, and $[c] = [\frac{n^2}{4}]$.

Interestingly, the following uniqueness statement is an immediate consequence of Observation 6.

4.11.3 Corollary. If $X^2 - [c] = [0]$ has a unique solution in $\mathbf{Z}_n$ for some $[c] \neq [0]$, then it is the only equation of that form having a unique solution in $\mathbf{Z}_n$. In particular, if $X^2 - [b] = [0]$ also has a unique solution in $\mathbf{Z}_n$ for some $[b] \neq [0]$, then $[c] = [b] = [\frac{n}{4}]$. (Justify.)

From rows 14 and 15 in Table 4.11.1, we see that if n is even, it is possible for an equation of the form $X^2 - [c] = [0]$ with $[c] \neq [0]$ to have more than one distinct solution in $\mathbf{Z}_n$. Furthermore, if n is even, it can happen that there is no integer c for which $[c] \neq [0]$ and the equation $X^2 - [c] = [0]$ has exactly one solution in $\mathbf{Z}_n$, e.g., $\mathbf{Z}_4$ has this property. (Verify.)

4.11.4 Question. For which even numbers $n > 1$ does there exist an integer c so that $[c] \neq [0]$ and the equation $X^2 - [c] = [0]$ has exactly one solution in $\mathbf{Z}_n$?

First, to answer Question 4.11.4, we show that certain even numbers can be eliminated. To reach this conclusion, it is instructive to examine a few examples.

From Observation 6 we know that if $X^2 - [c] = [0]$ has a unique solution in $\mathbf{Z}_n$ for some $[c] \neq [0]$, then $X = [\frac{n}{2}]$ is that unique solution. Thus, the element $[\frac{8}{2}] = [4] = -[4]$ in $\mathbf{Z}_8$ is the only possible element in $\mathbf{Z}_8$ that can be the unique solution of an equation $X^2 - [c] = [0]$ with $[c] \neq [0]$. Note that $[4]^2 = [0]$ in $\mathbf{Z}_8$, so there does not exist an integer c, and $[c] \neq [0]$, and the equation $X^2 - [c] = [0]$ has a unique solution in $\mathbf{Z}_8$. (Explain.)

We can draw the same conclusion about $\mathbf{Z}_{12}$. Specifically, the element $[\frac{12}{2}] = [6] = -[6]$ in $\mathbf{Z}_{12}$ has the property that $[6]^2 = [0]$, and so there does not exist an integer c such that $[c] \neq [0]$, and the equation $X^2 - [c] = [0]$ has a unique solution in $\mathbf{Z}_{12}$.

These examples are the template for the next observation.

Observation 7. If $4|n$ $(n > 1)$, then there does not exist an integer c so that $[c] \neq [0]$, and the equation $X^2 - [c] = [0]$ has a unique solution in $\mathbf{Z}_n$.

Justification. We proceed as in the motivating examples. If $4|n$, then $n = 4t$ for some integer t, and so $\frac{n}{2} = 2t$. Hence, $[\frac{n}{2}]^2 = [2t]^2 = [4t][t] = [0]$ in $\mathbf{Z}_{4t}$, therefore, there does not exist an integer c where $[c] \neq [0]$, and the equation $X^2 - [c] = [0]$ has a unique solution in $\mathbf{Z}_{4t}$.

If $4|n$ $(n > 1)$, it follows from Observation 4 that $X^2 = [0]$ has more than one distinct solution in $\mathbf{Z}_n$, thus, by Observation 7, there are no integers c having the property that the equation $X^2 - [c] = [0]$ has a unique solution in $\mathbf{Z}_n$.

Observation 7 does not apply when $n = 6$ or $n = 10$ (even numbers not divisible by 4), in particular, $[\frac{6}{2}]^2 = [3]^2 = [3]$ in $\mathbf{Z}_6$, and $[\frac{10}{2}]^2 = [5]^2 = [5]$ in $\mathbf{Z}_{10}$. Furthermore, it follows that the equation $X^2 - [3] = [0]$ has the unique solution $X = [3]$ in $\mathbf{Z}_6$ (check), and the equation $X^2 - [5] = [0]$ has the unique solution $X = [5]$ in $\mathbf{Z}_{10}$ (check).

In both $\mathbf{Z}_6$ and $\mathbf{Z}_{10}$, where n is even but not divisible by 4, the equation $X^2 - [\frac{n}{2}] = [0]$ has the unique solution $X = [\frac{n}{2}]$. However, these conditions on n are not strong enough in general to guarantee the existence of an integer c, so that

$[c] \neq [0]$ and the equation $X^2 - [c] = [0]$ has a unique solution in $\mathbf{Z}_n$ (which by Observation 6 is $X = [\frac{n}{2}]$).

For example, 18 is even and not divisible by 4, and the equation $X^2 - [9] = [0]$ has the three distinct solutions $X = [3]$, $X = -[3] = [15]$, and $X = [9]$. Hence, since $[9]^2 = [9]$, there does not exist an integer c such that $[c] \neq [0]$, and the equation $X^2 - [c] = [0]$ has a unique solution in $\mathbf{Z}_{18}$. (Explain.)

The following observation captures an important point from the previous three examples.

Observation 8. If n is even and not divisible by 4, then $X^2 - [\frac{n}{2}] = [0]$ has the (not necessarily unique) solution $X = [\frac{n}{2}]$ in $\mathbf{Z}_n$.

Justification. Assume $2|n$ and 4 does not divide n ($n > 1$), and thus $\frac{n}{2} = 2t + 1$ for some positive integer t. Calculating in $\mathbf{Z}_{(4t+2)}$ we have,

$$[\frac{n}{2}]^2 = [2t+1]^2 = [4t^2 + 4t + 1] = [t(4t+2) + (2t+1)]$$

$$= [t][4t+2] + [2t+1] = [2t+1] = [\frac{n}{2}].$$

The next two observations combine to completely answer Question 4.11.4. We first isolate a set of even integers n that are not divisible by 4 and that give rise to integers c where $[c] \neq [0]$, and the equation $X^2 - [c] = [0]$ has exactly one solution in $\mathbf{Z}_n$. We then show in Observation 10 that these are the only possible even integers that produce such equations having unique solutions.

Observation 9. If the canonical prime factorization of $n = 2p_2 \ldots p_s$, ($2 < p_2 < \ldots < p_s$), then the equation $X^2 - [p_2 \cdots p_s] = [0]$ has the unique nonzero solution $X = [p_2 \cdots p_s]$ in $\mathbf{Z}_n$ (and by Corollary 4.11.3, it is the only equation of the form $X^2 - [c] = [0]$ with $[c] \neq [0]$ that has a unique solution in $\mathbf{Z}_n$).

Justification. Assume that the canonical prime factorization of $n = 2p_2 \cdots p_s$, $(2 < p_2 < \ldots < p_s)$, and note that $[p_2 \cdots p_s] \neq [0]$. (Why?) We know from Observation 8 that $X = [p_2 \cdots p_s]$ is a solution of $X^2 - [p_2 \cdots p_s] = [0]$ in $\mathbf{Z}_n$. To prove that this is the only solution, suppose that $X = [a]$ is a solution of $X^2 - [p_2 \cdots p_s] = [0]$, where $0 < a < n$. We show that this assumption implies that $[a] = [p_2 \cdots p_s]$.

Since $n|(a^2 - (p_2 \cdots p_s))$ (why?), then $a^2 - (p_2 \cdots p_s) = (2p_2 \cdots p_s)t$ for some integer t. Hence, $a^2 = (p_2 \cdots p_s)(2t+1)$, so a^2 is odd, which means a is odd too. Thus, by the Fundamental Theorem of Arithmetic and the previous deduction, $a = (p_2 \cdots p_s)r$, for some **odd** integer r. (Why?) Therefore,

$$\left[a - \frac{n}{2}\right] = [(p_2 \cdots p_s)r - (p_2 \cdots p_s)] = [(p_2 \cdots p_s)(r-1)] = [0] \text{ in } \mathbf{Z}_{2p_2 \cdots p_s},$$

as $(2p_2 \cdots p_s)|(p_2 \cdots p_s)(r-1)$, since $r - 1$ is even. Therefore, $[a] = [\frac{n}{2}] = [p_2 \cdots p_s]$, and this establishes that $X^2 - [p_2 \cdots p_s] = [0]$ has the unique solution $X = [p_2 \cdots p_s]$ in $\mathbf{Z}_n$.

We next show that the n's appearing in the previous observation are the only such integers for which there is an equation of the form $X^2 - [c] = [0]$ with $[c] \neq [0]$ having a unique solution in $\mathbf{Z}_n$. First, an instructive example.

Example. If $n = 2 \cdot 3^4 \cdot 5 \cdot 11^3$, then the equation $X^2 - [\frac{n}{2}] = X^2 - [3^4 \cdot 5 \cdot 11^3] = [0]$ has more than one distinct solution. Moreover, there are no integers c, where $[c] \neq [0]$, and the equation $X^2 - [c] = [0]$ has a unique solution in $\mathbf{Z}_n$.

Analysis. To see the first assertion, note that Observation 8 shows that $X = [\frac{n}{2}] = [3^4 \cdot 5 \cdot 11^3]$ is a solution of the equation $X^2 - [\frac{n}{2}] = X^2 - [3^4 \cdot 5 \cdot 11^3] = [0]$ in $\mathbf{Z}_n$. We claim that $X = [3^3 \cdot 5 \cdot 11^2]$ is also a solution and that it is distinct from the first solution, i.e., $[3^3 \cdot 5 \cdot 11^2] \neq [3^4 \cdot 5 \cdot 11^3]$ in $\mathbf{Z}_n$.

To see that $X = [3^3 \cdot 5 \cdot 11^2]$ is a solution, we demonstrate that $[3^3 \cdot 5 \cdot 11^2]^2 = [3^4 \cdot 5 \cdot 11^3]$ in $\mathbf{Z}_n$. Note that,

$$(3^6 \cdot 5^2 \cdot 11^4) - (3^4 \cdot 5 \cdot 11^3) = (3^4 \cdot 5 \cdot 11^3)((3^2 \cdot 5 \cdot 11) - 1), \text{ and so}$$

$$(2 \cdot 3^4 \cdot 5 \cdot 11^3) | ((3^6 \cdot 5^2 \cdot 11^4) - (3^4 \cdot 5 \cdot 11^3)),$$

since $(3^2 \cdot 5 \cdot 11) - 1$ is even. (Justify entire calculation.) Thus the claim is verified.

Hence, $X = [3^3 \cdot 5 \cdot 11^2]$ is a solution to $X^2 - [3^4 \cdot 5 \cdot 11^3] = [0]$ in $\mathbf{Z}_n$ that is distinct from $X = [3^4 \cdot 5 \cdot 11^3]$, since $0 < (3^3 \cdot 5 \cdot 11^2) < (3^4 \cdot 5 \cdot 11^3) < 2 \cdot 3^4 \cdot 5 \cdot 11^3 = n$.

Let's analyze the second statement indirectly, i.e., suppose there is an integer c, where $[c] \neq [0]$ and the equation $X^2 - [c] = [0]$ has a unique solution in $\mathbf{Z}_n$. Hence, by Observation 6, $X = [\frac{n}{2}]$ is the unique solution. However, we know by Observation 8 that $[\frac{n}{2}]^2 = [\frac{n}{2}]$, so $[\frac{n}{2}]^2 - [c] = [\frac{n}{2}] - [c] = [0]$. Therefore, $[c] = [\frac{n}{2}]$, which means that the equation $X^2 - [\frac{n}{2}] = [0]$ has a unique solution. This contradicts the assertion established in the first part of the proof, and thus there are no integers c, where $[c] \neq [0]$ and the equation $X^2 - [c] = [0]$ has a unique solution in $\mathbf{Z}_n$.

The ideas and reasoning just considered serve to motivate the following observation. Although we will not include a formal proof, we could construct one that is modeled after the last example.

Observation 10. If $n > 1$ has the canonical prime factorization $n = 2p_2{}^{m_2} \cdots p_s{}^{m_s}$, where at least one of the $m_i > 1$, then there are no equations of the form $X^2 - [c] = [0]$ with $[c] \neq [0]$ and having a unique solution in $\mathbf{Z}_n$.

The consolidation of Observations 9 and 10 provide a characterization guaranteeing the existence of equations of the form $X^2 - [c] = [0]$, where $[c] \neq [0]$ and that have a unique solution in $\mathbf{Z}_n$.

Summary. There exists an equation of the form $X^2 - [c] = [0]$, where $[c] \neq [0]$ and having a unique solution in $\mathbf{Z}_n$ if and only if the canonical prime factorization of $n = 2p_2 \cdots p_s$. Moreover, for such n, $X^2 - [\frac{n}{2}] = [0]$ has the unique solution $X = [\frac{n}{2}]$ in $\mathbf{Z}_n$, and $X^2 - [\frac{n}{2}] = [0]$ is the only equation of the form $X^2 - [c] = [0]$, where $[c] \neq [0]$ and that has a unique solution in $\mathbf{Z}_n$. (Explain how these statements follow from Observations 9 and 10.)

Most of the material in this section focused on some solution possibilities in $\mathbf{Z}_n$ of equations having the form $X^2 - [c] = [0]$. The results obtained illustrated major differences between algebra over $\mathbf{Z}$ and algebra over $\mathbf{Z}_n$. However, if the modulus is an odd prime, then the differences we've encountered disappear. This point is expressed as the final observation of this section.

Observation 11. The equation $X^2 - [c] = [0]$ with $[c] \neq [0]$, where $p > 2$ is a prime, either has no solutions in $\mathbf{Z}_p$ or has *exactly two distinct solutions* in $\mathbf{Z}_p$. (Recall that the equation $X^2 - [1] = [0]$ defined over $\mathbf{Z}_2$ has exactly one solution in $\mathbf{Z}_2$.)

Justification. Suppose that $X^2 - [c] = [0]$ with $[c] \neq [0]$ has a solution $X = [a]$ in $\mathbf{Z}_p$, where $p > 2$ is a prime. We know that $-[a]$ is also a solution and that $[a] \neq -[a]$ (Observation 5). Since $[a]^2 = [c]$, the following factorization holds:

$$(X^2 - [c]) = (X^2 - [a]^2) = (X - [a])(X + [a]) = [0];$$

as $\mathbf{Z}_p$ has no divisors of zero, it follows that $X - [a] = [0]$ or $X + [a] = [0]$. Therefore, $X = [a]$ and $X = -[a]$ are the only distinct solutions of $X^2 - [c] = [0]$ with $[c] \neq [0]$.

Our work in this section uncovered a variety of results and highlighted some of the inherent complexities involved in algebra defined over $\mathbf{Z}_n$. There are many interesting and important questions concerning such equations and more general quadratic and polynomial equations that we have not considered in these materials; the mathematical literature abounds with their analyses.

EXERCISES 4.11

1. Find all the distinct solutions in the given $\mathbf{Z}_n$ (or indicate that there are none) to the following quadratic equations:
    a. $X^2 - [9]X + [8] = [0]$ in $\mathbf{Z}_{12}$ (Direction: $[8] = [20]$ in $\mathbf{Z}_{12}$)
    b. $X^2 - [2]X - [8] = [0]$ in $\mathbf{Z}_{11}$
    c. $X^2 + [3]X = [0]$ in $\mathbf{Z}_9$
    d. $X^2 + X + [1] = [0]$ in $\mathbf{Z}_5$
    e. $X^2 - X = [0]$ in $\mathbf{Z}_{10}$
    f. $X^2 = [0]$ in $\mathbf{Z}_{25}$
    g. $X^2 = [0]$ in $\mathbf{Z}_{27}$
    h. $X^2 = [0]$ in $\mathbf{Z}_{1331}$
    i. $X^2 = [0]$ in $\mathbf{Z}_{253}$
    j. $X^2 - [1] = [0]$ in $\mathbf{Z}_{16}$
    k. $X^2 - [119] = [0]$ in $\mathbf{Z}_{238}$
    l. $X^2 = -[1]$ in $\mathbf{Z}_{17}$
2. a. Show that the equation $X^2 - [1] = [0]$ has the unique solution $X = [1]$ in $\mathbf{Z}_n$ if and only if $n = 2$.
    b. Justify: If $p > 2$ is a prime number, then the equation $X^2 - [1] = [0]$ has exactly two distinct solutions $X = [1]$ and $X = -[1] = [p - 1]$ in $\mathbf{Z}_p$.
    c. Give an example showing that the converse of part b is not true in general.

d. Give an example showing that the equation $X^2 - [1] = [0]$ can have more than two distinct solutions in some $\mathbf{Z}_n$.

e. Prove: If $[a]$ is a solution of $X^2 - [1] = [0]$ in $\mathbf{Z}_n$, then $\gcd(a,n) = 1$. Direction: Assume $\gcd(a,n) > 1$, and derive a contradiction by considering the linear equation $[a]X = [1]$ in $\mathbf{Z}_n$.

3. Give an example of a quadratic equation of the form $X^2 - [c] = [0]$ having an odd number $m > 3$ distinct solutions in $\mathbf{Z}_n$, for some integer $n > 1$.

Algebraic Modeling in Geometry: The Pythagorean Theorem and More

CHAPTER 5

5.1 THE SIGNIFICANCE OF DARYL'S MEASUREMENTS AND RELATED GEOMETRY
5.2 CLASSROOM CONNECTIONS: THE PYTHAGOREAN THEOREM
5.3 REFLECTIONS ON CLASSROOM CONNECTIONS: THE PYTHAGOREAN THEOREM AND ITS CONVERSE
5.4 COMPUTING DISTANCE IN TWO-DIMENSIONAL AND THREE-DIMENSIONAL EUCLIDEAN SPACE: THE DISTANCE FORMULA
5.5 AN EXTENSION OF THE PYTHAGOREAN THEOREM: THE LAW OF COSINES
5.6 INTEGER DISTANCES IN THE PLANE
5.7 PYTHAGOREAN TRIPLES: POSITIVE INTEGER SOLUTIONS TO $x^2 + y^2 = z^2$
5.8 EXTENDED STUDIES: FURTHER INVESTIGATIONS INTO INTEGER DISTANCE POINT SETS—A THEOREM OF ERDÖS
5.9 EXTENDED STUDIES: ADDITIONAL QUESTIONS CONCERNING PYTHAGOREAN TRIPLES
5.10 FERMAT'S LAST THEOREM

In this chapter, we explore a variety of interesting geometric and algebraic questions. In the process, we uncover many important and fascinating aspects of the Pythagorean Theorem, its converse, and related results. The algebraic foundations that we laid in the previous chapters are integral to this work, and we employ them throughout this chapter. First, a story about my favorite carpenter.

Daryl the Carpenter. There is a general homeowner's rule concerning the hiring (and rehiring) of skilled individuals to work on your home: Once you find an honest, skilled, and dependable worker (electrician, plumber, carpenter, painter, exorcist, etc.), treat them right, hire them often, and never complain about their work.

Daryl is my carpenter and what a great carpenter he is. I have recommended his services to many of my friends, but this has turned out to be a really big mistake. My friends have now discovered how skilled he is at his job, and consequently it is nearly impossible to find an open time to hire him.

216 Chapter 5 Algebraic Modeling in Geometry: The Pythagorean Theorem

Anyway, what does this story have to do with mathematics? Daryl was installing a skylight in our kitchen, which involved cutting rectangular holes in my roof and in my kitchen ceiling. He also had to build a chute through the attic connecting the roof to the ceiling. As I watched him prepare for the cutting, I noticed he marked out the area to be cut with a straightedge and a square (right-angle ruler). He then commenced to perform several additional measurements. I asked him what purpose the additional measurements served. I assumed he was using some basic geometry to check whether his figure was actually a rectangle, but I was curious what his response would be. He looked down at me from his ladder, and after a bit of chuckling, he said, "Cutting a hole in a man's roof is serious business." Since Daryl is a man of few words, I decided not to press him any further on the geometric rationale behind his measurements.

FIGURE 5.1 Kitchen skylight.

5.1 THE SIGNIFICANCE OF DARYL'S MEASUREMENTS AND RELATED GEOMETRY

1. He first measured the diagonals of the penciled-out quadrilateral (which looked a lot like a parallelogram). Why would this information be useful to Daryl?
2. He also measured the lengths of the sides of a triangle formed by a diagonal and two adjacent sides of the parallelogram, and the scratched out some

computations on a crumpled coffee-stained pad of paper. Again, how would these actions be helpful to Daryl in double-checking his work before cutting?

Carpenters' Geometry: Some Geometric Rationale behind Daryl's Measurements of the Diagonals. Using ideas from Euclidean geometry, *establish the validity of the following statements* and indicate how some of these might be useful in carpentry. In particular, notice that Problem 5 of Exercises 5.1 provides a reasonable answer to Question 1 above. In Section 5.3, we consider some other mathematics that help us understand Daryl's further measurements and computations in Question 2 above.

EXERCISES 5.1

1. If the diagonals of a quadrilateral bisect each other, then the quadrilateral is a parallelogram. Is the converse of this statement valid? If so, prove it; if not, provide a counterexample.
2. If the opposite sides of a quadrilateral are equal in length, then the quadrilateral is a parallelogram.
3. If at least one angle of a parallelogram is a right angle, then the parallelogram is a rectangle.
4. If a quadrilateral is a parallelogram, then its opposite angles have the same measure, and its adjacent angles are supplementary.
5. If the diagonals of a parallelogram have equal length, then the parallelogram is a rectangle (see Question 1 in Section 5.1). Give an example to demonstrate that the parallelogram hypothesis cannot be replaced with a quadrilateral hypothesis.

5.2 CLASSROOM CONNECTIONS: THE PYTHAGOREAN THEOREM

In this section, we work through part of an eighth grade unit from Book 3 of *MathThematics*, focusing on the Pythagorean Theorem. See Figures 5.2.1 to 5.2.4 on pages 218–221. Our goal is to explore the mathematical content of these pages and to show how it relates to other (sometimes more general) mathematical concepts and results (such as some of the previous exercises). By the way, next time I see Daryl, I will be sure to show him Exercise 7 in Figure 5.2.4!

5.3 REFLECTIONS ON CLASSROOM CONNECTIONS: THE PYTHAGOREAN THEOREM AND ITS CONVERSE

What are some important geometric foundations needed for a middle-grade mathematics teacher to understand and effectively teach this lesson? The Pythagorean Theorem's statement and its applications to calculating certain distances is certainly a minimal requirement, but more background is needed to fully comprehend and utilize this result and related ones in varied situations. For example, Exercises 1–7 in Figure 5.2.4 are not consequences of the Pythagorean Theorem but follow instead from its converse.

A deeper understanding and appreciation of the mathematics supporting the Pythagorean Theorem and its converse can be reached by studying the arguments establishing their validity (proof!). There are known proofs that justify the

218 Chapter 5 Algebraic Modeling in Geometry: The Pythagorean Theorem

Pythagorean theorem's statement and that significantly contribute to a more concrete understanding of the result. Furthermore, the techniques and tools used in one proof of a theorem often can be adapted or modified for use in other arguments (see Exercises 5.3, no. 9). We now consider a few such proofs.

Internet Resource. See **www.cut-the-knot.com/pythagoras** for thirty-eight different proofs of the Pythagorean theorem.

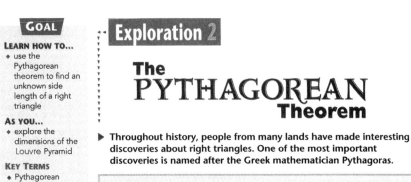

Reproduced from page 294 of Book 3 in *MathThematics*.

FIGURE 5.2.1

5.3.1 Theorem: Pythagorean Theorem. If $\triangle ABC$ is a right triangle, then the square of the length of the hypotenuse is equal to the sum of the squares of the lengths of the legs.

Proof 1. Consider the right $\triangle ABC$, where $\angle ACB$ is a right angle and sides $\overline{AB}, \overline{AC},$ and $\overline{BC}$ have lengths $AB = c, AC = b,$ and $BC = a$.

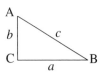

Section 5.3   Reflections on Classroom Connections   219

▶ The Louvre pyramid has a square base and four triangular faces. The *slant height* of the pyramid is the height of one of the triangular faces.

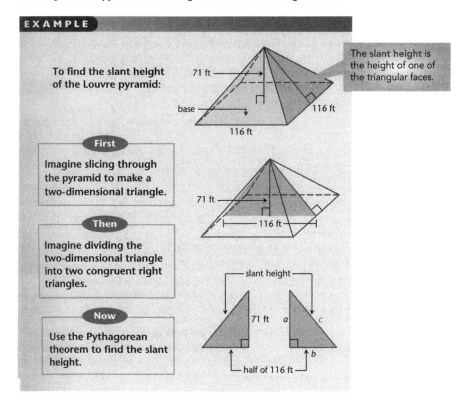

10 **Try This as a Class** Use the Example above.

   a. Use your equation from Question 9(a) to find the length of the hypotenuse of one of the congruent right triangles. $\sqrt{8405}$

   b. What is the slant height of the Louvre pyramid? Give your answer to the nearest foot. about 92 ft

   c. Now that you know the slant height, explain how you can find the area of one triangular face of the pyramid.
   Find $\frac{1}{2}bh$ where $b = 116$ and $h = 92$.

▶ You can use the Pythagorean theorem to find the length of one side of a right triangle if you know the lengths of the other two sides. This is shown in the Example on the next page.

Section 5   Working with Triangles   295

Reproduced from page 295 of Book 3 in *MathThematics*.

FIGURE 5.2.2

220 Chapter 5 Algebraic Modeling in Geometry: The Pythagorean Theorem

CLOSURE QUESTION

How can you use the lengths of the sides of a triangle to find out if the triangle is acute, right, or obtuse? When can you use the Pythagorean theorem and what does it tell you?

Sample Response: Look at the sum of the squares of the lengths of the shorter sides of the triangle. If this sum is less than the square of the length of the longer side, the triangle is obtuse. If the sum and the square of the length of the longer side are equal, the triangle is a right triangle. If the sum is greater than the square of the length of the longer side, the triangle is acute. If a triangle is a right triangle, you can use the Pythagorean Theorem to find the length of one of the sides if you know the lengths of the other two sides.

✓ QUESTION 14

...checks that you can use the Pythagorean theorem to find an unknown side length.

EXAMPLE

Use the Pythagorean theorem to find the unknown side length of the triangle below.

SAMPLE RESPONSE

The hypotenuse is given, so you need to find the length of one leg.

Let a = the unknown side length.

$$a^2 + 9^2 = 17^2$$
$$a^2 + 81 = 289$$
$$a^2 + 81 - 81 = 289 - 81$$
$$a^2 = 208$$
$$\sqrt{a^2} = \sqrt{208}$$
$$a \approx 14.42$$

The length of the unknown side is about 14.4 cm.

Use the Example to answer Questions 11–13.

11 How can you tell from looking at the triangle that the unknown side length is a leg of the triangle, and not the hypotenuse of the triangle? *Sample Response: It is not opposite the right angle.*

12 In the equations in the Example, why was 81 subtracted from both sides before the square roots of both sides were found? *The expression $\sqrt{a^2 + 81}$ cannot be simplified.*

13 Why is the length of the third side not an exact measurement? *The sum, 208, is not a perfect square.*

14 ✓ **CHECKPOINT** For each triangle, find the unknown side length.

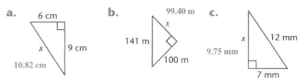

a. 6 cm, x, 9 cm, 10.82 cm
b. 99.40 m, x, 141 m, 100 m
c. x, 12 mm, 9.75 mm, 7 mm

15 Suppose the length of the longest side of a right triangle is $\sqrt{90}$. The other two side lengths are equal. Give the lengths of the shorter sides. Round your answer to the nearest tenth.
$\frac{\sqrt{90}}{2} < x < 45$ or $4.74 < x < 6.71$

HOMEWORK EXERCISES ▶ See Exs. 8–18 on pp. 298–299.

 Module 4 Patterns and Discoveries

Reproduced from page 296 of Book 3 in *MathThematics*.

FIGURE 5.2.3

Section 5.3   Reflections on Classroom Connections   221

Section 5 Practice & Application Exercises

YOU WILL NEED
For Ex. 14:
♦ graph paper

Tell whether a triangle with the given side lengths is *acute*, *right*, or *obtuse*.

1. 5 cm, 12 cm, 13 cm   right
2. 5 mm, 9 mm, 7 mm   obtuse
3. 8 in., 10 in., 9 in.   acute
4. 11.5 m, 6.2 m, 7 m   obtuse
5. 16 cm, 20 cm, 17 cm   acute
6. 15 mm, 12 mm, 9 mm   right

7. **Carpentry** Carpenters can use a method like the one used by the rope stretchers of ancient Egypt to check whether a corner is "square." For example, a carpenter took the measurements shown to check a right angle on a table. Is the angle opposite the 21 in. diagonal a right angle? Explain your thinking. No; The square root of the sum of the squares of the legs is 20, not 21.

12 in. 21 in. 17 in.

For each triangle, find the unknown side length. Give each answer to the nearest tenth.

8. 7 in., x, 10 in.   12.2 in.
9. 3 mm, 15 mm, x 14.7 mm
10. 14 ft, 13 ft, x, 5.2 ft
11. 15 cm, 17 cm, x 8 cm
12. 9.8 in., x, 10.2 in.   14.1 in.
13. 20 mm, x, 16 mm, 12 mm

14.
about 121 mi

14. **Visual Thinking** Sharon Ramirez receives directions for a party. Use graph paper to sketch Sharon's route. Draw a segment connecting Sharon's house and her friend's house. Find the distance represented by the segment.

Hey Sharon!
Here are the directions to the party! Hope you can make it!
First, drive 70 mi north on Highway 9.
Then drive 80 mi west on Route 16.
Then drive 40 mi north on Highway 7.
Then drive 30 mi east on Route 14.

15. Can a circular trampoline with a diameter of 16 ft fit through a doorway that is 10 ft high and 8 ft wide? (Assume that the legs of the trampoline can be removed.) Explain your answer. No; The hypotenuse is only about 12.8 ft.

Module 4   Patterns and Discoveries

Reproduced from page 298 of Book 3 in *MathThematics*.

FIGURE 5.2.4

We will prove that $c^2 = a^2 + b^2$. First construct a square of dimensions $a + b$ by $a + b$, and form four right triangles with each having a corner of the square as its right angle and with legs of length a and b.

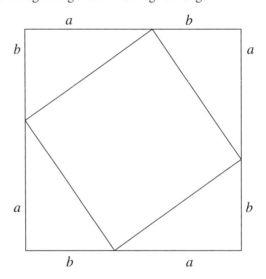

Notice that each triangle's hypotenuse has length c, since each is congruent to $\triangle ABC$ by SAS. In addition, the quadrilateral inscribed in the square is also a square, since the measure of each of its angles is equal to $180° - (m\angle CAB + m\angle CBA) = 90°$. Hence, $(a + b)^2 = 4(ab/2) + c^2$, so $a^2 + 2ab + b^2 = 2ab + c^2$. Finally, $a^2 + b^2 = c^2$.

Proof 2. Consider the right $\triangle ABC$, where $\angle ACB$ is a right angle and sides $\overline{AB}, \overline{AC}$, and $\overline{BC}$ have lengths $AB = c$, $AC = b$, and $BC = a$.

Construct the perpendicular (altitude) from C to D on side $\overline{AB}$.

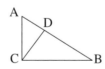

Recall that $\triangle ACD \sim \triangle ABC \sim \triangle BCD$. (Why?) Hence, $\frac{BC}{BD} = \frac{AB}{BC}$ and $\frac{AC}{AB} = \frac{AD}{AC}$, and so $a^2 = (AB)(BD)$ and $b^2 = (AB)(AD)$. Thus, $a^2 + b^2 = AB(BD + AD) = (AB)(AB) = c^2$.

Back to Daryl. If Daryl's penciled-out quadrilateral was actually a rectangle, then we know that his measurements of a diagonal and two adjacent legs would satisfy the Pythagorean Theorem. The problem with this reasoning is that Daryl did not know he had drawn a rectangle, and in fact, this is what he was really trying to prove. So what was his purpose for these measurements and computations? Certainly if his dimensions did not satisfy the Pythagorean Theorem, then he could conclude that the triangle was not a right triangle (and so the parallelogram was not a rectangle). But what if they did satisfy the Pythagorean Theorem—could he then conclude the triangle was a right triangle? It seems to me that this was Daryl's belief; and lucky for Daryl (and my roof), this is actually a valid deduction. In fact, this is precisely the converse of the Pythagorean Theorem. Here is a short proof of the converse.

5.3.2 Theorem: Converse of the Pythagorean Theorem. If the square of the length of a side of $\triangle ABC$ equals the sum of the squares of the lengths of the other two sides, then $\triangle ABC$ is a right triangle with a right angle opposite the longest side.

Proof. Assume $\triangle ABC$ satisfies $a^2 + b^2 = c^2$, where $AC = b$, $BC = a$, and $AB = c$. Construct a right $\triangle EFG$ with m$\angle EGF = 90°$, $EG = b$, and $FG = a$. We know that $(EF)^2 = a^2 + b^2 = c^2$, so $EF = c$. Hence, $\triangle ABC \cong \triangle EFG$ by SSS, and thus m$\angle ACB$ = m$\angle EGF = 90°$.

Back to the Classroom Lesson. In view of the converse of the Pythagorean Theorem, it is now clear how Daryl's measurements and computations helped him double-check his work before cutting. In addition, we know that if his calculations did not obey the Pythagorean Theorem, then the triangle he was checking would not be a right triangle. But then what kind of triangle would it be, and could he determine from his calculations whether it was an acute or obtuse triangle? This is precisely the situation in Exercise 7 in Figure 5.2.4. In fact, here is a diagram of that problem.

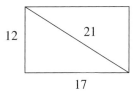

Notice that $12^2 + 17^2 = 433 < 441 = 21^2$, so the angle opposite the side of length 21 units is not a right angle (why?), consequently, the sketched quadrilateral is not a rectangle (even though it looks like one in the drawing). What kind of angle is it, and how would you justify your answer to this problem and to Exercises 1–6 in Figure 5.2.4? Direction: If the diagonal had length $\sqrt{433}$, then the angle opposite it would be a right angle. (Why?) However, the length of the given diagonal is greater than $\sqrt{433}$, so the angle opposite it must be greater than $90°$. (Why?)

The next proposition formally records the precise delineation of the acute, obtuse, and right angle cases, and its proof does not depend on the previous proof of the converse. This gives us another (more general) validation of the converse.

5.3.3 Theorem: A Generalized Converse of the Pythagorean Theorem. Consider $\triangle ABC$, with $AC = b$, $BC = a$, and $AB = c$.

  a. If $a^2 + b^2 > c^2$, then the angle opposite c is an acute angle.

b. If $a^2 + b^2 < c^2$, then the angle opposite c is an obtuse angle.
c. If $a^2 + b^2 = c^2$, then the angle opposite c is a right angle.

Proof. **a.** Assume $a^2 + b^2 > c^2$, and suppose that the angle opposite c is not an acute angle. Hence, this angle must be obtuse. (Why?) Construct a perpendicular from B to a point D on the extension of $\overline{AC}$, and let $CD = s$ and $BD = t$.

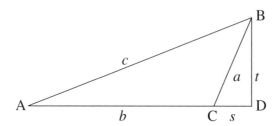

Note that $a^2 = s^2 + t^2$, so

$$[b^2 + (s^2 + t^2)] > c^2 = (b + s)^2 + t^2 = (b^2 + 2bs + s^2) + t^2.$$

Therefore, $0 > 2bs$, which is a contradiction.

b. Assume $a^2 + b^2 < c^2$, and suppose that the angle opposite c is not an obtuse angle. Thus, this angle must be acute. Construct a perpendicular from B to a point D on the side $\overline{AC}$, and let $CD = s$ and $BD = t$.

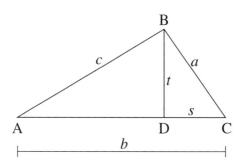

Using the same approach as in a, we have

$$[b^2 + (s^2 + t^2)] < c^2 = (b - s)^2 + t^2 = (b^2 - 2bs + s^2) + t^2,$$

so $0 < -2bs$, which is also a contradiction.

c. Assume $a^2 + b^2 = c^2$, and suppose the angle opposite c is not a right angle. Then it is either acute or obtuse. Using the ideas in a and b, it is routine to complete this part, and we leave this as Exercise 1 of Exercises 5.3.

EXERCISES 5.3

1. **a.** Complete part c of the Generalized Converse of the Pythagorean Theorem proof.
   b. Develop a coordinate proof for the converse of the Pythagorean Theorem. Direction: Position the triangle at the origin so that the legs have nonzero finite slope, and then show that the product of their slopes is -1.

2. Find three different proofs of the Pythagorean Theorem and write them out in a complete and precise manner.

3. If $\triangle ABC$ has sides of lengths $AB = 13$, $AC = 7$, and $BC = 9$, then find the length of the altitude on AB.

4. Show that the following triangles, which have sides of the given lengths, are each right triangles:
   a. $a = 2r + 1$, $b = 2r^2 + 2r$, and $c = 2r^2 + 2r + 1$, where $r > 0$ is a real number
   b. $a = r^2 - 1$, $b = 2r$, and $c = r^2 + 1$, where $r > 1$ is a real number
   c. $a = r^2 - s^2$, $b = 2rs$, $c = r^2 + s^2$, where $0 < s < r$ are real numbers

5. Suppose a right triangle has sides of lengths a, b, and c, where $c^2 = a^2 + b^2$ and a, b, and c are positive real numbers. Then a triangle with sides of lengths ra, rb, and rc, with $r > 0$ as a real number, is also a right triangle. (Give two different proofs of this statement.)

6. A triangle having sides of lengths 3, 4, and 5 is a right triangle. (Why?) Using Exercises 5.3 nos. 4 and 5 give several other examples of right triangles having integer length sides. A right triangle having integer length sides is called a **Pythagorean triangle**. Show that there are infinitely many Pythagorean triangles having sides of lengths a, b, and c, where $\gcd(a, b, c) = 1$. (Direction: Use Exercises 5.3 no. 4.) Pythagorean triangles having this property are called **primitive Pythagorean triangles**. (In Section 5.7, we conduct a thorough investigation of Pythagorean triangles and primitive Pythagorean triangles.)

7. In each part, either provide an example or explain why no such example exists:
   a. A Pythagorean triangle having a side of length 2 units
   b. A Pythagorean triangle having a leg of length 7 units
   c. A Pythagorean triangle with a hypotenuse of length 7 units
   d. A Pythagorean triangle with a hypotenuse of length 29 units
   e. A Pythagorean triangle having a perimeter of 30 units
   f. A Pythagorean triangle having an area of 84 square units
   g. A Pythagorean triangle with its perimeter equal to its area
   h. A Pythagorean triangle having two sides of even lengths and one side of odd length
   i. A Pythagorean triangle having all sides of odd lengths

8. Show that there are no isosceles Pythagorean triangles. Direction: Assume otherwise and seek a contradiction.

9. **The arithmetic mean—geometric mean inequality**: Prove that $(a + b)/2 \geq \sqrt{ab}$, where a and b are positive real numbers. Also show that the expressions are equal if and only if $a = b$. Recall that the arithmetic mean (average) of two real numbers a and b is defined as $(a + b)/2$, and the geometric mean of two positive real numbers a and b is defined to be a real number x so that $\frac{x}{a} = \frac{b}{x}$. Direction: Consider the following modification of the diagram used in Proof 1 of the Pythagorean Theorem 5.3.1 and show that $(a + b)^2 \geq 4ab$.

226 Chapter 5 Algebraic Modeling in Geometry: The Pythagorean Theorem

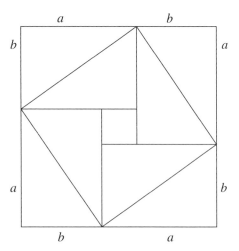

5.4 COMPUTING DISTANCE IN TWO-DIMENSIONAL AND THREE-DIMENSIONAL EUCLIDEAN SPACE: THE DISTANCE FORMULA

We have made several numerical computations when calculating the length of one side of a right triangle in terms of the lengths of the other two sides. This fundamental application of the Pythagorean Theorem can be utilized to find the distance between any two points in two- and three-dimensional Euclidean space.

Two-Dimensional Space. Let $A = (x_1, y_1)$ and $B = (x_2, y_2)$ be two distinct points in the plane. If $x_1 = x_2$ or $y_1 = y_2$, then the distance between the two points is $|y_2 - y_1|$ or $|x_2 - x_1|$, respectively, so we may as well consider the situation when $x_1 \neq x_2$ and $y_1 \neq y_2$. Construct a right $\triangle ABC$, with hypotenuse of length AB and legs of lengths $|y_2 - y_1|$ and $|x_2 - x_1|$.

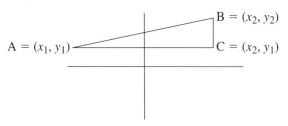

The Pythagorean Theorem gives us $(AB)^2 = (x_2 - x_1)^2 + (y_2 - y_1)^2$ (no need for absolute value signs here—why?), and then taking square roots establishes the **distance formula**:

$$AB = \sqrt{(x_2 - x_1)^2 + (y_2 - y_1)^2}. \quad \textbf{5.4.1}$$

Classroom Problems. Suppose you are given five distinct points in or on a unit square (a square of dimensions 1 unit by 1 unit). Show that the distance between at least two of these points is not greater than $\sqrt{2}/2$ units. Direction: Subdivide the square into four squares, each with a dimension of 1/2 unit by 1/2 unit. Argue that

Section 5.5   An Extension of the Pythagorean Theorem: The Law of Cosines   227

at least one of the subsquares must contain two of the given points, and then use the distance formula to complete the proof. ◆

Three-Dimensional Space. The distance formula for points $A = (x_1, y_1, z_1)$ and $B = (x_2, y_2, z_2)$ in three-dimensional Euclidean space can be derived similarly by applying the Pythagorean Theorem twice.

To see this, label $C = (x_2, y_2, z_1)$ and $D = (x_1, y_2, z_1)$ on the following diagram, then use the Pythagorean Theorem on $\triangle ACD$ to get:

$$(AC)^2 = (x_2 - x_1)^2 + (y_2 - y_1)^2.$$

Now apply the Pythagorean Theorem to $\triangle ABC$ to get:

$$(AB)^2 = (AC)^2 + (BC)^2 = (x_2 - x_1)^2 + (y_2 - y_1)^2 + (z_2 - z_1)^2, \text{ so}$$
$$AB = \sqrt{(x_2 - x_1)^2 + (y_2 - y_1)^2 + (z_2 - z_1)^2}.$$

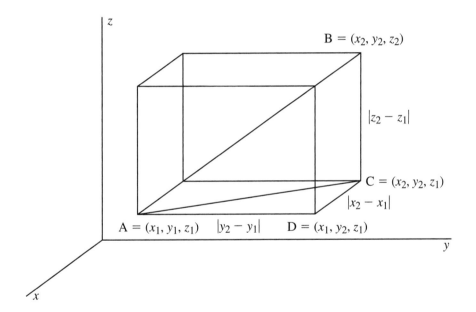

5.5 AN EXTENSION OF THE PYTHAGOREAN THEOREM: THE LAW OF COSINES

Is it possible to extend the Pythagorean Theorem to arbitrary triangles so that the length of any side can be determined just in terms of the lengths of the other two sides? (Note that a theorem Y is called an extension of a theorem X if theorem X can be realized as a special case of theorem Y.) To see that it is not possible in general, just notice that there are infinitely many noncongruent triangles having sides of lengths a and b, so the length of a third side would not be uniquely determined by the lengths of the other two sides (unless additional information was given). For example, suppose $a = 2$ and $b = 3$. Find several noncongruent triangles having these particular sides.

228 Chapter 5 Algebraic Modeling in Geometry: The Pythagorean Theorem

Let's look at an important extension of the Pythagorean Theorem that allows us to deduce the length of a side of a triangle if we know the lengths of the other two sides and the measure of the angle between them. Additionally, the angles of a triangle can be determined if we know the lengths of the sides. We will approach the problem from two points of view: Coordinate free approach and Coordinate approach. Compare the different approaches and notice that the coordinate free analysis involves two cases, whereas in the coordinate analysis it is possible to handle both cases simultaneously.

Coordinate Free Approach. Suppose we are given $\triangle ABC$ with sides of lengths $AB = c$ and $AC = b$ and $\angle BAC = \alpha$. (Note that it is common to denote the angle and the measure of the angle by the same symbol, α). How can we determine the length $BC = a$ in terms of the given data?

Analysis. It is necessary to look at two cases: 1. α acute; 2. α obtuse.

Case 1. Construct the perpendicular $\overline{BD}$ from vertex B to the point D on the side $\overline{AC}$; set $BD = h$ and $AD = s$ (thus $CD = b - s$).

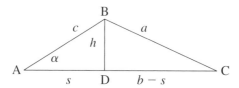

Since $\overline{BD}$ is a common leg to the right $\triangle ABD$ and the right $\triangle CBD$, an application of the Pythagorean Theorem gives $a^2 - (b - s)^2 = h^2 = c^2 - s^2$, so $a^2 = b^2 + c^2 - 2bs$. Furthermore, $\cos \alpha = s/c$, so

$$a^2 = b^2 + c^2 - 2bc \cos \alpha. \qquad \textbf{5.5.1}$$

Case 2. Construct the perpendicular $\overline{BD}$ from vertex B to the point D on the extension of side $\overline{AC}$; set $BD = h$, $AD = s$ (thus $CD = b + s$) and $\angle DAB = \overline{\alpha}$, the reference angle associated to α.

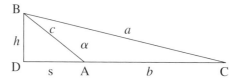

Using reasoning similar to Case 1 gives $a^2 - (b + s)^2 = h^2 = c^2 - s^2$; thus, $a^2 = b^2 + c^2 + 2bs$. Note that $\cos \alpha = -\cos \overline{\alpha}$, so, $s = c \cos \overline{\alpha} = -c \cos \alpha$, which in turn gives $a^2 = b^2 + c^2 - 2bc \cos \alpha$, as in Case 1.

Coordinate Approach. Again, assume we are given $\triangle ABC$ with sides of length $AB = c$ and $AC = b$ and $\angle BAC = \alpha$. Can the distance formula be utilized to determine the length $BC = a$ in terms of the given data?

Analysis. Let's position the given triangle so that vertex A is at $(0, 0)$ and vertex C is at $(b, 0)$. What are the coordinates (s, t) of vertex B? Trigonometry to the rescue! The definitions of sine and cosine give $s = c \cos \alpha$ and $t = c \sin \alpha$ (no need to consider two cases).

Calculating the distance $BC = a$ between points $B = (c \cos \alpha, c \sin \alpha)$ and $C = (b, 0)$ is now a matter of arithmetic and the trigonometric identity $\sin^2 \alpha + \cos^2 \alpha = 1$ (which is really a consequence of the Pythagorean Theorem, explain!).

$$a^2 = (c \cos \alpha - b)^2 + (c \sin \alpha)^2 = b^2 + c^2(\sin^2 \alpha + \cos^2 \alpha) - 2bc \cos \alpha$$
$$= b^2 + c^2 - 2bc \cos \alpha.$$

EXERCISES 5.5

1. The methods we used (in either approach) only depended on knowing the lengths of two sides of a triangle and the measure of the angle between them. If $\angle ABC = \beta$ and $\angle ACB = \gamma$, determine analogous expressions for b^2 and c^2.
2. Using the same notation as in the previous derivations, find the values of a, β, and γ in $\triangle ABC$ if $b = 7$, $c = 5$, and $\alpha = 50°$.
3. Find the values of b, α, and γ in $\triangle ABC$ if $a = 16.3$, $c = 12.1$, and $\beta = 120°$.
4. Explain why the law of cosines is an **extension** of the Pythagorean Theorem.
5. Given the length of one diagonal of a parallelogram, the length of one of its sides and the measure of the angle between these segments, outline a strategy to find the length of the parallelogram's other diagonal. Illustrate your method with a numerical example.

5.6 INTEGER DISTANCES IN THE PLANE

Question. How many points in the plane can be found with the property that the distance between any two of them is always an integer (not necessarily the same integer)?

Easy Answer. Infinitely many; for example, any collection of integer points on the x-axis has this property. More generally, it is geometrically obvious that any line contains infinitely many points that are integer distances apart (simply lay out integer distances on any line).

Modified Question. How many noncollinear points in the plane can be found that have the property that the distance between any two of them is an integer?

The vertices of a triangle having integer length sides comprise a set of three such points, and thus a set of four points can be found as the vertices of an appropriate rectangle.

A Simple Approach: "Almost Collinear" Points. Is it possible to find points $(0, P)$ and $(P_i, 0)$ for finitely many or infinitely many i, such that the distances between each of these points are integers? In particular, can a set of right triangles having a common leg and integer length sides be constructed with any finite (or possibly infinite) set of vertices?

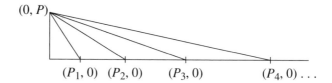

EXERCISES 5.6

1. Show how to explicitly find the coordinates of infinitely many points on the line $y = 3x + 2$ that are integer distances apart. Direction: For points 1 unit apart, apply the distance formula to get $1 = \sqrt{(x_2 - x_1)^2 + [(3x_2 + 2) - (3x_1 + 2)]^2}$.

   Does your method work for any line $y = mx + b$? If so, prove it; if not, develop a strategy that does apply to any line.

2. Find a rectangle having its vertices and the point of intersection of its diagonals all integer distances apart. Direction: Modify a 3, 4, 5 right triangle.

3. Show that it is not possible to find a set of four points that are the vertices of a square and are integer distances apart.

4. Using the Pythagorean Theorem, show that the following figure of two right triangles with four vertices is possible. Note that by reflecting about the *x*-axis and *y*-axis, this diagram can be used to find a set of seven points in the plane that are integer distances apart.

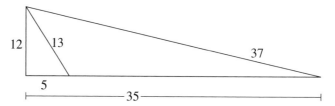

5. Find sets that contain eight (nine, ten, ...) points that are integer distances apart. Direction: In a fashion related to the previous problem, use a 5, 12, 13 right triangle and multiples of a 3, 4, 5 right triangle to build a set with nine points. This set includes the points $(0, 12)$, $(0, -12)$, $(16, 0)$, $(-16, 0)$, and five others. Extend this method to produce larger sets of points an integer distance apart, e.g., construct a set that includes the points $(0, 60)$ and $(0, -60)$. If we use right triangles with sides 3, 4, and 5; 5, 12, and 13; 7, 24, and 25, what is the largest set of points that we can construct (using the previous method) that are an integer distance apart?

5.7 PYTHAGOREAN TRIPLES: POSITIVE INTEGER SOLUTIONS TO $x^2 + y^2 = z^2$

In an effort to construct a collection of right triangles as previously described, it would be useful to find all possible right triangles having integer length sides, i.e., all positive integer solutions of the equation $x^2 + y^2 = z^2$.

A right triangle with integer length sides x, y, z is termed a **Pythagorean triangle** (see Exercises 5.3, no. 6); the set $\{x, y, z\}$ is called a **Pythagorean triple (PT)**.

Section 5.7   Pythagorean Triples: Positive Integer Solutions to $x^2 + y^2 = z^2$ 231

In Problem 4 of Exercises 5.6, we verified that $\{5, 12, 13\}$ and $\{35, 12, 37\}$ are PTs, and in Problems 4 and 5 of Exercises 5.3 we considered some templates that generate infinitely many PTs.

Historical Note. The question of expressing any square as the sum of two squares was considered by Euclid (about 325–265 BC in his book *Elements*, Book X, Lemma 1, following Proposition 28 of Heath's English edition [He 1]) and later as a problem of Diophantus (about AD 200–284) in his book *Arithmetic* (translated into Latin by Bachet and later into English by Heath [He 2]). In fact, Diophantus presented a solution to this problem in a section of the book that focused on right triangles. The mathematical language and notation of his treatment differs significantly from contemporary mathematical vernacular, but his underlying arguments have transcended time.

Internet Resource. See **www.perseus.tufts.edu/cgi-bin/ptext?lookup=Euc.+toc** for an online version of Heath's edition of Euclid's *Elements*.

Solving the Pythagorean Problem. Our search for all positive integer solutions of the equation $x^2 + y^2 = z^2$ proceeds through a sequence of elementary observations. As in most modern textbooks, the development follows the fundamental ideas of Diophantus.

Observation 1. $\{x, y, z\}$ is a PT if and only if $\{cx, cy, cz\}$ is a PT for any positive integer c. (Justify.)

This observation provides an easy method for generating infinitely many PTs. It also implies that if $\{x, y, z\}$ is a PT and d is a common divisor of x, y, and z, then $\{x/d, y/d, z/d\}$ is a PT. (Why?) Thus the members of any PT are multiples of the members of a PT having the greatest common divisor 1 (take $d = \gcd(x, y, z)$). Hence, to determine all PTs, it is enough to find those PTs having $\gcd(x, y, z) = 1$. (Why?)

A PT $\{x, y, z\}$ with $\gcd(x, y, z) = 1$ is called a **primitive Pythagorean triple** (**PPT**); a right triangle whose sides form a PPT is called a **primitive Pythagorean triangle** (see Exercises 5.3, no. 6). Notice that $\{12, 35, 37\}$ is a PPT, while $\{21, 28, 35\}$ is a PT but not a PPT.

5.7.1 Classroom Problems. Before moving to the next observation, let's work through some problems concerning PTs and PPTs.

1. **Pythagoras's PTs.** Verify that $\{2n + 1, 2n^2 + 2n, 2n^2 + 2n + 1\}$ is a PT for any positive integer n (see Exercises 5.3, no. 4a). The discovery of this infinite collection of PTs has been attributed to Pythagoras. Not all PTs can be generated by this formula. For example $\{35, 12, 37\}$ is a PT (see Exercises 5.3, no. 4b) that is not of this form. (Why?)
2. Demonstrate that $\{n^2 - 1, 2n, n^2 + 1\}$ is a PT for each positive integer $n \geq 2$ (see Exercises 5.3, no. 4b). Hence, each even number greater than or equal to 4 is the length of a side of some Pythagorean triangle. For example, 10 is

a member of the PT $\{24, 10, 26\}$. Using the collection of PTs considered in Problem 1, show how each odd number greater than or equal to 3 is the length of a side of some Pythagorean triangle. For example, 15 is a member of the PT $\{15, 112, 113\}$.

3. **Pythagoras's PPTs.** Show that Pythagoras's PTs (Problem 1) are always PPTs. Direction: For a positive integer n, set $d = \gcd(2n + 1, 2n^2 + 2n, 2n^2 + 2n + 1)$ and show $d|1$.

4. Are the PTs of Problem 2 always PPTs? Direction: Consider when n is even and when n is odd. ◆

Observation 2. If $\{x, y, z\}$ is a PPT, then $\gcd(x, y) = \gcd(x, z) = \gcd(y, z) = 1$.

Justification. Suppose $\gcd(x, z) > 1$, and thus x and z have a common prime factor p. Hence p divides $z^2 - x^2$, which implies that p divides y^2. Finally, since p is a prime number, p divides y, which contradicts the assumption that $\gcd(x, y, z) = 1$. The other cases follow similarly.

Note that for three positive integers u, v, and w with $\gcd(u, v, w) = 1$, it is possible that $\gcd(u, v) > 1$ or $\gcd(u, w) > 1$ or $\gcd(v, w) > 1$. Give examples to demonstrate that any combination of these conditions can occur. Also note that if $\gcd(u, v) = 1$ or $\gcd(u, w) = 1$ or $\gcd(v, w) = 1$, then $\gcd(u, v, w) = 1$. (Why?)

Observation 3. If $\{x, y, z\}$ is a PPT, then exactly one of its members is even.

Justification. Since no two of the terms can be even, assume all are odd. Then $x^2 + y^2$ is even, and z^2 is odd, a contradiction.

Observation 4. The even member of a PPT $\{x, y, z\}$ is either x or y.

Justification. Suppose z is even and thus x and y are odd. Write $x = 2n + 1$ and $y = 2m + 1$. Notice that z^2 is divisible by 4, which implies $x^2 + y^2$ is as well. This is impossible, since $x^2 + y^2 = 2 + 4(n + n^2 + m + m^2)$.

Convention. For $\{x, y, z\}$ a PPT, we choose **y to be even** and both **x and z to be odd**. Write $y^2 = z^2 - x^2 = (z + x)(z - x)$, and note that $y, z + x$ and $z - x$ are all even positive integers, so $(y/2)^2 = [(z + x)/2][(z - x)/2]$.

Observation 5. If the product uv of two positive integers u and v is a square and $\gcd(u, v) = 1$, then both u and v are squares (see Section 2.6). Clearly this statement can fail to be true if $\gcd(u, v) > 1$. (Why?)

Justification. Recall that a positive integer is a square if and only if each exponent in its prime factorization is even. Therefore, since uv is a square and u and v have no prime factors in common, then the prime factors of u and v must each have even exponents.

Section 5.7   Pythagorean Triples: Positive Integer Solutions to $x^2 + y^2 = z^2$

Observation 6. $\gcd((z + x)/2, (z - x)/2) = 1$, and thus $(z + x)/2$ and $(z - x)/2$ are squares (see Observation 5).

Justification. Arguing indirectly, suppose that the two integers in question have a common divisor $d > 1$. Then d is a divisor of $(z + x)/2 + (z - x)/2 = z$ and a divisor of $(z + x)/2 - (z - x)/2 = x$, which is impossible. (Why?)

Notation. Set $\sqrt{(z + x)/2} = m$ and $\sqrt{(z - x)/2} = n$, so

$$m^2 = (z + x)/2, n^2 = (z - x)/2, m^2 n^2 = y^2/4.$$

Note that m and n are positive integers with $\gcd(m, n) = 1$. (Explain.)

Summary of Observations: Necessary Conditions of a PPT. We have shown that if $\{x, y, z\}$ is a PPT with y being even, then there exists relatively prime positive integers $m > n$, with one even and the other odd, such that

$$x = m^2 - n^2, y = 2mn, z = m^2 + n^2.$$

To obtain a complete characterization of PPTs, we demonstrate in the next theorem that the previous conditions on m and n are in fact sufficient to guarantee that $\{m^2 - n^2, 2mn, m^2 + n^2\}$ is a PPT.

5.7.2 Theorem: Characterizing Primitive Pythagorean Triples. A set $\{x, y, z\}$ of positive integers with y being an even integer is a PPT if and only if there exists positive integers m and n such that:

1. $x = m^2 - n^2, y = 2mn, z = m^2 + n^2$;
2. $m > n$;
3. $\gcd(m, n) = 1$; and
4. of m and n, one is even and the other is odd.

Proof. ($\Rightarrow$) This direction of the proof follows from the previous observations and remarks. (Indicate precisely how each part follows from previous work.)

($\Leftarrow$) Assume Conditions 1 through 4 are satisfied. We wish to show that $\{x, y, z\}$ is a PPT. Conditions 1 and 2 guarantee that x, y, z are positive integers, and we have previously checked that $\{x, y, z\}$ is a PT.

It remains to show that $\gcd(x, y, z) = 1$. Suppose this is not the case and let p be a prime divisor of x, y, and z. Since p divides $y = 2mn$, then p divides either 2, m, or n. (Why?) If p divides 2, then $p = 2$, and this can't be, since Condition 4 implies that x and z are odd. Hence, we may suppose p is an odd prime divisor of m or n. If p divides m, then p divides $z - m^2 = n^2$, and thus p divides n, a violation of Condition 3. Similarly, another violation of Condition 3 occurs if p divides n. Therefore, $\gcd(x, y, z) = 1$.

Classroom Discussion: The PPTs Discovered by Pythagoras. Before returning to the problem concerning the construction of sets of points that are an integer distance apart in the plane, we can use Theorem 5.7.2 to find several examples of PPTs that are not members of the infinite collection $\{\{2n + 1, 2n^2 + 2n, 2n^2 + 2n + 1\}: n$ is a positive integer$\}$ of PPTs discovered by Pythagoras (see Classroom Problems 5.7.1, no. 3).

In particular, the problem at hand is to find PPTs $\{x, y, z\}$ with y being even, where y and z are not consecutive positive integers. One way to find such examples is to simply use Theorem 5.7.2 to search for all PPTs that are not of the given form. A systematic list helps in our search and may be useful for raising and answering other questions concerning the nature of PPTs.

A List of Some Primitive Pythagorean Triples

m	n	$m^2 - n^2$	$2mn$	$m^2 + n^2$	Pythagoras PPT
2	1	3	4	5	yes
3	2	5	12	13	yes
4	1	15	8	17	no
4	3	7	24	25	yes
5	2	21	20	29	no
5	4	9	40	41	yes
6	1	35	12	37	no
6	5	11	60	61	yes
7	2	45	28	53	no
7	4	33	56	65	no
7	6	13	84	85	yes
8	1	63	16	65	no
8	3	55	48	73	no

◆

5.7.3 Question. The PPTs discovered by Pythagoras have the distinctive feature that y and z are consecutive positive integers (here we are assuming that y is even). Could it be that all PPT's of this form turn out to be the ones discovered by Pythagoras? Let's analyze this question by using the characterization of PPT's described in Theorem 5.7.2.

Analysis. First assume that $\{x, y, y + 1\}$ is a PPT with y even. By Theorem 5.7.2, there exist integers m and n where one is even and the other is odd such that

Section 5.7   Pythagorean Triples: Positive Integer Solutions to $x^2 + y^2 = z^2$

$\gcd(m, n) = 1$, $m > n$, and $x = m^2 - n^2$, $y = 2mn$, and $y + 1 = m^2 + n^2$. Thus $(m^2 + n^2) - 2mn = 1$, and so by rearranging and factoring we get $(m - n)^2 = 1$. Since $m > n$, then $m - n = 1$, i.e., $m = n + 1$. The desired form (discovered by Pythagoras) is now achieved by replacing m with $n + 1$ in the expressions for $x, y, y + 1$. (Verify this.)

Back to the question concerning points that are an integer distance apart in the plane:

Method 1. Recall that our analysis of PPTs was initially motivated by a particular approach used to determine collections of noncollinear points that are an integer distance apart in the plane. In particular, we are looking for collections of points $(0, P)$ and $(P_i, 0)$ that are integer distances apart. In other words, it suffices to find collections of primitive Pythagorean triangles that all share the same vertical leg and have different hypotenuses. The characterization of PPTs developed in this chapter is an important tool in the search for such collections.

In the following diagram, label the leg on the y-axis with y (so $P = y$ in the diagram), where y is assumed to be a positive integer.

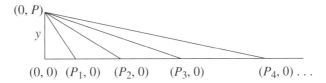

Question. Which *even* integers y can arise as the length of a primitive Pythagorean triangle's leg (i.e., y is the even member of a PPT)? Furthermore, is there a method to determine all primitive Pythagorean triangles with a given even length side?

Observation. First note that $y = 2mn$ must be divisible by 4, since either m or n is even.

Example. Let's determine all PPTs with $y = 56$, i.e., find all positive integers m and n, where $56 = 2mn$ and where m and n satisfy the conditions of Theorem 5.7.2. In particular, we are looking for all relatively prime pairs of positive integers $m > n$, where one is even and the other is odd, such that $56 = 2mn$ (i.e., $28 = mn$).

The factorizations $28 = 7 \cdot 4$ and $28 = 28 \cdot 1$ are the only ones satisfying the given criteria, so $m = 7$ and $n = 4$ produces the PPT $\{33, 56, 65\}$, while $m = 28$ and $n = 1$ generates the PPT $\{783, 56, 785\}$.

Notice that if we choose a factorization other than one of these, say $28 = 14 \cdot 2$, then $m = 14$ and $n = 2$ yields the PT $\{192, 56, 200\}$, which is not a PPT. (What is the PPT that can be derived from the PT $\{192, 56, 200\}$?)

5.7.4 Summary: Finding the Even Leg of a Primitive Pythagorean Triangle.

All positive even integers y that are divisible by 4 can occur as the length of the even side of a primitive Pythagorean triangle. Moreover, all primitive Pythagorean triangles that have a common leg of length y can be obtained through the factorizations

$y/2 = mn$, where $m > n$ and $\gcd(m,n) = 1$. Notice that under these conditions, one of the factors must be odd and the other even. (Why?) The following table lists some particular examples of PPTs generated by this procedure.

y	y/2	m	n	$x = m^2 - n^2$	$y = 2mn$	$z = m^2 + n^2$
4	2	2	1	3	4	5
12	6	6	1	35	12	37
12	6	3	2	5	12	13
20	10	10	1	99	20	101
20	10	5	2	21	20	29
36	18	18	1	323	36	325
36	18	9	2	79	36	85

Example: 9 Points in the Plane, Each an Integer Distance Apart. Using the ideas developed in the previous discussion, it is sufficient to find an integer y that is divisible by 4, so that $y/2$ can be factored into 8 different relatively prime pairs. This can be easily accomplished if we choose y so that $y/2$ has exactly 4 distinct prime factors (each prime may occur a multiple number of times). For example, let $y = 420 = 2^2 \cdot 3 \cdot 5 \cdot 7$, and consider the following factorizations:

y/2	m	n	$x = m^2 - n^2$	$y = 2mn$	$z = m^2 + n^2$
210	$(2 \cdot 3 \cdot 5 \cdot 7) = 210$	1	44,099	420	44,101
210	$(3 \cdot 5 \cdot 7) = 105$	2	11,021	420	11,029
210	$(2 \cdot 3 \cdot 5) = 30$	7	851	420	949
210	$(2 \cdot 3 \cdot 7) = 42$	5	1,739	420	1,789
210	$(2 \cdot 5 \cdot 7) = 70$	3	4,891	420	4,909
210	$(5 \cdot 7) = 35$	$(2 \cdot 3) = 6$	1,189	420	1,261
210	$(3 \cdot 7) = 21$	$(2 \cdot 5) = 10$	341	420	451
210	$(3 \cdot 5) = 15$	$(2 \cdot 7) = 14$	29	420	421

Set $P = 420$, $P_1 = 29$, $P_2 = 341$, $P_3 = 851$, $P_4 = 1{,}189$, $P_5 = 1{,}739$, $P_6 = 4{,}891$, $P_7 = 11{,}021$, $P_8 = 44{,}099$, and consider the points: $(0, P)$ and the $(P_i, 0)$. This is a set of 9 points in the plane where the distance between any two of these points is an integer.

General Solution. Explain how the previous example can be generalized to show that for any integer $n \geq 2$, there are n (noncollinear) points that are an integer distance apart in the plane. (Exercise.)

Another approach to a general solution:

Method 2 [8]. For any positive integer n, let $\{a_i, b_i, c_i\}$ be distinct PPTs (with b_i even) where $i = 1, 2, 3, \ldots, n - 1$. Let $P = \prod_1^{n-1} a_i$, and consider the following noncollinear points (with positive integer components): $(0, P)$ and $(b_i P/a_i, 0)$, where $i = 1, 2, 3, \ldots, n - 1$.

Observation 1. There are n distinct points in this collection.

Justification. It is enough to show that if $i \neq j$, then $b_i P/a_i \neq b_j P/a_j$. Suppose otherwise (i.e., assume that for some $i \neq j$, $b_i P/a_i = b_j P/a_j$, so $b_i a_j = b_j a_i$). Since $\gcd(a_i, b_i) = 1$, then $a_i | a_j$. (Why?) Likewise, $a_j | a_i$, as $\gcd(a_j, b_j) = 1$. Therefore, $a_i = a_j$, and similarly $b_i = b_j$. This contradicts the fact that $\{a_i, b_i, c_i\} \neq \{a_j, b_j, c_j\}$; hence, the $(b_i P/a_i, 0)$ are all distinct.

Observation 2. The distance between any two of the points in the given collection is an integer.

Justification. Since $b_i P/a_i$ is an integer for each $1 \leq i \leq n - 1$, it is clear that the distances between the $(b_i P/a_i, 0)$s are integers. Moreover, the distance between $(0, P)$ and any $(b_i P/a_i, 0)$ is

$$\sqrt{(b_i^2 P^2/a_i^2) + P^2} = P\sqrt{(b_i^2/a_i^2) + 1} = P\sqrt{(b_i^2 + a_i^2)/a_i^2} = P\sqrt{c_i^2/a_i^2} = Pc_i/a_i,$$

which is an integer.

EXERCISES 5.7

1. Find all PPTs $\{x, y, z\}$ with $y =$
   a. 68
   b. 240
   c. 148
   d. 256
2. Construct ten incongruent primitive Pythagorean triangles that share an even leg.

5.8 EXTENDED STUDIES: FURTHER INVESTIGATIONS INTO INTEGER DISTANCE POINT SETS—A THEOREM OF ERDÖS

Question. Is it possible to find for any integer $n \geq 3$ a set of n points in the plane such that no three are collinear and the distance between any two is an integer?

Although it is not entirely obvious how to produce such sets of points, they do exist. In fact, we use a bit of trigonometry and some theory of complex numbers to produce an example. We shall provide a brief description of a construction of these sets for those readers possessing the prerequisites.

Construction Sketch [8]. Let θ be the angle such that $\cos\theta = 4/5$ and $\sin\theta = 3/5$. The points on the unit circle, $P_n = e^{i(2n\theta)} = \cos(2n\theta) + i\sin(2n\theta), n = 0, 1, 2, 3, \ldots$ are all at rational distances from one another. (To show this, some trigonometry is needed as is an application of the Binomial Theorem and de Moivre's Theorem.) Finally, a change of scale can make any finite set of these points have the property that the distance between any two of them is a natural number.

Question. Is it possible to find an *infinite* collection of noncollinear points in the plane such that the distance between any two of them is an integer?

The surprising answer is NO, and a theorem of Anning and Erdös deals with this explicitly. Their proof was a bit messy, and Irving Kaplansky asked Erdös (1913–1996) if he could find a simpler proof. What Erdös discovered was elementary and amazingly elegant, and we shall reproduce it here.

A brief review of two elementary concepts provides the necessary background to understand Erdös's argument. As motivation for the first one, let's consider the following story from Book 1 of *MathThematics*.

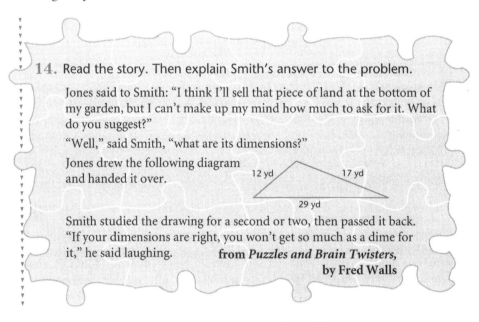

Reproduced from page 24 of Book I in *MathThematics*.

FIGURE 5.8.1

Why was Smith laughing? The next statement provides a suitable explanation.

5.8.1 Theorem: Triangle Inequality. The length of a triangle's side is always less than the sum of the lengths of the other two sides.

Proof. Given $\triangle ABC$, we know that the lengths AB, AC, and BC have some ordering, say, $AB \leq BC \leq AC$. Thus, $AB < AC + BC$ and $BC <$

$AC + AB$. To show that $AC < AB + BC$, construct the perpendicular from C to a point D on the line determined by A and C, and note that D must lie between A and C, since $AB \leq BC \leq AC$. (Justify this assertion.) So $AC = AD + DC, AD < AB, DC < BC$, and consequently, $AD + DC < AB + BC$. Therefore, $AC < AB + BC$.

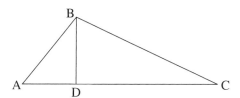

Defining the Hyperbola. Let F_1 and F_2 be two fixed points in the plane. A hyperbola H with foci at F_1 and F_2 is defined to be a set of points in the plane

$$H = \{P : |PF_1 - PF_2| = k\},$$

where k is a fixed real number. The midpoint of the segment $\overline{F_1F_2}$ is called the "center" of the hyperbola. The **standard form** of the equation of the hyperbola with center at $(0,0)$, foci $F_1 = (-c, 0)$, $F_2 = (c, 0)$, where $c > 0$ and $k = 2a$ is

$$\frac{x^2}{a^2} - \frac{y^2}{b^2} = 1, \text{ where } b = \sqrt{c^2 - a^2}.$$

5.8.2 Theorem [8]. If an infinite set S of points in the plane has the property that the distance between any two of them is an integer, then all of the points of S lie on some line.

Proof. Assume S is an infinite set of points in the plane where the distance between any two of them is an integer, and suppose there is no line containing all the points of S. Consequently, S must contain at least three noncollinear points A, B, and C. Form $\triangle ABC$ and set $AB = c$, $BC = a$, and $AC = b$. Let D be any other point of S, and note that D cannot lie on more than one side of $\triangle ABC$. Accordingly, suppose that D does not lie on sides AC and BC.

We know that the lengths of both AD and CD are positive integers, and thus the difference $|AD - CD| = m$ is a positive integer. Hence, utilizing the triangle inequality on $\triangle ADC$ gives, $m = |AD - CD| < AC = b$, so m is an integer with $0 \leq m \leq b - 1$. Therefore, D is a point on the hyperbola that has foci at A and C and has constant difference m between its focal radii. Note that there is a family of "b" hyperbolas with foci at A and C and constant difference k, where $k = 0, 1, 2, \ldots, b - 1$; D lies on one of these.

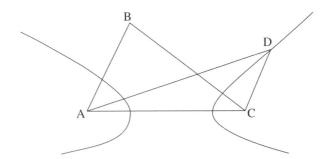

Analogously, since D does not lie on BC, it must lie on one of the "a" hyperbolas that has foci at B and C and, constant difference k, where $k = 0, 1, 2, \ldots, a - 1$.

Since two distinct hyperbolas can intersect in at most 4 points, D must be one of the possible $4ab$ points of intersection of the hyperbolas of these two families. Hence, there are only a finite number of possible locations for points of S other than A, B, C, and this contradicts the fact that S is infinite.

EXERCISES 5.8

1. In $\triangle ABC$, show that the length of any side is less than $(AB + BC + AC)/2$.
2. Consider a rectangle $ABCD$ in the plane and any point P in the plane. Show that the distance from P to any vertex of the rectangle is less than the sum of the distances from P to the other vertices.
3. Consider $\triangle ABC$, with $AC = a$, $BC = b$, and $AB = c$. True or false: Since $c < a + b$, then $c^2 < a^2 + b^2$. Justify your answer (see Theorem 5.3.3).

5.9 EXTENDED STUDIES: ADDITIONAL QUESTIONS CONCERNING PYTHAGOREAN TRIPLES

The Method 1 approach in Section 5.7 used to find finite sets of points that are an integer distance apart in the plane sought to determine which *even* positive integers y can arise as the length of a primitive Pythagorean triangle's leg (i.e., y is the even member of a PPT). The useful aspect of this questions analysis involved in how such even integers y can be expressed as $y = 2mn$, where m and n are positive integers satisfying the conditions of Theorem 5.7.2.

Question. Which *odd* positive integers x can arise as the length of a primitive Pythagorean triangle's leg and how can such x be expressed as $x = m^2 - n^2$, where m and n are positive integers satisfying the conditions of Theorem 5.7.2? Moreover, is there a process to determine all primitive Pythagorean triangles with a given odd length side?

Note that the PPTs of Pythagoras provide a partial answer to this question (see Problems 5.7.1, no. 3), but they do not provide a means to find all primitive Pythagorean triangles with a fixed odd length side.

Example. Let's find all PPTs with $x = 45$ (i.e., find all pairs of positive integers m and n, where $45 = m^2 - n^2$ and m and n satisfy the conditions of Theorem

Section 5.9 Additional Questions Concerning Pythagorean Triples

5.7.2). More specifically, we are searching for all relatively prime pairs of positive integers $m > n$, where one is even and the other is odd, such that $45 = m^2 - n^2 = (m+n)(m-n)$.

Notice that if such an m and n did exist, then both $m+n$ and $m-n$ would have to be odd and relatively prime. (Why?) Therefore, the only choices for the terms $m+n$ and $m-n$ that can possibly result in a PPT are those odd, relatively prime factors of 45.

The factorizations $45 = 9 \cdot 5$ and $45 = 45 \cdot 1$ are the only ones satisfying the necessary conditions and result in the following systems of linear equations:

System 1	System 2
$m + n = 9$	$m + n = 45$
$m - n = 5$	$m - n = 1$

System 1 has solutions $m = 7$ and $n = 2$, and System 2 has solutions $m = 23$ and $n = 22$. Both pairs of solutions satisfy the conditions of Theorem 5.7.2 and produce the PPTs $\{45, 28, 53\}$ and $\{45, 1{,}012, 1{,}013\}$.

5.9.1 Summary: Finding the Odd Leg of a Primitive Pythagorean Triangle.

All odd integers $x \geq 3$ can arise as the length of the odd leg of a primitive Pythagorean triangle. In particular, by factoring $x = rs$, where $r > s$ and $\gcd(r, s) = 1$, then positive integers m and n with $x = m^2 - n^2$ can be determined by setting $r = m + n$ and $s = m - n$ (note that r and s are odd since x is odd).

Justification. The rational numbers $m = (r+s)/2$ and $n = (r-s)/2$ are positive integers that satisfy the conditions of Theorem 5.7.2. To see this, note that m and n are positive integers with $m > n$, since r and s are odd and $r > s$. Also, $\gcd(m, n) = 1$, since $\gcd(r, s) = 1$. (Why?) Hence, m and n are not both even. Moreover, they are not both odd, since $m + n = r$ is odd. Therefore, $x = rs = m^2 - n^2$, where m and n are integers that satisfy the conditions of Theorem 5.7.2.

Some numerical examples help illustrate the process described in Summary 5.9.1.

x	r	s	$m = (r+s)/2$	$n = (r-s)/2$	$x = m^2 - n^2$	$y = 2mn$	$z = m^2 + n^2$
7	7	1	4	3	7	24	25
9	9	1	5	4	9	40	41
15	15	1	8	7	15	112	113
15	5	3	4	1	15	8	17
17	17	1	9	8	17	144	145
21	21	1	11	10	21	220	221
21	7	3	5	2	21	20	29
25	5^2	1	13	12	25	312	313
27	3^3	1	14	13	27	364	365

5.9.2 Question. Which *odd* positive integers z can arise as the length of a primitive Pythagorean triangle's hypotenuse, and how can such z be expressed as $z = m^2 + n^2$, where m and n are positive integers that satisfy the conditions of Theorem 5.7.2?

For an odd positive integer x and a positive integer y divisible by 4, the factorizations $x = m^2 - n^2 = (m + n)(m - n)$ and $y/2 = mn$, in combination with the prime factorizations of x and y, played a key role in determining how to find integers m and n that satisfy the conditions of Theorem 5.7.2. Although z certainly has a prime factorization, no comparable algebraic factorization exists for $z = m^2 + n^2$ over the integers, and thus a different approach is necessary.

Example. To help illustrate some of the challenges in resolving Question 5.9.2, let's find all PPTs $\{x, y, z\}$ (or show that there are none) for three different possibilities: $z = 21$, $z = 45$, and $z = 65$.

A few calculations will quickly convince you that 21 cannot be expressed as $m^2 + n^2$ for positive integers m and n satisfying Theorem 5.7.2 (actually, it can't be written as the sum of any two squares), so there are no PPTs with $z = 21$.

Even though 45 is the sum of two squares ($45 = 6^2 + 3^2$), it still cannot be represented this way with positive integers m and n satisfying Theorem 5.7.2. (Verify.)

Finally, $65 = 7^2 + 4^2 = 8^2 + 1^2$. Consequently, there are two different PPTs with $z = 65$ ($m = 7$ and $n = 4$ generate the PPT $\{33, 56, 65\}$, and $m = 8$ and $n = 1$ generate the PPT $\{63, 16, 65\}$).

As we continue to analyze this question in more depth, the mathematics underlying these examples become more apparent.

A Necessary Condition for z to Be a Member of a PPT $\{x, y, z\}$. If z is a member of a PPT $\{x, y, z\}$, then $z = 4t + 1$ for some positive integer $t \geq 1$, but not conversely. For example, $z = 21 = 4 \cdot 5 + 1$ is not a member of a PPT $\{x, y, z\}$. (Why?)

Justification. Recall that if z is a member of a PPT $\{x, y, z\}$, then there exists relatively prime positive integers $m > n$, one even and the other odd, such that $z = m^2 + n^2$. If m is odd and n is even, then $z = (2r + 1)^2 + (2s)^2 = 4(r^2 + s^2 + r) + 1 = 4t + 1$, likewise if m is even and n is odd.

The next table lists some positive integers of the form $4t + 1$ that can be expressed as $m^2 + n^2$, where m and n are positive integers satisfying the conditions of Theorem 5.7.2.

$z = 4t + 1$	$z = m^2 + n^2$
5	$2^2 + 1^2$
13	$3^2 + 2^2$
17	$4^2 + 1^2$
$25 = 5 \cdot 5$	$4^2 + 3^2$

29	$5^2 + 2^2$
37	$6^2 + 1^2$
41	$5^2 + 4^2$
53	$7^2 + 2^2$
$65 = 5 \cdot 13$	$7^2 + 4^2$
$65 = 5 \cdot 13$	$8^2 + 1^2$
$85 = 5 \cdot 17$	$7^2 + 6^2$
$85 = 5 \cdot 17$	$9^2 + 1^2$
$221 = 13 \cdot 17$	$11^2 + 10^2$
$221 = 13 \cdot 17$	$14^2 + 5^2$

Notice that the prime factors of each z in this list are also of the form $4t + 1$ and can be expressed as $m^2 + n^2$, where m and n are positive integers satisfying the conditions of Theorem 5.7.2. We further discuss this phenomenon later in this section.

Some positive integers of the form $4t + 1$ can be written as the sum of two squares, but not in the configuration we have stipulated (satisfying the conditions of Theorem 5.7.2), e.g., $9 = 3^2 + 0^2$ and $45 = 6^2 + 3^2$. Moreover, there are also positive integers of the form $4t + 1$ that cannot be written as the sum of two squares, such as 21 and 33.

Reformulation of Question 5.9.2. Which positive integers of the form $z = 4t + 1$ can be represented as $z = m^2 + n^2$, where m and n are positive integers satisfying the conditions of Theorem 5.7.2? Is there some general procedure for explicitly finding m and n?

Although the mathematics needed to fully explore this reformulation is beyond the scope of this book, we indicate an important reduction used in the analysis. Once we have accomplished this, we then state a general theorem that completely settles the question of which positive integers can arise as the length of a primitive Pythagorean triangle's hypotenuse. First, a general observation is helpful in establishing a reduction of reformulated Question 5.9.2.

Observation. If $u = a^2 + b^2$ and $v = c^2 + d^2$, then $uv = (ad + bc)^2 + (ac - bd)^2$, where a, b, c, d are integers. (Verify.)

A Reduction of Reformulated Question 5.9.2 to Prime Factors. If each prime factor p of a positive integer $z = 4t + 1$ can be represented as $a^2 + b^2$, where a and b are positive integers satisfying the conditions of Theorem 5.7.2, then z can be represented by a similar expression.

244   Chapter 5   Algebraic Modeling in Geometry: The Pythagorean Theorem

Justification. We consider the special case when $z = 4t + 1 = pq$, where p and q are prime numbers. An argument for the general case can be constructed by using mathematical induction and the analysis from this special case.

Assume $p = a^2 + b^2$ and $q = c^2 + d^2$, where a, b and c, d are respectively positive integers that satisfy the conditions of Theorem 5.7.2. Two possibilities can occur: $p = q$ and $p \neq q$.

a. Assume $p = q$. Then (by the previous Observation),

$$z = p^2 = (a^2 + b^2)(a^2 + b^2) = (2ab)^2 + (a^2 - b^2)^2.$$

We claim that $\gcd(2ab, a^2 - b^2) = 1$. Let g be a *prime* divisor of $2ab$ and $a^2 - b^2$, and so g divides either 2, a, or b. Since $a^2 - b^2$ is odd (why?), then g must divide either a or b (why?). If $g|a$, then $g|a^2$, and thus $g|[a^2 - (a^2 - b^2)] = b^2$. Hence, $g|b$, which contradicts the fact that $\gcd(a, b) = 1$. A similar argument handles the case if initially $g|b$. Therefore, $\gcd(2ab, a^2 - b^2) = 1$.

From this we can conclude that either $2ab < a^2 - b^2$ or $a^2 - b^2 < 2ab$ (see the next paragraph for examples illustrating these possibilities). Let m be the largest and n the smallest of these two numbers, and observe that one is even and the other is odd. (Why?) Finally, $z = m^2 + n^2$, where m and n satisfy the conditions of Theorem 5.7.2.

In part a, if $z = 25 = (2^2 + 1^2)(2^2 + 1^2)$, then $2ab > a^2 - b^2$ $(4 > 3)$; whereas, if $z = 17 \cdot 17 = (4^2 + 1^2)(4^2 + 1^2)$, then $a^2 - b^2 > 2ab$ $(15 > 8)$.

b. Assume $p \neq q$. Then,

$$z = pq = (a^2 + b^2)(c^2 + d^2) = (ad + bc)^2 + (ac - bd)^2.$$

As above, we claim that $\gcd(ad + bc, ac - bd) = 1$. Let g be a *prime* divisor of $ad + bc$ and $ac - bd$. Hence, $g|[(ad + bc)^2 + (ac - bd)^2] = pq$, so $g = p$ or $g = q$. Say $g = q$. Note that

$$g|[a(ad + bc) - b(ac - bd)] = (a^2d + b^2d) = (a^2 + b^2)d,$$

so $g|(a^2 + b^2)$ or $g|d$.

If $g|d$, then by assumption, $g|(ad + bc)$, so $g|[(ad + bc) - ad] = bc$. Therefore, $g|b$ or $g|c$, and since $\gcd(c, d) = 1$, then $g|b$. (Why?) Thus, $g|[(ac - bd) + bd] = ac$, which gives that $g|a$ or $g|c$.

We have observed that g does not divide c, and since $\gcd(a, b) = 1$, then g cannot divide a. This contradiction implies that g does not divide d and $g|(a^2 + b^2) = p$. This can't be, since p and q are distinct primes and therefore $g \neq q$. A similar argument shows that $g \neq p$, and therefore $\gcd(ad + bc, ac - bd) = 1$.

Conclusion. Although we have not completely answered Question 5.9.2 (which *odd* positive integers z can arise as the length of the hypotenuse of a primitive Pythagorean triangle, and how can such z be expressed as $z = m^2 + n^2$, where m and n are positive integers satisfying the conditions of Theorem 5.7.2?), we have uncovered several important components that are essential to the total characterization. The following two theorems (whose proofs we do not include), which build on the foundation we have developed, completely solve the problem.

5.9.3 Theorem. An odd prime p can be expressed as the sum of two squares if and only if $p = 4t + 1$.

5.9.4 Theorem: Finding the Hypotenuse of a Primitive Pythagorean Triangle. A positive integer z can be expressed as $z = m^2 + n^2$, where m and n are positive integers satisfying the conditions of Theorem 5.7.2 if and only if each prime factor of z is of the form $4t + 1$. In addition, there is a general procedure for explicitly finding the m and n.

Question. Which positive integers P can arise as the perimeter of a primitive Pythagorean triangle? Given such P, how can the sides $x, y,$ and z be determined?

Before considering a few specific examples, it is advantageous to consider some necessary conditions. In fact, such conditions will steer us away from unnecessary computations and will substantially reduce the amount of work needed to complete the job at hand.

Necessary Conditions for a Positive Integer P to Be the Perimeter of a Primitive Pythagorean Triangle. If $P = x + y + z$, where $\{x, y, z\}$ is a PPT, then

$$P = (m^2 - n^2) + 2mn + (m^2 + n^2) = 2m(m + n)$$

for some integers m and n, satisfying the conditions of Theorem 5.7.2. In this circumstance,

a. P is even;
b. $m < (m + n) < 2m$ (why?);
c. $\gcd(m + n, m) = 1$ (why?); and
d. $(m + n)$ is odd (why?).

Example.

1. With the previous constraints in mind, let's find all PPTs $\{x, y, z\}$ (or show there are none) that have $P = x + y + z = 90$. To accomplish this task, we must find all pairs of positive integers m and n that satisfy the conditions of Theorem 5.7.2 and that have the property that $90 = 2m(m + n)$, i.e., $45 = m(m + n)$.

   The only possible ways to express 45 as the product of two distinct positive integers are $45 = 45 \cdot 1 = 15 \cdot 3 = 9 \cdot 5$, and it remains to determine which of these factorizations (if any) give rise to PPTs.

   Notice that each of the factorizations $45 = (m + n)m = 45 \cdot 1 = 15 \cdot 3$ satisfy the necessary conditions of a, c, and d but fail for b. (Verify.) Hence, these factorizations do not give rise to pairs of positive integers m and n that satisfy the conditions of Theorem 5.7.2.

246   Chapter 5   Algebraic Modeling in Geometry: The Pythagorean Theorem

The factorization $45 = 9 \cdot 5 = (m + n)m$ is consistent with all of the listed necessary conditions and implies that $m = 5$ and $n = 4$. These values produce the PPT $\{9, 40, 41\}$ that has the perimeter $P = 90$.

2. Again, let's discover all PPTs $\{x, y, z\}$ (or show there are none) that have $P = x + y + z = 204$.

The factorizations of $P/2 = 102$ as the product of two distinct positive integers are $102 = 102 \cdot 1 = 51 \cdot 2 = 34 \cdot 3 = 17 \cdot 6$. It follows that none of these satisfy all of the necessary conditions previously detailed (verify), so there are no primitive Pythagorean triangles having perimeter $P = 204$.

Question.   Are the listed necessary conditions also sufficient conditions? The next result answers this in the affirmative.

5.9.5 Characterizing Perimeters of Primitive Pythagorean Triangles. A positive integer P can arise as the perimeter of a primitive Pythagorean triangle if and only if P is even and $P/2 = rs$, where r is odd, $s < r < 2s$, and $\gcd(r, s) = 1$. If such a factorization is possible, then positive integers m and n that satisfy the conditions of Theorem 5.7.2 can be determined by setting $r = m + n$ and $s = m$, and the PPT $\{x, y, z\}$ generated by m and n satisfies $x + y + z = P$.

Justification.   ($\Rightarrow$) The necessary conditions determined earlier handle this direction.

($\Leftarrow$) Assume that P is even and factors as specified. We claim that the integers $m = s$ and $n = (r - s)$ are positive integers satisfying the conditions of Theorem 5.7.2. First note that m and n are positive integers, since $r > s$; secondly, observe that $m > n$, as $r < 2s$. (Why?) Also, since $r = m + n$ is odd, then of m and n, one is even and the other is odd. (Why?) Finally, $\gcd(m, n) = 1$, since $\gcd(r, s) = 1$. (Why?) Therefore, $P/2 = rs = m(m + n)$, i.e., $P = 2m(m + n) = x + y + z$.

Some numerical examples further illustrate the process described in the preceding characterization. The following table contains some even positive integers P that factor as $P/2 = rs$, where r is odd, $s < r < 2s$, and $\gcd(r, s) = 1$.

$P/2$	r	s	$m = s$	$n = (r - s)$	$x = m^2 - n^2$	$y = 2mn$	$z = m^2 + n^2$
6	3	2	2	1	3	4	5
15	5	3	3	2	5	12	13
20	5	4	4	1	15	8	17
28	7	4	4	3	7	24	25
35	7	5	5	2	21	20	29
42	7	6	6	1	35	12	37
45	9	5	5	4	9	40	41
63	9	7	7	2	45	28	53
72	9	8	8	1	63	16	65

Section 5.9   Additional Questions Concerning Pythagorean Triples   247

Question.   Which positive integers A can occur as the area of a primitive Pythagorean triangle? Given such A, how can the sides x, y, and z be determined?

We first show that A must be a multiple of 6.

A Necessary Condition on the Area of a Primitive Pythagorean Triangle.   If $\{x, y, z\}$ is a PPT, then xy is a multiple of 12, so the area A of a primitive Pythagorean triangle must be a multiple of 6.

Justification.   We may write $xy = (m^2 - n^2)2mn$, where m and n are positive integers that satisfy the conditions of Theorem 5.7.2. Since we know that $4|y$, it is sufficient to show $3|xy$. (Why?)

If either m or n is divisible by 3, then $3|xy$, so suppose that 3 does not divide m and n. By Fermat's Little Theorem (Theorem 4.8.2), we know that

$$[m]^2 = [1] \text{ and } [n]^2 = [1] \text{ in } \mathbf{Z}_3.$$

Hence, $[m^2 - n^2] = [0]$ in $\mathbf{Z}_3$ (i.e., $3|y$, so $3|xy$). Therefore, $12|xy$, so $6|A$, where $A = xy/2$.

Notice that if $\{x, y, z\}$ is a PPT, then $3|x$ or $3|y$ (proof of previous result).

Corollary: Necessary Condition on the Product of the Sides of a Primitive Pythagorean Triangle.   If $\{x, y, z\}$ is a PPT, then xyz is a multiple of 60.

Justification.   In view of the necessary condition on the area of a primitive Pythagorean triangle, it is enough to show that $5|xyz$. (Why?) We may assume 5 does not divide m and n, otherwise we would have the desired conclusion. (Why?) Again, by Fermat's Little Theorem (Theorem 4.8.2),

$$[m]^4 = [0] \text{ and } [n]^4 = 0 \text{ in } \mathbf{Z}_5;$$

consequently, $5|(m^4 - n^4)$. Hence, $5|xyz$, since $xz = m^4 - n^4$.

Question.   Do there exist distinct PPTs $\{x, y, z\}$ and $\{u, v, w\}$ such that $xyz = uvw$? Equivalently, do there exist pairs of positive integers m, n, and r, s that satisfy the conditions of Theorem 5.7.2 so that

$$(m^2 - n^2)(2mn)(m^2 + n^2) = (r^2 - s^2)(2rs)(r^2 + s^2),$$

i.e.,

$$mn(m^4 - n^4) = rs(r^4 - s^4)?$$

It is somewhat surprising that an answer to this question is not yet known by anyone!

Necessary Conditions for a Positive Integer A to Be the Area of a PPT.   If $A = xy/2$, where $\{x, y, z\}$ is a PPT, then

$$A = mn(m^2 - n^2) = mn(m + n)(m - n)$$

for some positive integers m and n that satisfy the conditions of Theorem 5.7.2. In this context,

a. $6|A$;
b. $\gcd(m,n) = \gcd(m, m - n) = \gcd(m, m + n) = \gcd(n, m - n) = \gcd(n, m + n) = \gcd(m - n, m + n) = 1$ (why?);
c. $m + n$ and $m - n$ are odd (why?); and
d. if $A > 6$, then $m + n, m, n, m - n$ are distinct.

To see why condition d is a necessary condition, note that since $m > n$, then the only possible terms that be could be equal are n and $m - n$. If $n = m - n$, then $m = 2n$. Furthermore, since $\gcd(m,n) = 1$, then $n = 1, m = 2$, and $A = 6$. (Why?)

Examples.

1. With the above necessary conditions in mind, let's find all PPTs $\{x, y, z\}$ with $A = xy/2 = 60$ (or show that there are none), i.e., determine all pairs of positive integers m and n that satisfy the conditions of Theorem 5.7.2 such that $60 = mn(m + n)(m - n)$.

   The only possible ways to express 60 as the product of four distinct positive integers are $60 = 10 \cdot 3 \cdot 2 \cdot 1 = 5 \cdot 4 \cdot 3 \cdot 1 = 6 \cdot 5 \cdot 2 \cdot 1$. The first and last factorizations can be eliminated, since $m + n$ must be odd and thus can't equal 10 or 6. The factorization $60 = 5 \cdot 4 \cdot 3 \cdot 1$ does not violate any of the necessary conditions, and thus might lead to a possible solution. In fact, $m = 4, n = 1$ produces the given factorization and leads to the PPT $\{15, 8, 17\}$.

2. For this example, let's find all PPTs $\{x, y, z\}$ with $A = xy/2 = 100$ (or show that there are none), i.e., ascertain all pairs of positive integers m and n that satisfy the conditions of Theorem 5.7.2 and that have the property that $100 = mn(m + n)(m - n)$.

   There is only one way to write 100 as the product of four distinct positive integers, namely $100 = 10 \cdot 5 \cdot 2 \cdot 1$. This is incompatible with the necessary condition b or c. Therefore, there are no primitive Pythagorean triangles with area $A = 100$.

5.9.6 Characterizing Areas of Primitive Pythagorean Triangles. A positive integer A can arise as the area of a primitive Pythagorean triangle if and only if A is divisible by 6 and if $A = rsuv$, where r, s, u, and v are positive integers satisfying:

a. $\gcd(r,s) = \gcd(r,u) = \gcd(r,v) = \gcd(s,u) = \gcd(s,v) = \gcd(u,v) = 1$;
b. r is greater than $s, u,$ and v;
c. $r, s,$ and u are odd and v is even;
d. $(r + s)/2$ equals either u or v; and
e. $(r - s)/2 = u$ if $(r + s)/2 = v$ or $(r - s)/2 = v$ if $(r + s)/2 = u$.

Section 5.9   Additional Questions Concerning Pythagorean Triples   249

If such a factorization is possible, then positive integers m and n that satisfy the conditions of Theorem 5.7.2 can be determined by setting $r = m + n$ and $s = m - n$, and the PPT $\{x, y, z\}$ generated by m and n satisfies $xy/2 = A$.

Justification.   ($\Rightarrow$) This part is handled by the previous discussion of necessary conditions.

($\Leftarrow$) Assume A is a multiple of 6 and factors as specified in parts a, ..., e. The rational numbers $m = (r + s)/2$ and $n = (r - s)/2$ are positive integers that satisfy the conditions of Theorem 5.7.2. To see this, notice that $m > n$ are positive integers, since $r > s$ are odd positive integers. Furthermore, $\gcd(m, n) = 1$ as $\gcd(r, s) = 1$. (Why?) Also, of the integers m and n, one is even and the other is odd (parts c, d, and e). (Why?) Therefore, $A = rsuv = (m + n)(m - n)mn = xy/2$.

Most of condition a is redundant. More specifically, if $\gcd(r, s) = 1$ and conditions d and e are satisfied, then the remaining parts of a are true. For example, we'll show $\gcd(r, u) = 1$. Suppose $\gcd(r, u) = d$, so $d|r$ and $d|u$. Consider the case when $u = (r + s)/2$. (The case when $u = (r - s)/2$ is similar.) It follows that $d|[2r - (r + s)]/2$ (why?), and thus $d|(r - s)/2$. Hence, $d|[(r + s)/2 - (r - s)/2]$, which gives $d|s$. Therefore, $d = 1$.

Example.   Some numerical examples help illustrate the process described in the characterization of result 5.9.6. The following table contains some positive multiples of 6 that can be factored as factor $A = rsuv$, where r, s, u, and v satisfy the conditions of characterization 5.9.6.

A	r	s	u	v	$m = (r + s)/2$	$n = (r - s)/2$	x	y	z
6	3	1	1	2	2	1	3	4	5
30	5	1	3	2	3	2	5	12	13
60	5	3	1	4	4	1	15	8	17
84	7	1	3	4	4	3	7	24	25
180	9	1	4	5	5	4	9	40	41
210	7	3	5	2	5	2	21	20	29
210	7	5	1	6	6	1	35	12	57

EXERCISES 5.9

1. Find all PPTs $\{x, y, z\}$ with $x =$

   a. 75
   b. 289
   c. 313
   d. 1,155

2. Recall from Question 5.7.3 that we further investigated the PPTs discovered by Pythagoras and characterized them as those PPTs $\{x, y, z\}$, where $z = y + 1$.

Show that if $\{x, y, z\}$ is a PPT, where x is an odd prime number, then $\{x, y, z\}$ is a PPT of the form Pythagoras discovered. Direction: Apply the factorization used in Summary 5.9.1 to an odd prime number x.

3. Find all PPTs $\{x, y, z\}$, or say why there are none, with $z =$

   a. 61
   b. 43
   c. 145
   d. 4,199

4. Determine all PPTs $\{x, y, z\}$ (or show that there are none) with $P = x + y + z$ for the following Ps:

   a. 300
   b. 154
   c. 140
   d. 1,260

5. Show that $P/2$ (1/2 of the perimeter of a primitive Pythagorean triangle) can never be a prime number.

6. Give an example of two distinct primitive Pythagorean triangles that have the same perimeter. Direction: Consider $P/2 = 17{,}160 = 2^3 \cdot 3 \cdot 5 \cdot 11 \cdot 13$.

7. Find an example of a positive integer $A \neq 210$ that is the area of exactly two different primitive Pythagorean triangles.

8. Find three different primitive Pythagorean triangles, each having area $A = 13{,}123{,}110 = 2 \cdot 3 \cdot 5 \cdot 7 \cdot 11 \cdot 13 \cdot 19 \cdot 23$.

9. Show that a primitive Pythagorean triangle has its area equal to its perimeter if and only if it is a 5, 12, 13 right triangle.

10. Let a, b, and c be positive integers and suppose that a^2, b^2, c^2 form an arithmetic sequence with common difference d. Show that $24|d$. Direction: We will sketch a proof in several steps and leave most of the details up to you.

    Step 1: $c^2 - a^2$ is even $[c^2 - a^2 = (c^2 - b^2) + (b^2 - a^2) = 2(c^2 - b^2)]$.
    Step 2: Since $(c^2 - a^2) = (c + a)(c - a)$, it follows that $(c + a)$ and $(c - a)$ are both even. Direction: (Indirect) Since $(c + a)$ and $(c - a)$ cannot both be odd, then assume one is even and the other is odd and show that this leads to a contradiction.
    Step 3: Set $r = (c + a)/2$ and $s = (c - a)/2$. Then, $r + s = c$ and $r - s = a$, so $c^2 = r^2 + 2rs + s^2$ and $a^2 = r^2 - 2rs + s^2$. Thus, $((c^2 + a^2)/2 = r^2 + s^2)$.
    Step 4: Since $(c^2 - a^2) = 2(c^2 - b^2) = 2c^2 - 2b^2$ (Step 1), then $(c^2 + a^2)/2 = b^2$, so $r^2 + s^2 = b^2$. Therefore, $\{r, s, b\}$ is a PT.
    Step 5: The common difference of the arithmetic sequence a^2, b^2, c^2 equals $(c^2 - b^2) = 2rs$, since $c^2 = r^2 + 2rs + s^2$ and $b^2 = r^2 + s^2$.
    Step 6: We know from Theorem 5.7.2 that $4|r$ or $4|s$, so $8|2rs$. Also, we have shown that $3|rs$ (see Section 5.9, a necessary condition on the area of a primitive Pythagorean triangle), so $3|r$ or $3|s$. Therefore, $24|rs$, i.e., $24|(c^2 - b^2)$, and the proof is now complete.

5.10 FERMAT'S LAST THEOREM

In Section 5.7 we found all positive integer solutions to the equation $x^2 + y^2 = z^2$. Analogously, it is reasonable to attempt to determine all positive integer solutions for $x^3 + y^3 = z^3$, $x^4 + y^4 = z^4$, and in general for $x^n + y^n = z^n$, where $n > 2$. A much less ambitious project might be to find examples of specific solutions to some of these equations. Some computational experimentation quickly shows that this task is much more difficult than finding PTs.

In the seventeenth century, a French lawyer named Pierre de Fermat (1601–1665), who loved thinking about mathematics in his leisure and happened to have a brilliant mathematical mind, became interested in these questions and inadvertently altered the course of mathematical history. While studying Bachet's translation (from Greek to Latin) of Diophantus's *Arithmetic*, Fermat wrote several comments in the margins of his book. Specifically, in the margin adjacent to the problem pertaining to representing a square as the sum of two squares (Pythagorean triples), Fermat noted that it was impossible to write a cube as a sum of two cubes, to write a fourth power as the sum of two fourth powers, and in general to write an nth power as the sum of two nth powers for all $n > 2$. He indicated that he had found a remarkable proof of this fact but that the margin was too narrow to include it.

Fermat's claim that the equation $x^n + y^n = z^n$ has no positive integer solutions for all integers $n > 2$ is called **Fermat's Last Theorem.** Although the source of this nomenclature is not explicitly known, it does not refer to Fermat's last mathematical effort, but rather to the fact that this was the last enduring statement of his mathematical work that had yet to be proved or refuted. In fact, for more than three centuries after Fermat's death, the discovery of a valid proof had escaped generations of eminent mathematicians and innumerable mathematical amateurs. Although many special cases (specific exponents) had been verified, and countless purported proofs periodically announced, a complete solution seemed intractable.

It is worth noting that if $x^d + y^d = z^d$ has no positive integer solutions for some $d > 2$, then $x^{cd} + y^{cd} = z^{cd}$ has no positive integer solutions for $c \geq 1$. (Why?) In particular, Fermat showed that $x^4 + y^4 = z^4$ has no positive integer solutions, and thus neither does $x^{4c} + y^{4c} = z^{4c}$ for $c \geq 1$. Therefore, since any integer $n > 2$ is divisible by either 4 or an odd prime, to prove Fermat's Last Theorem, it suffices to prove it for all odd prime numbers p.

The nineteenth century was an especially active and fruitful period in the history of Fermat's Last Theorem. A particularly compelling story occurred in March of 1847, at the meeting of the Paris Academy of Sciences, where Lamé (a French mathematician) announced that he had found a proof of Fermat's Last Theorem. His "proof" was based on a method used in proofs of other known cases (e.g., $n = 3, 4, 5, 7$) and involved factoring $x^n + y^n$ (where n is an odd prime) into lower-degree terms, and in his case into linear terms. This could be accomplished by using a complex number $r \neq 1$, where $r^n = 1$ and writing $x^n + y^n = (x + y)(x + ry)(x + r^2 y) \cdots (x + r^{n-1} y) = z^n$. Using this factorization, Lamé reasoned that if the factors were relatively prime, then each one of the factors must be an nth power. If they were not relatively prime, then he would divide the factors by their greatest common divisor and proceed as in the initial relatively prime case. (Recall in the PPT case that we used the Fundamental Theorem of Arithmetic to conclude that if a product of relatively prime numbers was a square, then the numbers themselves must be squares; see Section 5.7, Observation 5). The whole point of this process was that if $\{x, y, z\}$ is a solution, then this method would produce a "smaller solution," and this would lead to a contradiction via Fermat's method of infinite descent (solutions cannot get smaller indefinitely).

Lamé attributed this fundamental idea of his proof to Liouville (a French mathematician), but Liouville dismissed recognition for himself and indicated that several other well-known mathematicians had used this same idea in their work. In connection to this idea, he noted that Lamé's proof had a serious problem, Lamé had concluded that if a product of relatively prime numbers in his new number system was an nth power, then each factor must be an nth power.

In May of 1847, Liouville received a letter from Kummer (a German mathematician) indicating that Liouville's suspicions concerning Lamé's proof were correct. In fact, 3 years earlier, Kummer had published a manuscript that addressed this point exactly. Namely, Lamé had implicitly used unique factorization (a critical part of the Fundamental Theorem of Arithmetic) in his proof, and it turns out that this property does not always hold in other number systems. More importantly, Kummer had introduced a new type of complex number (cyclotomic integers) to regain some form of unique factorization, and with these he was able to establish the validity of Fermat's Last Theorem for a large class of prime exponents. (The events previously described are an abbreviated version of Section 4.1, **The events of 1847**, from Harold Edwards's outstanding book, *Fermat's Last Theorem: A Genetic Introduction to Algebraic Number Theory* [Ed].)

Would the twentieth century see a solution, and if not, how many more centuries would pass without one? What the world did not know but found out in 1993 was that Andrew Wiles, a mathematician at Princeton University, had been developing a proof of Fermat's Last Theorem for about 7 years. In June of 1993, while giving the third lecture of a three-part lecture series at the Issac Newton Institute in Cambridge, England, Wiles ended his presentation by noting that, as a consequence of his work, Fermat's Last Theorem was now proved.

Unfortunately, the "last theorem" saga did not end there. During the reviewing process, a serious gap in the proof was discovered, and Wiles worked very hard to repair this hole. He solicited help from Richard Taylor, a former student, and together they attempted to mend the proof. In October of 1994, they announced their success and distributed two manuscripts containing the completed proof. In May of 1995, the papers "Modular Elliptic Curves and Fermat's Last Theorem" by Andrew Wiles and "Ring Theoretic Properties of Certain Hecke Algebras" by Andrew Wiles and Richard Taylor appeared in the *Annals of Mathematics*. At last Fermat's Last Theorem is really a theorem!

Arithmetic and Algebra of Matrices

CHAPTER 6

6.1 CLASSROOM CONNECTIONS: SYSTEMS OF LINEAR EQUATIONS
6.2 REFLECTIONS ON CLASSROOM CONNECTIONS: SYSTEMS OF LINEAR EQUATIONS
6.3 RATIONAL AND IRRATIONAL NUMBERS
6.4 SYSTEMS OF LINEAR EQUATIONS
6.5 POLYNOMIAL CURVE FITTING: AN APPLICATION OF SYSTEMS OF LINEAR EQUATIONS
6.6 MATRIX ARITHMETIC AND MATRIX ALGEBRA
6.7 MULTIPLICATIVE INVERSES: SOLVING THE MATRIX EQUATION $AX = B$
6.8 CODING WITH MATRICES

In Chapters 1 through 4, we introduced and investigated many new ideas and tools concerning the integers and the related system of integers modulo n. Our studies in Chapter 5 applied this work to some interesting geometrical considerations, and in this present chapter, we continue to advance our knowledge through the exploration of additional mathematical questions.

Golf and Mathematics. There are days when even the most ardent golfers would agree with Mark Twain's assessment that "Golf is a good walk spoiled." The game of golf is a difficult and often frustrating sport, but it is still my number-one recreational diversion. The beauty and tranquility of the golf course provides a wonderful setting to clear my mind of everyday matters (and yes, that includes mathematical theorems) and, as Winston Churchill put it, to "... hit a very small ball into an even smaller hole with weapons singularly ill-designed for the purpose."

During a typical round of golf (usually eighteen holes), most comments and discussions involve one and only one topic: golf! I was totally surprised one recent Sunday morning when a fellow golfer in my foursome (a group of four golfers) somewhat broke tradition and asked me a golf-related mathematical question. Usually my mathematical expertise is only called upon when it is time to add scorecards and divide wagers. He said quite proudly, "Last week I played eighteen holes and shot a seventy-eight, one of my best rounds. Unfortunately, I lost the scorecard, but I do remember only making fours or fives on any one hole. Is it possible to figure out how many holes I played in four strokes and how many I played in five strokes?"

Before I had a chance to respond, another golfer in our group said that it would not be possible to determine this answer, furthermore, he questioned whether the 78 might actually be an 88 (golfers, like fisherman, tend to exaggerate). A situation like this, where golf egos are involved, can quickly escalate out of control, so I immediately asked if we could concentrate on today's game and continue this discussion in the clubhouse. After a bit of grumbling (one of them insisted I could not figure out the answer and that I was merely stalling), both agreed to play on and to settle this afterward.

Not a word was spoken of the infamous problem until the round was complete and all wagers had been paid. At this time, I asked for a clean scorecard and a fresh pencil. I said, "I hope you boys are ready for a little algebra, because this is precisely what is needed to solve the problem." As I started to write some equations on the blank scorecard, I could sense that my friends were becoming a bit uneasy. "Don't be frightened, algebra is your friend," I said, but this proclamation did not seem to diminish their anxiety. A few algebraic manipulations stimulated one of them to announce that he sort of remembered doing this "stuff" in high school but never thought it ever had any actual use in everyday life.

Rather than lecturing my friends on the beauty and utility of algebra, I simply said to them, "Six of the holes were played in five strokes each, and the remaining twelve holes were played in four strokes each." To convince them of this answer, I showed them the computation $(5 \cdot 6) + (4 \cdot 12) = 30 + 48 = 78$ and then told them that I'd be happy to explain how I arrived at this answer. In perfect unison, they shouted, "NO WAY!" and hastily disappeared from the clubhouse.

Question. How can algebra be used to arrive at the answer I gave my golfing companions, and is this the only solution to the problem?

6.1 CLASSROOM CONNECTIONS: SYSTEMS OF LINEAR EQUATIONS

The golfing problem could have been solved by trial and error, but it was much more efficient to tackle the problem using algebra. Moreover, an algebraic approach has the added benefit of showing that there is one and only one solution to the problem. In some similar simple problems, it might be easier to reach a solution by methods other than algebraic ones (e.g., guess and check), however, this is not a reliable method for more complicated situations.

We begin our work in this section by looking at some related problems from the sixth grade unit *Comparing Quantities* from the middle-grade curriculum *Mathematics in Context* (see Figures 6.1.1 to 6.1.3 on pages 255–257). The students confronted with these problems would not yet have formal algebraic methods to call upon but will be developing such methods in the process of solving the problems. It is especially important that teachers understand various student solutions and be able to clearly compare them and discuss their relative merits.

Section 6.1   Classroom Connections: Systems of Linear Equations   255

Furthermore, the ultimate goal is to incorporate some of the student solution strategies into the development of a general algebraic framework for solving problems of this type.

We initially approach these problems as middle-school students might. Later we interpret the solutions in a general algebraic context. The teacher's edition of this unit (some sample pages are included at the end of Section 6.2) discusses several different strategies used by students to solve these problems and is an extremely helpful resource for teachers first using the materials.

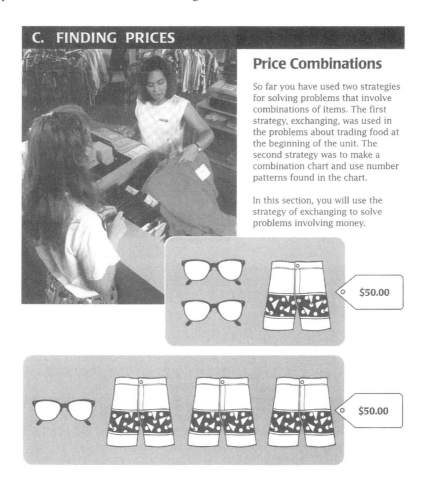

1. Without knowing the prices of a pair of glasses or a pair of shorts, you can determine which item is more expensive. Explain how.

2. How many pairs of shorts can you buy for $50?

3. What is the price of one pair of glasses? Explain your reasoning.

Reproduced from page 15 of *Comparing Quantities* in *Mathematics in Context*.

FIGURE 6.1.1

256 Chapter 6 Arithmetic and Algebra of Matrices

4. Which is more expensive, a cap or an umbrella? How much more expensive is it?

5. Use the two pictures above to make a new combination of umbrellas and caps. Write down the cost of the combination.

6. Make a group of only caps or only umbrellas. Then find its price.

7. What is the price of one umbrella? one cap?

Reproduced from page 16 of *Comparing Quantities* in *Mathematics in Context*.

FIGURE 6.1.2

Section 6.2 Reflections on Classroom Connections: Systems of Linear Equations 257

Sean bought two T-shirts and one sweatshirt for a total of $30. When he got home, he regretted his purchase. He decided to exchange one T-shirt for another sweatshirt.

Sean was able to do this, but he had to pay $6 more because the sweatshirt was more expensive than the T-shirt.

8. What was the price of each item? Explain your reasoning.

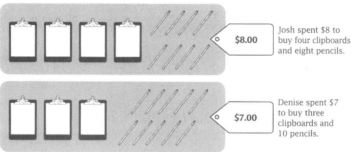

Josh spent $8 to buy four clipboards and eight pencils.

Denise spent $7 to buy three clipboards and 10 pencils.

Denise wants to trade Josh two pencils for a clipboard.

9. Is that a fair exchange? If not, who has to pay the difference, and how much is it?

10. What is the price of a pencil? What is the price of a clipboard? Explain your reasoning.

Reproduced from page 17 of *Comparing Quantities* in *Mathematics in Context*.

FIGURE 6.1.3

6.2 REFLECTIONS ON CLASSROOM CONNECTIONS: SYSTEMS OF LINEAR EQUATIONS

What responses would middle-grade students give for Problems 1, 2, and 3? Guess and check is always a strategy for some students, but try to think of other strategies that sixth graders might use in solving these problems. Keep in mind that most sixth grade students would not approach these problems with formal algebraic equations, but in fact many will use algebraic methods without explicitly knowing it.

Problem 1 (in Figure 6.1.1). Without knowing the prices of a pair of glasses or a pair of shorts, determine which item is more expensive. Explain your answer.

Understanding the Meaning of the Pictures. Before attempting to solve the problems, it is important to make sure that students understand the meaning of the price-combination pictures. To accomplish this, have students explain aloud or in writing what they think the pictures' statements represent.

Answer. Glasses cost more than shorts.

Justification. Since two pairs of glasses and one pair of shorts costs the same as one pair of glasses and three pairs of shorts, a pair of glasses and a pair of shorts can be removed from each picture and the resulting items must have the same price, i.e., a pair of glasses is the same price as two pairs of shorts (students often use their hands or a piece of paper to cover up a pair of glasses and a pair of shorts from each picture). Therefore, glasses cost more than shorts.

Problem 2 (in Figure 6.1.1). How many pairs of shorts can you buy for $50?

Answer. Five pairs.

Justification. In Problem 1, it was concluded that a pair of glasses and two pairs of shorts were the same price, so by exchanging in either picture two pairs of shorts for each pair of glasses, we see that five pairs of shorts can be purchased for $50.

Problem 3 (in Figure 6.1.1). What is the price of one pair of glasses? Explain your reasoning.

Answer. $20.

Justification. We know from Problem 2 that each pair of shorts costs $10. (Why?) Since one pair of glasses and three pairs of shorts cost $50 (second picture), then a pair of glasses must cost $20. Employing similar reasoning, the same conclusion can be reached by using the first picture and concluding that two pairs of glasses must cost $40 (middle-grade students usually communicate these same solutions by using pictures or their own words).

The exchanging strategies used in the previous problems are quite effective, and we employ them in the next sequence of problems. However, the students must slightly modify their thinking to handle these problems, since the total prices in each picture are not the same (as they were in Problems 1–3).

Problem 4 (in Figure 6.1.2). Which is more expensive, a cap or an umbrella, and by how much?

Answer. An umbrella costs $4 more than a cap.

Justification. Since two umbrellas and one cap costs $4 more than one umbrella and two caps, the removal of an umbrella and a cap from each picture shows that an umbrella must be $4 more than a cap (again, students frequently use their hands or a piece of paper to cover up an umbrella and a cap from each picture).

Problem 5 (in Figure 6.1.2). Use the given pictures to make a new combination of the umbrellas and caps. Write down the cost of the combination.

Some Answers.

   a. Doubling the first picture: four umbrellas and two caps costing $160.

Section 6.2   Reflections on Classroom Connections: Systems of Linear Equations   259

   b. Doubling the second picture: two umbrellas and four caps costing $152.
   c. Adding the original combinations: three umbrellas and three caps costing $156.

Problem 6 (in Figure 6.1.2).   Make a group of only caps or umbrellas and then find its price.

Some Answers.

   a. Exchanging an umbrella for a cap in the first picture and increasing the price by $4 to $84 gives a group with three umbrellas.
   b. Similarly, swapping a cap for an umbrella in the second picture and reducing the price by $4 to $72 results in a combination of three caps.

Problem 7 (in Figure 6.1.2).   What is the price of one umbrella and one cap?

Answer.   A cap costs $24, and an umbrella costs $28.

Justification.   In Problem 6 it was determined that three umbrellas cost $84, so one umbrella costs $28. Since an umbrella costs $4 more than a cap, a cap must cost $24. Similarly, since three caps cost $72, one costs $24, and thus an umbrella costs $28.

Problems 1–7 were designed to stimulate algebraic thinking and to encourage students to develop problem-solving strategies other than "guess and check." The picture formatting of the problems is very helpful for students in constructing their solutions. However, some students prefer to develop their own shorthand notations rather than drawing additional pictures.

Using words or letters to represent the pictures is common, and it is important to make sure that the students' representations make sense. For example, a student may let U stand for umbrella and C for cap and then write $2U + C = \$80$. At this point, it is crucial that the teacher ask the student to explain his or her notation and to have him or her realize that their descriptions for U and C need modification. In particular, when U stands for the *price* of an umbrella and C stands for the *price* of a cap, then the statement $2U + C = \$80$ is meaningful and does correspond to the cost combination given in the first picture.

Problem 8 is similar in nature to the previous problems, except that it is presented without pictures and is stated entirely in words. This format is intended to encourage students to find their own representations of these problems and then use them to find solutions. Moreover, these student representations and solution strategies lead naturally to a general algebraic framework involving systems of linear equations and a highly efficient algorithm for solving them.

Problem 8 (in Figure 6.1.3).   What price did Sean pay for a T-shirt and for a sweatshirt? Explain your reasoning.

Modeling the Problem with Pictures or Letters.   Since this problem was not presented with accompanying pictures as the other problems were, some students draw their own pictures, and others use letters to represent the problems details.

As before, it is vital to stress to students the importance of accurately defining their symbols so that they convey the correct meaning when translating the problems details into symbolic form. For example, using T to represent the price of a T-shirt and S to represent the price of a sweatshirt, the student can then express in symbolic form the fact that Sean bought two T-shirts and one sweatshirt for $30. Namely,

$$2T + 1S = \$30.$$

After arriving home, Sean decided he wanted to exchange one T-shirt for an additional sweatshirt; however, the cost of a sweatshirt was $6 more than the cost of a T-shirt. This cost differential can be expressed as

$$1S = 1T + \$6.$$

The expression $1T + \$6$ (cost of one T-shirt plus $6) can now be exchanged for the expression $1S$ (cost of one sweatshirt) in the original cost combination equality to give

$$3T + 6 = \$30, \text{ or equivalently, } 3T = \$24.$$

Therefore, $T = \$8$ and $S = \$14$.

In the next section, we view the exchanging strategies used in the previous problems in terms of algebraic operations (i.e., algebraic adjustments of the given equations that do not affect the solutions). This perspective justifies the methods used in the previous problems and provides an approach that applies to more complicated situations.

Classroom Problems. Complete Problems 9 and 10 from Figure 6.1.3 by using middle-grade strategies similar to those used in Problems 1–8. Discuss how middle-grade students might approach these problems and potential difficulties that might arise.

An important distinction between these two problems and the others is that the resulting answers are no longer integers, thus it is necessary to expand our number system beyond the integers in order to find a solution. This new complexity makes "guess and check" strategies somewhat ineffective and forces students to find other more viable approaches. ◆

Solution Strategy Summary. A common solution strategy that emerged in the previous problems involved an exchanging process that led to expressions with a single unknown quantity. Once such expressions were found, then the remaining unknown quantity was easy to determine. This elimination procedure is a simple but powerful process and can be applied in a variety of general contexts. We will investigate it further in the next section.

Sample Teacher Pages. We have included some sample pages from the Teacher's Edition of *Comparing Quantities* (see Figures 6.2.1 and 6.2.2 on pages 261–262) to further illustrate various student strategies and to stimulate discussion on the mathematical concepts underlying the lesson.

Solutions and Samples
of student work

1. Glasses are more expensive than shorts. Explanations will vary. Sample explanation:

   Covering up one pair of glasses and one pair of shorts from each picture shows that one pair of glasses costs the same as two pairs of shorts. So, a pair of glasses costs more than a pair of shorts.

2. You can buy five pairs of shorts for $50. Explanations will vary. Sample explanation:

   Since two pairs of shorts cost the same as one pair of glasses, exchange the glasses for two pairs of shorts in either picture. The result is that five pairs of shorts cost $50.

3. One pair of glasses costs $20. Explanations will vary. Sample strategy:

   Some students may determine the cost of one pair of shorts and then draw pictures.

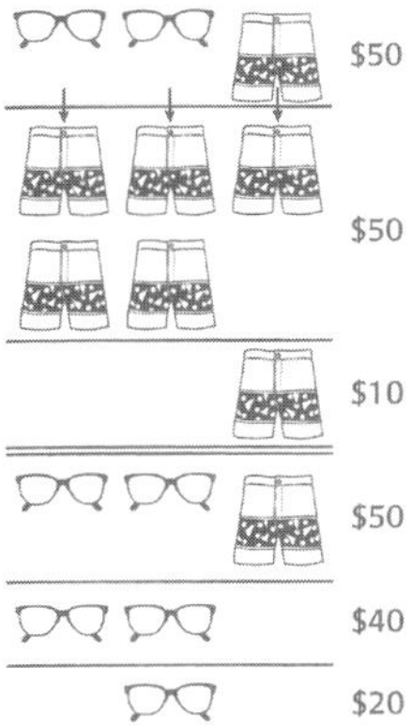

   [Diagram showing: 2 glasses + 1 shorts = $50; 5 shorts = $50; 1 shorts = $10; 2 glasses + 1 shorts = $50; 2 glasses = $40; 1 glasses = $20]

   Since five pairs of shorts cost $50, one pair of shorts costs $10. Then, using the first picture, two pairs of glasses cost $40, so one pair of glasses costs $20.

Hints and Comments

Overview Given the costs of two combinations of glasses and shorts, students use an exchanging strategy to determine the price of each item.

About the Mathematics In this section students use the costs of some combinations of items to find the price of each item. Students build on their experience with exchanging in this section. Most problems are presented in pictures. Since making drawings can be cumbersome, students may employ symbols such as letters. Make sure students understand that the letters represent the items.

Comments about the Problems

1–3. Encourage students to use exchanging to solve these problems, rather than a combination chart.

1. Tell students to try to solve the problem without finding the price of each item. If they do, ask them to try solving the problem in a different way.

2. Students may solve this problem by imagining a picture that contains only shorts. Drawing the picture is not necessary.

3. Some students may use a guess-and-check strategy. Other students may double the first picture to get four pairs of glasses and two pairs of shorts, then exchange two pairs of shorts for one pair of glasses, to end up with five pairs of glasses that total $100.

Reproduced from the Teachers Edition of *Comparing Quantities* in *Mathematics in Context*.

FIGURE 6.2.1

262 Chapter 6 Arithmetic and Algebra of Matrices

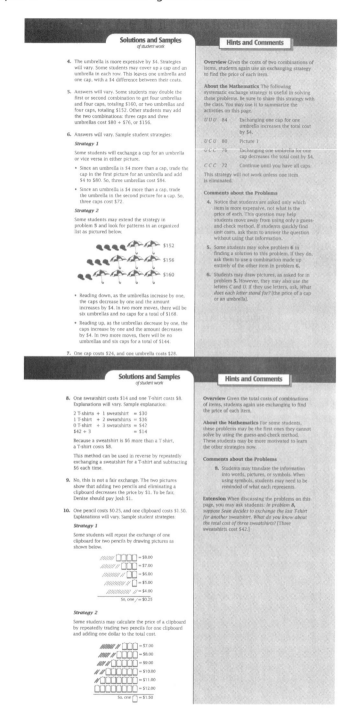

Reproduced from the Teachers Edition of *Comparing Quantities* in *Mathematics in Context*.

FIGURE 6.2.2

6.3 RATIONAL AND IRRATIONAL NUMBERS

Problems 1–10 in the previous section provided a gradual introduction to linear systems of equations and some solution strategies. These types of systems and their associated solution methods are part of a more general theory that we begin to examine in the next section. Our studies consolidate various student solution approaches and lead to a highly effective algorithm for solving such systems. This work provides us with new and deeper understandings of this fundamental subject matter. Before undertaking this study, it is necessary to expand our number universe.

Expanding Our Number Universe. The focal point of our work has been the arithmetic and algebra of the integers. The allied system of integers modulo n has been an important tool in this study, and we have thoroughly investigated its properties and applications. In some instances, our work has taken us to places where we did use numbers on the number line other than integers, but for the most part, we have concentrated on the integers. Such a sharp focus has helped us reach a deeper understanding of the integers, and to continue this effort, it is advantageous to enlarge our mathematical universe to include all the **real numbers** (i.e., the rational numbers and the irrational numbers). Working in this enriched number context affords us greater power and versatility in problem solving.

Rational Numbers. Recall that a number is called **rational** if and only if it can be expressed as the quotient $\frac{a}{b}$ of integers a and b, where $b \neq 0$. The set of all rational numbers is usually denoted by **Q**.

In terms of the number line, rational numbers correspond to division points of a specified fixed scale. For example, given the integers representing equally spaced points on the number line,

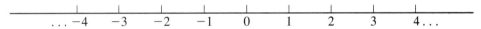

each segment can be subdivided into four congruent subsegments.

In particular, starting with a blown-up view of the subdivided segment from 0 to 1, we subdivide the segment into four congruent parts and label the numbers determined by the endpoints of the segments with the following notation.

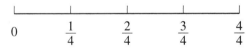

The number 1 is also represented by the symbol $\frac{4}{4}$, and to be uniform, we represent the number 0 as $\frac{0}{4}$. The notation for the numbers representing the points determined in the next subdivision of the interval between 1 and 2 would be $\frac{5}{4}, \frac{6}{4}, \frac{7}{4}, \frac{8}{4} = 2$.

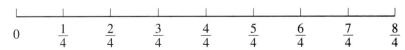

Continuing this subdivision process in the positive and negative directions produces the numbers $\frac{a}{4}$, where a is any integer. Furthermore, since there is nothing special about subdividing by 4, we could subdivide the intervals into $b \neq 0$ congruent pieces (b is a positive integer) and represent the subdivision points by numbers of the form $\frac{a}{b}$, i.e., rational numbers (ratios of integers).

Decimal Representation of Rational Numbers. We know from our previous studies (K–12 and college mathematics courses) that another important representation of rational numbers is in terms of their decimal expansions. In particular, a number is **rational** if and only if it has a repeating decimal representation (terminating decimal expansions are considered repeating, since they are zero from some finite point on).

For example, the decimal representation of the rational number $\frac{2}{7}$ is $0.\overline{285714}$ (the bar represents the repeated block of digits), which was determined by the following long division calculation:

$$
\begin{array}{r}
.285714\ldots \\
7\overline{)2.000000\ldots} \\
\underline{14} \\
60 \\
\underline{56} \\
40 \\
\underline{35} \\
50 \\
\underline{49} \\
10 \\
\underline{7} \\
30 \\
\underline{28} \\
2
\end{array}
$$

In this calculation, note (by the division algorithm) that there are only seven possible remainders, namely $0, 1, 2, \ldots, 6$, so one remainder must be repeated on or before an eighth division step. Using this same reasoning, it is easy to see why any rational number $\frac{a}{b}$ has a repeating decimal expansion. (Explain.)

Converting Repeating Decimals to Quotients of Integers. Conversely, given a repeating decimal $x = .72\overline{413}$, it is always possible to express $x = \frac{a}{b}$, for some integers a, $b \neq 0$. In fact, there is a simple procedure for accomplishing this conversion.

Step 1. Move the decimal point to the end of the first repeating block by multiplying the initial equality by an appropriate power of 10. In the example given,

$$100{,}000x = 72{,}413.413413\ldots$$

Step 2. Move the decimal point to the start of the first repeating block by multiplying the initial equality by an appropriate power of 10 (if the first block starts immediately, then multiply by $10^0 = 1$). Thus,

$$100x = 72.413413\ldots$$

Section 6.3 Rational and Irrational Numbers 265

Step 3. Subtract the second equation from the first equation. Hence,
$$99{,}900x = 72{,}413.$$
Step 4. Solve the equation obtained in Step 3 for x. Finally,
$$x = \frac{72{,}413}{99{,}900} \text{ (check this answer by using long division or a calculator).}$$

Using these steps, we see that any repeating decimal can be expressed as the quotient of integers. It is interesting to note that Steps 1, 2, and 3 involve multiplying and subtracting numbers with infinite decimal representations, and concepts from calculus (arithmetic of convergent infinite series) are needed to completely justify these calculations.

Classroom Problems.

1. Convert the following decimals into quotients of integers:

   a. $x = .\overline{35}$ **b.** $x = .444\overline{152}$ **c.** $x = .4\overline{9}$

2. Find a rational number of the form $\frac{a}{b}$, where $a, b \neq 0$ are integers, that starts repeating at the eighth digit to the right of the decimal point. ◆

The study of rational numbers and their decimal representations is a significant part of elementary and middle-school mathematics curricula. It is essential that teachers of these grade levels have a broad practical and conceptual understanding of these important numbers. This knowledge is a necessary prerequisite for these materials, and it is assumed that the reader is intimately familiar with these numbers and their arithmetic.

Irrational Numbers. We have noted that the rational numbers correspond to points on the number line that arise as division points in some positive integer subdivision of a given unit scale. To the novice number-line observer, it might seem that all points on the number line emerge this way; however, students of mathematics know differently. In fact, any nonrepeating decimal, such as
$$x = .212112111211112\ldots,$$
represents a point on the number line that is not a division point in some subdivision of a given unit scale. (Why?) The numbers corresponding to these nonrational points on the number line are called **irrational numbers.** Hence, the rational numbers are numbers having a repeating decimal expansion, and the irrational numbers are numbers with a nonrepeating decimal expansion.

As we just saw, it is easy to construct infinitely many irrational numbers using decimals (explain), but in fact the Greek mathematicians of antiquity studied irrational numbers centuries before the notion of decimal representation. In particular, Pythagoras (or one of his followers) showed that the length $c = \sqrt{2}$ of a unit square's diagonal is irrational. Even if Pythagoras had decimals to work with, a direct computational approach would not be feasible, since it would be impossible to verify through computation that $\sqrt{2}$ has a nonrepeating decimal expansion.

Realizing that calculation would not be useful in this endeavor, Pythagoras justified the assertion through pure mathematical reasoning. His elegant proof, with its stunning simplicity, vividly demonstrates the power and beauty of mathematical reasoning.

6.3.1 Theorem. $\sqrt{2}$ is irrational.

Proof **(Indirect).** Assume that $\sqrt{2}$ is rational, i.e., $\sqrt{2} = \frac{a}{b}$ for some integers $a, b \neq 0$. There is no harm in assuming that $\gcd(a,b) = 1$, because if this is not the case, we may replace $\frac{a}{b}$ with an equivalent rational number $\frac{c}{d}$, where $\gcd(c,d) = 1$.

Since $\sqrt{2} = \frac{a}{b}$, then $2 = \frac{a^2}{b^2}$, so $2b^2 = a^2$. Hence, a^2 is even, and thus a must be even, i.e., $a = 2t$ for some integer t. Whence, $2b^2 = a^2 = 4t^2$, which means that $b^2 = 2t^2$. Thus, b^2 is even, which implies that b is even. The original assumption has led to the conclusion that both a and b are even, which is a contradiction of the fact that $\gcd(a,b) = 1$. Therefore, $\sqrt{2}$ is irrational.

Classroom Problems. Use the previous proof to show that $\sqrt[n]{2}$ is irrational for $n > 2$. ◆

Killing an Ant with an Atom Bomb. Recall that Fermat's Last Theorem was finally proved in 1995 (after more than three centuries) by Andrew Wiles and asserts that the equation $x^n + y^n = z^n$ has no positive integer solutions for all integers $n > 2$ (see Section 5.10). Wiles's proof is extremely complicated, and the details involve several hundred pages of sophisticated mathematical ideas and techniques.

It is the nature of mathematics to use known theorems to prove new theorems, but it is not good practice to establish simple results (having elementary proofs) by using other results with complex proofs (circular reasoning is a common pitfall of this approach). An especially absurd and comical example of this phenomenon is the proof (noticed by several mathematicians) showing that Fermat's Last Theorem implies $\sqrt[n]{2}$ is irrational for each integer $n > 2$.

"Proof." As before, assume that $\sqrt[n]{2}$ is rational for some integer $n > 2$, i.e., $\sqrt[n]{2} = \frac{a}{b}$ for some integers $a, b \neq 0$. Hence, $2b^n = a^n$, so $b^n + b^n = a^n$. Therefore, the equation $x^n + y^n = z^n$ has a positive integer solution for some integer $n > 2$, and this contradicts Fermat's Last Theorem (killing an ant with an atom bomb). Therefore, $\sqrt[n]{2}$ is irrational for each integer $n > 2$.

Section 6.3   Rational and Irrational Numbers   267

In view of Theorem 6.3.1, it is reasonable to ask the following question.

Question. If an integer $a > 0$ is not a square, is $\sqrt{a}$ irrational?
For example, is $\sqrt{15}$ an irrational number? Although Pythagoras's argument does not directly apply to this case (try it), let's try to develop an argument that proves $\sqrt{15}$ is irrational.

Analysis. Proceeding as above, assume $\sqrt{15}$ is a rational number, i.e., $\sqrt{15} = \frac{a}{b}$ for some integers $a, b \neq 0$. Hence, $15 = \frac{a^2}{b^2}$, so $15b^2 = a^2$. We know (from our work in Chapter 2, result 2.6.8) that the distinct prime factors of a^2 appear an even number of times (i.e., the exponents in the canonical prime factorization are even), and the same is true for the prime factors of $15b^2$ by the Fundamental Theorem of Arithmetic. However, this is impossible, since the prime 3 (and the prime 5) appears an odd number of times in the prime factorization of $15b^2$. (Why?) This contradiction shows that $\sqrt{15}$ is an irrational number.

Classroom Problems. Write a proof showing that $\sqrt{a}$ is irrational for any integer $a > 0$ that is not a square. Direction: If a is not a square, then it has at least one prime factor that appears an odd number of times in its prime factorization. ◆

It is not always possible to show that a given number is irrational by means of an elementary argument. In fact, advanced mathematics is needed to establish the irrationality of the number π. Even though it is sometimes difficult to prove that certain numbers are irrational, we have observed (using decimals) that there are infinitely many irrational numbers. In fact, the next set of Classroom Problems helps illustrate how densely packed the rationals and irrationals are on the number line.

Classroom Problems. We provide a solution for part b and leave the other parts for classroom group or individual investigation. General statements concerning these problems appear in Exercises 6.3, no. 6.

  a. Find a rational number r with $\frac{25}{100} < r < \frac{25}{99}$.
  b. Find an irrational number t with $\frac{25}{100} < t < \frac{25}{99}$.
  c. Find a rational number r with $\sqrt{2} < r < \sqrt{2.1}$.
  d. Find an irrational number t with $\sqrt{2} < t < \sqrt{2.1}$.

Solution b. First note that $\frac{25}{100} = .25\overline{0}$ and that $\frac{25}{99} = .\overline{25}$. Thus, if we choose $t = .25202002000200002\ldots$, it satisfies the inequality and is irrational. What are some other choices for t that will also solve this problem? ◆

EXERCISES 6.3

  1. Convert the following decimals into quotients of integers:
    a. $x = .516\overline{277}$ **b.** $x = .\overline{9}$ **c.** $x = .22222\overline{273}$

2. In each part, indicate whether the given statement concerning integers is true or false. If it is true, provide a sound argument demonstrating the statement's truth or indicate how the statement follows from some other known statement(s). If it is false, provide a concrete counterexample with a complete explanation of why your example shows the statement is false.
   a. The sum of two rational numbers is rational.
   b. The product of two rational numbers is rational.
   c. The sum of a rational number and an irrational number is irrational.
   d. The product of a rational number and an irrational number is irrational.
   e. The sum of two irrational numbers is irrational.
   f. The product of two irrational numbers is irrational.
3. Prove that $\sqrt[3]{2}$ is irrational.
4. Show that $\sqrt{3} + \sqrt{5}$ is irrational.
5. Show that $\sqrt{\frac{2}{3}}$ is irrational.
6. a. Show that (strictly) between any two distinct rational numbers there is at least one rational number.
   b. Show that (strictly) between any two distinct rational numbers there is at least one irrational number.
   c. Show that (strictly) between any two distinct irrational numbers there is at least one rational number.
   d. Show that (strictly) between any two distinct irrational numbers there is at least one irrational number.

6.4 SYSTEMS OF LINEAR EQUATIONS

The most prevalent approach that middle-school students used (besides "guess and check") to solve the problems in Section 6.2 involved manipulating and exchanging product cost statements until a single product statement resulted. Once they accomplished this, they determined the cost of that product and then used this to ascertain the cost of the remaining item. In terms of algebra, this process amounts to eliminating one of the variables and then solving for the remaining one, and it is referred to as performing elementary row operations on the linear equations of the system.

Elementary Row Operations. To interpret the middle-school exchanging strategies in terms of algebra, it is important to recall from high school and college algebra that the solution set of a system of linear equations can be determined algebraically by performing certain permissible operations on the equations called **elementary row operations** (i.e., operations that do not affect the solution set of the system). Two systems of linear equations having the same solution set are called **equivalent systems**. The following elementary row operations applied to a system always result in an equivalent system.

1. Interchange any two equations.
2. Multiply an equation by a nonzero constant.
3. Add a multiple of an equation to another equation.

It is clear that Row Operations 1 and 2 would not alter the solution set of a linear system (explain), but it is not so obvious why Row Operation 3 also has this property. To see this, let's consider the effect of Row Operation 3 on the system of equations in Problems 9 and 10 in *Comparing Quantities* (see Figure 6.1.3). The argument given in terms of this specific example is presented in a general manner and can easily be adapted to handle any system of linear equations.

Let c = the cost of a clipboard and p = the cost of a pencil; consider the system (labeled S_1) of two linear equations in two unknowns:

S_1:
$$4c + 8p = 8,$$
$$3c + 10p = 7.$$

Before applying Row Operation 3, let's simplify the system by multiplying the first equation by $\frac{1}{4}$ (Row Operation 2) to get the equivalent system (labeled S_2):

S_2:
$$1c + 2p = 2,$$
$$3c + 10p = 7.$$

Performing Row Operation 3. Multiply the first equation by -3 and then add it to the second equation. The system of equations (labeled S_3) resulting from this row operation is

S_3:
$$1c + 2p = 2,$$
$$0c + 4p = 1.$$

We would like to show that systems S_2 and S_3 have the same solution set. To accomplish this goal, we demonstrate that the solution set of S_2 is a subset of the solution set of S_3 and that the solution set of S_3 is a subset of the solution set of S_2. Note that in this specific example, the solution sets consist of exactly one solution, but this is not always the case in other linear systems.

To this end, suppose that (c_1, p_1) is any solution to system S_2, where c_1 and p_1 are real numbers. Thus, we have the following equalities involving real numbers:

$$1c_1 + 2p_1 = 2,$$
$$3c_1 + 10p_1 = 7.$$

Multiplying both sides of the first equality by -3 results in the equality

$$-3c_1 + (-6)p_1 = -6.$$

We may now add $-3c_1 + (-6)p_1$ to the left-hand side of the second number equality and add -6 to the right-hand side of that same equality, i.e.,

$$[-3c_1 + (-6)p_1] + (3c_1 + 10p_1) = (-6) + 7,$$

and obtain the equality
$$0c_1 + 4p_1 = 1.$$

We have shown that the following equalities of real numbers hold:

$$1c_1 + 2p_1 = 2,$$
$$0c_1 + 4p_1 = 1,$$

so (c_1, p_1) is a solution to system S_3. Therefore, the solution set of system S_2 is a subset of the solution set of system S_3.

For the reverse inclusion, assume that (c_1, p_1) is any solution to system S_3, where c_1 and p_1 are real numbers, i.e.,

$$1c_1 + 2p_1 = 2,$$
$$0c_1 + 4p_1 = 1.$$

Multiplying both sides of the first equality by 3 results in the equality

$$3c_1 + 6p_1 = 6.$$

Adding $3c_1 + 6p_1$ to the left-hand side of the second number equality and adding 6 to the right-hand side of that same equality gives

$$(3c_1 + 6p_1) + (0c_1 + 4p_1) = 6 + 1,$$

thus,

$$3c_1 + 10p_1 = 7.$$

We have shown that the following equalities hold:

$$1c_1 + 2p_1 = 2,$$
$$3c_1 + 10p_1 = 7,$$

so (c_1, p_1) is a solution to system S_2. Hence the solution set of system S_3 is a subset of the solution set of system S_2, and thus we have successfully demonstrated that systems S_2 and S_3 have the same solution sets, i.e., they are equivalent systems.

Therefore, finding the solutions of system S_1 is equivalent to finding the solutions to the simpler system S_3

$$1c + 2p = 2,$$
$$0c + 4p = 1.$$

In particular, substituting the value from the second equation $p = \frac{1}{4}$ into the first equation yields

$$c = 2 - \frac{1}{2} = \frac{3}{2}.$$

A quick check verifies that $(\frac{3}{2}, \frac{1}{4})$ is a solution (actually the only solution) to systems S_1 and S_2.

Section 6.4 Systems of Linear Equations

Classroom Problems. Revisit Problems 1–10 from Figures 6.1.1–6.1.3 in Section 6.2 and interpret the solutions we discussed in terms of elementary row operations. ◆

Representing Linear Systems of Equations with Matrices. The systems in the middle-grade examples were comprised of two linear equations in two unknowns, but our studies are not limited exclusively to systems of this size. In particular, some of our investigations involve m linear equations in n unknowns.

Such a general system can be represented as

$$a_{11}x_1 + a_{12}x_2 + \cdots + a_{1n}x_n = b_1,$$
$$a_{21}x_1 + a_{22}x_2 + \cdots + a_{2n}x_n = b_2,$$
$$\vdots$$
$$a_{m1}x_m + a_{m2}x_2 + \cdots + a_{mn}x_n = b_m,$$

where a_{ij} is the coefficient of the jth variable of the ith equation counting equations from top to bottom.

When simplifying a system of linear equations by means of elementary row operations, it is cumbersome to write the variables after each operation, but it is crucial to keep track of which coefficients belong to which variables and which constants belong to which equations. The employment of **matrices (rectangular arrays of real numbers)** helps address this problem.

For example, the coefficients of system S_1

$$4c + 8p = 8$$
$$3c + 10p = 7$$

can be represented by the rectangular array $\begin{bmatrix} 4 & 8 \\ 3 & 10 \end{bmatrix}$, called the **coefficient matrix** of the system, where the rows correspond to the coefficients and the columns correspond to the variables. A related matrix $\begin{bmatrix} 4 & 8 & 8 \\ 3 & 10 & 7 \end{bmatrix}$, called the **augmented matrix**, represents the coefficients and the constant terms.

In terms of a general system of m linear equations in n unknowns, the coefficient matrix consisting of m rows and n columns is given by

$$A = \begin{bmatrix} a_{11} & a_{12} & \cdots & a_{1n} \\ a_{21} & a_{22} & \cdots & a_{2n} \\ & & \vdots & \\ a_{m1} & a_{m2} & \cdots & a_{mn} \end{bmatrix}$$

and has **order** or **size m by n**, denoted by $m \times n$ (m and n are integers with $m \geq 1$ and $n \geq 1$). The entry in the ith row and j^{th} column of matrix A is called the $(i, j)^{\text{th}}$ **entry** and is denoted by a_{ij}.

If $m = n$, then the matrix A is called a **square matrix of size $n \times n$**.

The **augmented matrix** of a general system of m linear equations in n unknowns is the $m \times (n + 1)$ matrix

$$\begin{bmatrix} a_{11} & a_{12} & \cdots & a_{1n} & b_1 \\ a_{21} & a_{22} & \cdots & a_{2n} & b_2 \\ & & \vdots & & \\ a_{m1} & a_{m2} & \cdots & a_{mn} & b_m \end{bmatrix}.$$

Set Notation. The **set of all $m \times n$ matrices with real number entries** is denoted by $M_{m \times n}(\mathbf{R})$. In the case when $m = n$ (square matrices), we write $M_n(\mathbf{R})$.

Equality of Matrices. Two matrices $A = [a_{ij}]$ and $B = [b_{ij}]$ in the set $M_{m \times n}(\mathbf{R})$ are said to be **equal** if $a_{ij} = b_{ij}$ for each i and j.

Elementary Row Operations on the Augmented Matrix. The elementary row operations were originally stated in terms of equations, but it is now convenient to restate them in terms of matrices. More specifically, the elementary row operations and the notation describing them are as follows:

Elementary row operations	Notation
1. Permute (interchange) rows i and j	$P(i, j)$
2. Multiply row i by a nonzero constant c	$(c)Ri$
3. Add c times row i to row j	$(c)Ri + Rj$

Performing elementary row operations on the augmented matrix of a system of linear equations results in an augmented matrix of an equivalent system of linear equations. For example, multiplying row 1 of the augmented matrix of system S_1 by $1/4$ produces the augmented matrix of the equivalent system S_2. This row operation is symbolized by

$$\begin{bmatrix} 4 & 8 & 8 \\ 3 & 10 & 7 \end{bmatrix} \xrightarrow{(1/4)R1} \begin{bmatrix} 1 & 2 & 2 \\ 3 & 10 & 7 \end{bmatrix}.$$

Multiplying the first row of the augmented matrix of System S_2 by -3 and then adding it to the second row produces the augmented matrix of the equivalent system S_3. This particular row operation is denoted by

$$\begin{bmatrix} 1 & 2 & 2 \\ 3 & 10 & 7 \end{bmatrix} \xrightarrow{(-3)R1+R2} \begin{bmatrix} 1 & 2 & 2 \\ 0 & 4 & 1 \end{bmatrix}.$$

Row Operations Are Reversible. Perhaps you have noted that any row operation is reversible. This is certainly clear for elementary Row Operations 1, and it is also evident for Row Operations 2. In particular, if a row in system S_1 is multiplied by a nonzero constant c to produce system S_2, then multiplying that altered equation in system S_2 by $1/c$ results in system S_1. Similarly, we have seen that Row Operations 3 are reversible, since if system S_2 is obtained from system S_1 by adding c times row

Section 6.4   Systems of Linear Equations   273

i to row j, then this can reversed by adding $-c$ times row i to row j of system S_2 to get system S_1.

Determining Solution Sets of Linear Systems. Using elementary row operations, the solutions of a linear system of equations can efficiently be determined in a systematic sequence of steps. We first illustrate such a process with a few examples and then formalize the procedure in general terms.

We start with system S_1, i.e.,

$$4c + 8p = 8$$
$$3c + 10p = 7.$$

Goal. Try to reduce the augmented matrix $\begin{bmatrix} 4 & 8 & 8 \\ 3 & 10 & 7 \end{bmatrix}$ of the given system to an augmented matrix of the form $\begin{bmatrix} 1 & 0 & a \\ 0 & 1 & b \end{bmatrix}$ of an equivalent system. If this can be accomplished, then translating back to variables gives the solution $c = a$ and $p = b$.

Process.

$\begin{bmatrix} 4 & 8 & 8 \\ 3 & 10 & 7 \end{bmatrix} \xrightarrow{(1/4)R1} \begin{bmatrix} 1 & 2 & 2 \\ 3 & 10 & 7 \end{bmatrix} \xrightarrow{(-3)R1+R2} \begin{bmatrix} 1 & 2 & 2 \\ 0 & 4 & 1 \end{bmatrix} \xrightarrow{(1/4)R2} \begin{bmatrix} 1 & 2 & 2 \\ 0 & 1 & 1/4 \end{bmatrix}$

$\xrightarrow{(-2)R2+R1} \begin{bmatrix} 1 & 0 & 3/2 \\ 0 & 1 & 1/4 \end{bmatrix}$. Therefore, $c = \frac{3}{2}$ and $p = \frac{1}{4}$ is the unique solution (check) to the system

$$4c + 8p = 8,$$
$$3c + 10p = 7.$$

Terminology: Row Equivalent Matrices. Two matrices $A, B \in M_{m \times n}(\mathbf{R})$ are said to be **row equivalent (denoted by $A \sim B$)** if one can be transformed into the other by a finite sequence of elementary row operations.

In the system just considered, the matrix $\begin{bmatrix} 4 & 8 & 8 \\ 3 & 10 & 7 \end{bmatrix}$ is row equivalent to the matrix $\begin{bmatrix} 1 & 0 & 3/2 \\ 0 & 1 & 1/4 \end{bmatrix}$, i.e., $\begin{bmatrix} 4 & 8 & 8 \\ 3 & 10 & 7 \end{bmatrix} \sim \begin{bmatrix} 1 & 0 & 3/2 \\ 0 & 1 & 1/4 \end{bmatrix}$.

Classroom Problems. Let $A, B, C \in M_{m \times n}(\mathbf{R})$. Justify the following statements:

a. If $A \sim B$, then $B \sim A$.
b. If $A \sim B$ and $B \sim C$, then $A \sim C$. ◆

Geometric Interpretation for Two Linear Equations in Two Unknowns. We know (from our studies in school and college algebra) that each of the equations in this system describes a line in the plane, and the solution determined is the point of intersection of the two lines. From this geometric perspective, it is clear that the only solution possibilities for two equations and two unknowns are:

a. **A unique solution (c_1, p_1)**: intersecting lines;
b. **No solution**: parallel lines; and
c. **Infinitely many solutions**: identical lines.

Let's see the outcome of applying elementary row operations to a system consisting of two parallel lines:

$$5x + 7y = 3,$$
$$5x + 7y = 4.$$

Goal. We will attempt to use the same strategy as in the previous example, i.e., to try to reduce the augmented matrix to a row equivalent matrix of the form

$$\begin{bmatrix} 1 & 0 & a \\ 0 & 1 & b \end{bmatrix}.$$

If this cannot be achieved, then we'll reduce it as close to this form as possible.

Process.

$$\begin{bmatrix} 5 & 7 & 3 \\ 5 & 7 & 4 \end{bmatrix} \xrightarrow{(-1)R1+R2} \begin{bmatrix} 5 & 7 & 3 \\ 0 & 0 & 1 \end{bmatrix} \xrightarrow{(1/5)R1} \begin{bmatrix} 1 & 7/5 & 3/5 \\ 0 & 0 & 1 \end{bmatrix}.$$

Note that the second row of the last simplified matrix gives the equation $0x + 0y = 1$, and since there are no real numbers x and y satisfying this equation, it follows that the system has no solutions.

Next we'll consider an example from the *Comparing Quantities* unit (Mario's Restaurant, exercise 5) of *Mathematics in Context* that involves a system of three equations and three unknowns (see Figure 6.4.1).

First, we define the variables. In particular, let t = the cost of a taco, s = the cost of a salad, and d = the cost of a drink. The information on Mario's pad can be expressed in terms of the following system of equations:

$$2t + 4s + 0d = 10$$
$$t + 2s + 3d = 8$$
$$3t + 0s + 3d = 9$$

Goal. Try to reduce the augmented matrix to a row equivalent matrix of the form

$$\begin{bmatrix} 1 & 0 & 0 & a \\ 0 & 1 & 0 & b \\ 0 & 0 & 1 & c \end{bmatrix}$$

As before, if this cannot be accomplished, then we'll reduce it as close to this form as possible.

Section 6.4 Systems of Linear Equations 275

Mario's Restaurant

Mario runs a Mexican restaurant, and he is very busy. He moves from one table to another, writing down all the orders. Below you see how he writes the orders on his order pad.

ORDER	TACO	SALAD	DRINK	TOTAL
1	2	4	--	$10
2	1	2	3	$8
3	3	--	3	$9
4	1	2	--	
5	1	--	1	
6	2	2	1	
7	4	2	3	
8				
9				
10				

3. Some of the orders do not have total prices indicated. What are the prices of these orders?

4. Make up two new orders and write them in your notebook. Fill in the prices of these orders.

5. What is the price of each individual item?

Reproduced from page 22 of *Comparing Quantities* in *Mathematics in Contact*.

FIGURE 6.4.1

Process.

$$\begin{bmatrix} 2 & 4 & 0 & 10 \\ 1 & 2 & 3 & 8 \\ 3 & 0 & 3 & 9 \end{bmatrix} \xrightarrow{P(1,2)} \begin{bmatrix} 1 & 2 & 3 & 8 \\ 2 & 4 & 0 & 10 \\ 3 & 0 & 3 & 9 \end{bmatrix} \xrightarrow{(-2)R1+R2} \begin{bmatrix} 1 & 2 & 3 & 8 \\ 0 & 0 & -6 & -6 \\ 3 & 0 & 3 & 9 \end{bmatrix} \xrightarrow{(-3)R1+R3}$$

$$\begin{bmatrix} 1 & 2 & 3 & 8 \\ 0 & 0 & -6 & -6 \\ 0 & -6 & -6 & -15 \end{bmatrix} \xrightarrow{P(2,3)} \begin{bmatrix} 1 & 2 & 3 & 8 \\ 0 & -6 & -6 & -15 \\ 0 & 0 & -6 & -6 \end{bmatrix} \xrightarrow{(-1/6)R2} \begin{bmatrix} 1 & 2 & 3 & 8 \\ 0 & 1 & 1 & 5/2 \\ 0 & 0 & -6 & -6 \end{bmatrix}$$

$$\xrightarrow{(-2)R2+R1} \begin{bmatrix} 1 & 0 & 1 & 3 \\ 0 & 1 & 1 & 5/2 \\ 0 & 0 & -6 & -6 \end{bmatrix} \xrightarrow{(-1/6)R3} \begin{bmatrix} 1 & 0 & 1 & 3 \\ 0 & 1 & 1 & 5/2 \\ 0 & 0 & 1 & 1 \end{bmatrix} \xrightarrow{(-1)R3+R1}$$

$$\begin{bmatrix} 1 & 0 & 0 & 2 \\ 0 & 1 & 1 & 5/2 \\ 0 & 0 & 1 & 1 \end{bmatrix} \xrightarrow{(-1)R3+R2} \begin{bmatrix} 1 & 0 & 0 & 2 \\ 0 & 1 & 0 & 3/2 \\ 0 & 0 & 1 & 1 \end{bmatrix}.$$

Therefore, $t = 2$, $s = \frac{3}{2}$, and $d = 1$ is the unique solution to the original system of equations (check). In regards to Mario's restaurant, a taco costs \$2, a salad costs \$1.50, and a drink costs \$1.

Classroom Problems. Using a similar elimination strategy, solve the same system of equations by completing the following initial steps:

$$\begin{bmatrix} 2 & 4 & 0 & 10 \\ 1 & 2 & 3 & 8 \\ 3 & 0 & 3 & 9 \end{bmatrix} \xrightarrow{(1/2)R1} \begin{bmatrix} 1 & 2 & 0 & 5 \\ 1 & 2 & 3 & 8 \\ 3 & 0 & 3 & 9 \end{bmatrix} \xrightarrow{?} \blacklozenge$$

Gauss-Jordan Elimination Method: Reduced Row Echelon Form. In the previous examples, we employed elementary row operations to determine the solution sets of given systems of linear equations. For each system, the process began by representing that system with an augmented matrix; then, after performing a finite number of elementary row operations to arrive at row equivalent augmented matrices of simpler form, the process concluded by reintroducing the variables from a "final" augmented matrix. This solution method is referred to as **Gauss-Jordan elimination.** A quintessential feature of this solution technique, especially for theoretical purposes (to be discussed later), is that the process terminates in a unique matrix that has a special form, known as the "reduced row echelon" form. In particular, an $m \times n$ matrix is said to be in **reduced row echelon form** if

    a. all rows consisting entirely of zeros (if there are any) are placed at the bottom of the matrix;

    b. the first nonzero entry of each (nonzero) row is 1 and is referred to as the **leading one** of that row;

    c. given two nonzero consecutive rows j and $j + 1$, the leading one of row $j + 1$ is to the right of the leading one of row j; and

Section 6.4   Systems of Linear Equations   277

   d. if a column contains a leading one, then all other entries of that column are zero.

Note that the following matrices, which were obtained by the Gauss-Jordan elimination method in the previous examples above, are in reduced row echelon form:

$$\begin{bmatrix} 1 & 0 & 3/2 \\ 0 & 1 & 1/4 \end{bmatrix}, \begin{bmatrix} 1 & 7/5 & 3/5 \\ 0 & 0 & 0 \end{bmatrix}, \begin{bmatrix} 1 & 0 & 0 & 2 \\ 0 & 1 & 0 & 3/2 \\ 0 & 0 & 1 & 1 \end{bmatrix}.$$

Examples. We find the solution sets of the following systems using Gauss-Jordan elimination.

1. $\begin{aligned} x_1 - 2x_2 + x_3 &= 3 \\ x_1 + 5x_2 - 4x_3 &= -1 \end{aligned}$
2. $\begin{aligned} x + y &= 2 \\ -2x + 3y &= -3 \\ x - y &= 5 \end{aligned}$

Analysis of System 1.

$$\begin{bmatrix} 1 & -2 & 1 & 3 \\ 1 & 5 & -4 & -1 \end{bmatrix} \xrightarrow{(-1)R1+R2} \begin{bmatrix} 1 & -2 & 1 & 3 \\ 0 & 7 & -5 & -4 \end{bmatrix} \xrightarrow{(1/7)R2} \begin{bmatrix} 1 & -2 & 1 & 3 \\ 0 & 1 & -5/7 & -4/7 \end{bmatrix}$$

$\xrightarrow{(2)R2+R1} \begin{bmatrix} 1 & 0 & -3/7 & 13/7 \\ 0 & 1 & -5/7 & -4/7 \end{bmatrix}$. The process is now complete, since the last matrix is in reduced row echelon form. Hence, the solution set is infinite and is given by

$$S = \left\{ (x_1, x_2, x_3): x_1 = \frac{3}{7}x_3 + \frac{13}{7}, x_2 = \frac{5}{7}x_3 - \frac{4}{7}, \text{ and } x_3 \text{ is any real number} \right\}.$$

Let's check a few solutions. We first choose values for x_3 to determine values for x_1 and x_2, and then we substitute those quantities back into the original equations. If $x_3 = 0$, then $x_1 = \frac{13}{7}$ and $x_2 = -\frac{4}{7}$. Thus,

$$\frac{13}{7} - 2\left(-\frac{4}{7}\right) + 0 = \frac{21}{7} = 3 \text{ and } \frac{13}{7} + 5\left(-\frac{4}{7}\right) + 0 = -\frac{7}{7} = -1.$$

If $x_3 = -\frac{1}{2}$, then $x_1 = \frac{3}{7}(-\frac{1}{2}) + \frac{13}{7} = \frac{23}{14}$ and $x_2 = \frac{5}{7}(-\frac{1}{2}) - \frac{4}{7} = -\frac{13}{14}$. Hence,

$$\frac{23}{14} - 2\left(-\frac{13}{14}\right) - \frac{1}{2} = \frac{42}{14} = 3 \text{ and } \frac{23}{14} + 5\left(-\frac{13}{14}\right) - 4\left(-\frac{1}{2}\right) = -\frac{14}{14} = -1.$$

Analysis of System 2.

$$\begin{bmatrix} 1 & 1 & 2 \\ -2 & 3 & -3 \\ 1 & -1 & 5 \end{bmatrix} \xrightarrow{(2)R1+R2} \begin{bmatrix} 1 & 1 & 2 \\ 0 & 5 & 1 \\ 1 & -1 & 5 \end{bmatrix} \xrightarrow{(-1)R1+R3} \begin{bmatrix} 1 & 1 & 2 \\ 0 & 5 & 1 \\ 0 & -2 & 3 \end{bmatrix} \xrightarrow{(1/5)R2}$$

$$\begin{bmatrix} 1 & 1 & 2 \\ 0 & 1 & 1/5 \\ 0 & -2 & 3 \end{bmatrix} \xrightarrow{(-1)R2+R1} \begin{bmatrix} 1 & 0 & 9/5 \\ 0 & 1 & 1/5 \\ 0 & -2 & 3 \end{bmatrix} \xrightarrow{(2)R2+R3} \begin{bmatrix} 1 & 0 & 9/5 \\ 0 & 1 & 1/5 \\ 0 & 0 & 17/5 \end{bmatrix} \xrightarrow{(5/17)R3}$$

$$\begin{bmatrix} 1 & 0 & 9/5 \\ 0 & 1 & 1/5 \\ 0 & 0 & 1 \end{bmatrix} \xrightarrow{-(1/5)R3+R2} \begin{bmatrix} 1 & 0 & 9/5 \\ 0 & 1 & 0 \\ 0 & 0 & 1 \end{bmatrix} \xrightarrow{-(9/5)R3+R1} \begin{bmatrix} 1 & 0 & 0 \\ 0 & 1 & 0 \\ 0 & 0 & 1 \end{bmatrix}.$$

Since the last matrix is in reduced row echelon form, we know that the Gauss-Jordan elimination process is complete. Notice that the third row of the reduced row echelon form gives the equation

$$0x + 0y = 1.$$

Since there are no real numbers x and y that satisfy this equation, it follows that the system has no solutions, i.e., the solution set of the original system is the empty set.

Classroom Problems. Determine which of the following matrices are in reduced row echelon form. For those matrices not in reduced row echelon form, perform the necessary row operations to put them in reduced row echelon form:

a. $\begin{bmatrix} 0 & 1 & 5 \\ 1 & 0 & 2 \end{bmatrix}$ **b.** $\begin{bmatrix} 1 & 1 & 1 \\ 0 & 0 & 0 \end{bmatrix}$ **c.** $\begin{bmatrix} 1 & 0 & 2 & 5 \\ 0 & 1 & 0 & 3 \\ 0 & 0 & 1 & 4 \end{bmatrix}$

d. $\begin{bmatrix} 0 & 1 & 0 & 2 \\ 0 & 0 & 1 & 9 \\ 0 & 0 & 0 & 0 \end{bmatrix}$ **e.** $\begin{bmatrix} 1 & 0 & -7 & 3 \\ 0 & 1 & 4 & 5 \end{bmatrix}$ **f.** $\begin{bmatrix} 1 & 0 & 0 & 7 \\ 0 & 0 & 1 & 4 \\ 0 & 0 & 0 & 0 \end{bmatrix}$ ◆

Solution Possibilities Summary. The systems of linear equations we've considered either have no solutions (called **inconsistent systems**), a unique solution, or infinitely many solutions (a system having at least one solution is called a **consistent system**). As we indicated earlier in this section, it is geometrically clear in the two equation–two unknown context that these three outcomes are the only solution possibilities. However, for general systems of linear equations, it is not obvious that these are the only solution sets that can arise. This circumstance motivates the following question.

6.4.1 Question. Are there systems of linear equations with exactly $t > 1$ solutions (t an integer)?

Analysis. Let's first attempt to analyze the two equation–two unknown case without relying on geometry and then determine if our reasoning can be extended to other cases. Our analysis employs Gauss-Jordan elimination from a theoretical perspective.

Two Equation–Two Unknown System. The only possible reduced row echelon forms of a two equation–two unknown system of linear equations are:

$$\begin{bmatrix} 1 & 0 & * \\ 0 & 1 & * \end{bmatrix}, \begin{bmatrix} 1 & * & * \\ 0 & 0 & 0 \end{bmatrix}, \begin{bmatrix} 1 & * & 0 \\ 0 & 0 & 1 \end{bmatrix}, \begin{bmatrix} 0 & 1 & * \\ 0 & 0 & 0 \end{bmatrix}, \begin{bmatrix} 0 & 1 & 0 \\ 0 & 0 & 1 \end{bmatrix}, \begin{bmatrix} 0 & 0 & 1 \\ 0 & 0 & 0 \end{bmatrix},$$ where $*$

represents (not necessarily the same) real numbers. (Explain.) Hence, such a system has exactly one solution given by the matrix $\begin{bmatrix} 1 & 0 & * \\ 0 & 1 & * \end{bmatrix}$, infinitely many solutions described by the matrix $\begin{bmatrix} 1 & * & * \\ 0 & 0 & 0 \end{bmatrix}$, or no solution corresponding to the matrix $\begin{bmatrix} 1 & * & 0 \\ 0 & 0 & 1 \end{bmatrix}$. Therefore, there does not exist a two equation–two unknown linear system with exactly $t > 1$ solutions.

Three Equation–Three Unknown System. As in the previous case, we consider all possible reduced row echelon forms for the augmented matrix of a three equation–three unknown system. We then interpret the solution outcomes corresponding to each one. (Check to see that these are all the possible 3×4 matrices in reduced row echelon form.)

Unique Solution:

$$\begin{bmatrix} 1 & 0 & 0 & * \\ 0 & 1 & 0 & * \\ 0 & 0 & 1 & * \end{bmatrix}.$$

Infinitely Many Solutions. (for each matrix, explain why there are infinitely many solutions):

$$\begin{bmatrix} 1 & 0 & * & * \\ 0 & 1 & * & * \\ 0 & 0 & 0 & 0 \end{bmatrix}, \begin{bmatrix} 1 & * & 0 & * \\ 0 & 0 & 1 & * \\ 0 & 0 & 0 & 0 \end{bmatrix}, \begin{bmatrix} 1 & * & * & * \\ 0 & 0 & 0 & 0 \\ 0 & 0 & 0 & 0 \end{bmatrix}, \begin{bmatrix} 0 & 1 & 0 & * \\ 0 & 0 & 1 & * \\ 0 & 0 & 0 & 0 \end{bmatrix}, \begin{bmatrix} 0 & 1 & * & * \\ 0 & 0 & 0 & 0 \\ 0 & 0 & 0 & 0 \end{bmatrix},$$

$$\begin{bmatrix} 0 & 0 & 1 & * \\ 0 & 0 & 0 & 0 \\ 0 & 0 & 0 & 0 \end{bmatrix}.$$

No Solution. (for each matrix, explain why there are no solutions):

$$\begin{bmatrix} 1 & 0 & * & 0 \\ 0 & 1 & * & 0 \\ 0 & 0 & 0 & 1 \end{bmatrix}, \begin{bmatrix} 1 & * & 0 & 0 \\ 0 & 0 & 1 & 0 \\ 0 & 0 & 0 & 1 \end{bmatrix}, \begin{bmatrix} 1 & * & * & 0 \\ 0 & 0 & 0 & 1 \\ 0 & 0 & 0 & 0 \end{bmatrix}, \begin{bmatrix} 0 & 1 & 0 & 0 \\ 0 & 0 & 1 & 0 \\ 0 & 0 & 0 & 1 \end{bmatrix}, \begin{bmatrix} 0 & 1 & * & 0 \\ 0 & 0 & 0 & 1 \\ 0 & 0 & 0 & 0 \end{bmatrix},$$

$$\begin{bmatrix} 0 & 0 & 1 & 0 \\ 0 & 0 & 0 & 1 \\ 0 & 0 & 0 & 0 \end{bmatrix}, \begin{bmatrix} 0 & 0 & 0 & 1 \\ 0 & 0 & 0 & 0 \\ 0 & 0 & 0 & 0 \end{bmatrix}.$$

Therefore, as in the two equation–two unknown case, there does not exist a three equation–three unknown linear system with exactly $t > 1$ solutions.

6.4.2 Classroom Problems.
Using the reduced row echelon form (as in the previous examples), determine the solution possibilities of a linear system consisting of two equations in three unknowns

$$a_{11}x + a_{12}y + a_{13}z = c_1,$$
$$a_{21}x + a_{22}y + a_{23}z = c_2. \quad \blacklozenge$$

Geometric Interpretation for Three Linear Equations in Three Unknowns. The solution possibilities for a three equation–three unknown system can be observed geometrically using the fact (studied in analytic geometry) that the set of points satisfying an equation $ax + by + cz = d$, where a, b, c, and d are real numbers, corresponds to a plane in three-dimensional (Euclidean) space (e.g., a wall, floor, or ceiling in a typical room).

Classroom Problems. Look around your room and describe all the solution set possibilities for three planes. Find examples that describe the solution outcomes we just determined by using reduced row echelon form. For example, a room's corner point is where two perpendicular walls meet the floor and is the unique point lying on those three planes. $\blacklozenge$

6.4.3 General Case: m Equation–n Unknown System.
For the systems we have analyzed, we concluded that Question 6.4.1 has a negative answer (i.e., those systems cannot have exactly $t > 1$ solutions). Our justifications involved inspecting reduced row echelon forms for the systems we considered, and from those forms it was straightforward to see that the systems either had no solution, exactly one solution, or infinitely many solutions. Using similar reasoning, it follows that we can reach the same conclusions for a general m equation–n unknown system (i.e., they either have no solutions, a unique solution, or infinitely many solutions). In Section 6.6, we construct a simple matrix algebra proof of this fact.

EXERCISES 6.4

1. A friend told me that he earned $1,200 last year from a $25,000 investment. He said that his stock fund paid 6% per year, and his fixed annuity paid 3% per year. How much of the $25,000 did my friend invest in stocks and how much in annuities?

2. More on golf! A golf-ball manufacturer produces three different kinds of golf balls that are designed for specific playing characteristics. Those golfers seeking maximum distance use a distance ball. Golfers who want to control the ball when it hits the green like to use balls that maximize spin. Finally, golfers who want a little of both features prefer a combination spin/distance ball.

   The manufacturing process involves three major steps: (1) Preparing the polymers (natural and synthetic rubbers) for the ball's core and cover, (2) shaping (also includes placing the dimple patterns on the covers); and (3) finishing (painting, drying, quality-control checks).

   The following table delineates the time (in minutes) each process takes for each kind of ball.

	Preparing polymers	Shaping	Finishing
Distance	2	3	2.5
Spin	3	3.5	3.5
Combination	2.5	3	3

Determine the total number of distance, spin, and combination balls that can be produced in a production run. A production run consists of running the polymer machine for 35 minutes, running the shaping machine for 45 minutes, and running the finishing machine for 42 minutes.

3. Determine the solution sets of the following systems of linear equations using Gauss-Jordan elimination:

a. $2x - 7y = 8$
    $3x + 4y = -6$

b. $-(1/5)x + (2/3)y = (1/30)$
    $-3x + 10y = (1/2)$

c. $x + y = 4$
    $-2x - 3y = 2$
    $5x + y = -1$

d. $2x_1 + 4x_2 - x_3 = 4$
    $-x_1 - 2x_2 + x_3 = -6$
    $2x_1 - 3x_2 + x_3 = -1$

e. $3x_1 - 0x_2 + 2x_3 = 0$
    $x_1 + 8x_2 - 4x_3 = 0$

f. $0x_1 - x_2 + 2x_3 = 0$
    $3x_1 + 9x_2 + 0x_3 = 0$
    $x_1 - x_2 - x_3 = 0$

4. A system of linear equations where each of the constant terms is 0 (as in e and f in Exercise 3) is called a **homogeneous** system of equations.
   a. Explain why a homogeneous system of equations either has a unique solution or infinitely many solutions.
   b. Show that the sum of two solutions of a homogeneous system of m linear equations in n unknowns is also a solution of the system. Direction: If $x_1 = u_1, x_2 = u_2, \ldots, x_n = u_n$ and $x_1 = w_1, x_2 = w_2, \ldots, x_n = w_n$ are (not necessarily distinct) solutions of the system

   $$a_{11}x_1 + a_{12}x_2 + \cdots + a_{1n}x_n = 0,$$
   $$a_{21}x_1 + a_{22}x_2 + \cdots + a_{2n}x_n = 0,$$
   $$\cdot$$
   $$\cdot$$
   $$\cdot$$
   $$a_{m1}x_1 + a_{m2}x_2 + \cdots + a_{mn}x_n = 0,$$

   show that $x_1 = u_1 + w_1, x_2 = u_2 + w_2, \ldots, x_n = u_n + w_n$ is also a solution to the system.
   c. Is the sum of two solutions of any system of m linear equations in n unknowns also a solution of the system? If yes, prove it; if no, give a counterexample.

5. Find all values of t so that the following systems have a unique solution. Direction: Determine those values of t that obstruct you from transforming the augmented matrix into the reduced row echelon form $\begin{bmatrix} 1 & 0 & * \\ 0 & 1 & * \end{bmatrix}$.

a. $x + ty = 3$
    $-x + 2y = 1$

b. $-x + ty = 2$
    $tx - 9y = 5$

6. Show that the linear system

$$a_{11}x + a_{12}y = c_1,$$
$$a_{21}x + a_{22}y = c_2$$

has a unique solution if and only if $a_{11}a_{22} - a_{12}a_{21} \neq 0$. Direction: Show that $\begin{bmatrix} a_{11} & a_{12} & c_1 \\ a_{21} & a_{22} & c_2 \end{bmatrix}$ is row equivalent to $\begin{bmatrix} 1 & 0 & * \\ 0 & 1 & * \end{bmatrix}$ if and only if $a_{11}a_{22} - a_{12}a_{21} \neq 0$.

6.5 POLYNOMIAL CURVE FITTING: AN APPLICATION OF SYSTEMS OF LINEAR EQUATIONS

We know that the points $(3, 9)$ and $(5, 13)$ determine a line in the plane that is described by a linear polynomial function $f(x) = a_1 x + a_0$, where $f(3) = 3a_1 + a_0 = 9$ and $f(5) = 5a_1 + a_0 = 13$. One way to find the coefficients of this polynomial is to solve the following system of linear equations:

$$3a_1 + a_0 = 9,$$
$$5a_1 + a_0 = 13.$$

In particular,

$$\begin{bmatrix} 3 & 1 & 9 \\ 5 & 1 & 13 \end{bmatrix} \xrightarrow{(-1)R1+R2} \begin{bmatrix} 3 & 1 & 9 \\ 2 & 0 & 4 \end{bmatrix} \xrightarrow{(1/2)R2} \begin{bmatrix} 3 & 1 & 9 \\ 1 & 0 & 2 \end{bmatrix} \xrightarrow{P(1,2)} \begin{bmatrix} 1 & 0 & 2 \\ 3 & 1 & 9 \end{bmatrix}$$

$$\xrightarrow{(-3)R1+R2} \begin{bmatrix} 1 & 0 & 2 \\ 0 & 1 & 3 \end{bmatrix}.$$ Therefore, $a_1 = 2$, $a_0 = 3$, and so $f(x) = 2x + 3$.

Therefore, the polynomial function $f(x) = 2x + 3$ is the only linear polynomial that satisfies $f(3) = 9$ and $f(5) = 13$.

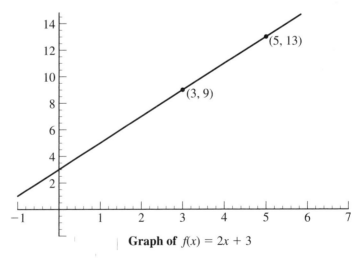

Graph of $f(x) = 2x + 3$

Curve Fitting for Two Arbitrary Points in the Plane. The reasoning used in the previous example easily extends to the general case. In particular, it follows that for

Section 6.5   Polynomial Curve Fitting: Systems of Linear Equations

two points in the plane (x_1, y_1) and (x_2, y_2), where $x_1 \neq x_2$, there exists a unique linear polynomial function $f(x) = a_1 x + a_0$ that satisfies $f(x_1) = y_1$ and $f(x_2) = y_2$. To see this, we must show that the following system (in the variables a_1 and a_0) has a unique solution:

$$a_1 x_1 + a_0 = y_1,$$

$$a_1 x_2 + a_0 = y_2.$$

Analysis.

$$\begin{bmatrix} x_1 & 1 & y_1 \\ x_2 & 1 & y_2 \end{bmatrix} \xrightarrow{(-1)R1+R2} \begin{bmatrix} x_1 & 1 & y_1 \\ x_2 - x_1 & 0 & y_2 - y_1 \end{bmatrix}$$

$$\xrightarrow{P(1,2)} \begin{bmatrix} x_2 - x_1 & 0 & y_2 - y_1 \\ x_1 & 1 & y_1 \end{bmatrix} \xrightarrow{[1/(x_2 - x_1)]R1} \begin{bmatrix} 1 & 0 & (y_2 - y_1)/(x_2 - x_1) \\ x_1 & 1 & y_1 \end{bmatrix}$$

$$\xrightarrow{(-x_1)R1+R2} \begin{bmatrix} 1 & 0 & (y_2 - y_1)/(x_2 - x_1) \\ 0 & 1 & y_1 - x_1[(y_2 - y_1)/(x_2 - x_1)] \end{bmatrix}.$$

Therefore, the system has a unique solution, and our goal is accomplished.

Note that it is possible for $a_1 = 0$. For example, if $(1, 1)$ and $(2, 1)$ are the given points, then the unique linear function whose graph passes through these points is $f(x) = 1$.

Hence, in terms of the degree of a polynomial function, the previous curve-fitting statement can be restated as follows: Given two points in the plane (x_1, y_1) and (x_2, y_2), where $x_1 \neq x_2$, there exists a unique polynomial function $f(x) = a_1 x + a_0$ of *degree less than or equal to 1* that satisfies $f(x_1) = y_1$ and $f(x_2) = y_2$.

Now that we have settled the two-point case, let's consider the three-point analogue. For example, given the points $(-1, 4)$, $(0, 1)$, and $(3, 4)$, does there exist a unique polynomial function $f(x) = a_2 x^2 + a_1 x + a_0$ of degree less than or equal to 2, such that $f(-1) = 4$, $f(0) = 1$, and $f(3) = 4$? Our analysis proceeds as in the previous example.

Analysis. As before, we apply Gauss-Jordan elimination to the augmented matrix of the system:

$$f(-1) = 1 a_2 - 1 a_1 + 1 a_0 = 4,$$

$$f(0) = 0 a_2 + 0 a_1 + 1 a_0 = 1,$$

$$f(3) = 9 a_2 + 3 a_1 + 1 a_0 = 4.$$

$$\begin{bmatrix} 1 & -1 & 1 & 4 \\ 0 & 0 & 1 & 1 \\ 9 & 3 & 1 & 4 \end{bmatrix} \xrightarrow{P(2,3)} \begin{bmatrix} 1 & -1 & 1 & 4 \\ 9 & 3 & 1 & 4 \\ 0 & 0 & 1 & 1 \end{bmatrix} \xrightarrow{(-9)R1+R2} \begin{bmatrix} 1 & -1 & 1 & 4 \\ 0 & 12 & -8 & -32 \\ 0 & 0 & 1 & 1 \end{bmatrix}$$

$$\xrightarrow{(1/12)R2} \begin{bmatrix} 1 & -1 & 1 & 4 \\ 0 & 1 & -2/3 & -8/3 \\ 0 & 0 & 1 & 1 \end{bmatrix} \xrightarrow{R2+R1} \begin{bmatrix} 1 & 0 & 1/3 & 4/3 \\ 0 & 1 & -2/3 & -8/3 \\ 0 & 0 & 1 & 1 \end{bmatrix}$$

$$\xrightarrow{(2/3)R3+R2} \begin{bmatrix} 1 & 0 & 1/3 & 4/3 \\ 0 & 1 & 0 & -2 \\ 0 & 0 & 1 & 1 \end{bmatrix} \xrightarrow{(-1/3)R3+R1} \begin{bmatrix} 1 & 0 & 0 & 1 \\ 0 & 1 & 0 & -2 \\ 0 & 0 & 1 & 1 \end{bmatrix}.$$

Thus, $a_2 = 1$, $a_1 = -2$, and $a_0 = 1$. Hence, $f(x) = x^2 - 2x + 1$ is the only polynomial function of degree less than or equal to 2 that satisfies $f(-1) = 4$, $f(0) = 1$, and $f(3) = 4$.

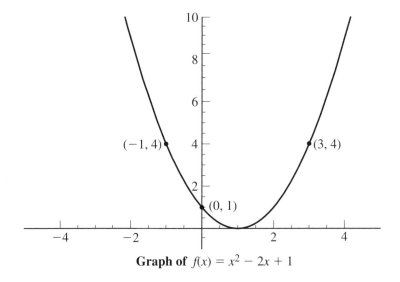

Graph of $f(x) = x^2 - 2x + 1$

Question. Given three points in the plane (x_1, y_1), (x_2, y_2), and (x_3, y_3), where $x_i \neq x_j$ if $i \neq j$, does there exist a unique polynomial function $f(x) = a_2 x^2 + a_1 x + a_0$ of degree less than or equal to 2 that satisfies $f(x_1) = y_1$, $f(x_2) = y_2$, and $f(x_3) = y_3$?

Classroom Problems. Using Gauss-Jordan elimination, show that the following linear system (in the variables a_2, a_1, and a_0) has a unique solution, and thus the previous question has an affirmative answer.

$$f(x_1) = a_2 x_1^2 + a_1 x_1 + a_0 = y_1$$
$$f(x_2) = a_2 x_2^2 + a_1 x_2 + a_0 = y_2$$
$$f(x_3) = a_2 x_3^2 + a_1 x_3 + a_0 = y_3 \quad \blacklozenge$$

The arguments for cases involving more than three points are similar to the previous verifications, and although we do not provide the details, we record the general statement.

Section 6.5 Polynomial Curve Fitting: Systems of Linear Equations

Polynomial Curve Fitting. Given n points in the plane (x_1, y_1), (x_2, y_2), ..., (x_n, y_n), where $x_i \neq x_j$ if $i \neq j$, there exists a unique polynomial function $f(x) = a_{n-1}x^{n-1} + \ldots + a_1 x + a_0$ of degree less than or equal to $n - 1$ that satisfies $f(x_i) = y_i$ for each $i = 1, 2, \ldots, n$.

This unique polynomial, whose graph passes through the specified points, is called the **interpolating polynomial of minimal degree**.

Example: (Not another golf problem!) The driving distances of professional golfers have dramatically increased over the last 25 years. Technology (new space-age clubs and balls), physical training, and golf-course maintenance are the main reasons for this increase. The following table lists the driving leaders and their average distances for the given years:

Year	Driving Distance Leader	Average Distances
1980	Dan Pohl	274.3 yards
1985	Andy Bean	278.2 yards
1995	John Daly	289.0 yards

Let's find the interpolating polynomial of minimal degree for this data and then use it to approximate the leading driving distances in 1990 and in 2000. To help make the calculations manageable, we'll view 1980 as the starting time 0, so 5 represents 1985, and 15 corresponds to 1995. Thus we are interested in determining the coefficients of the polynomial $f(x) = a_2 x^2 + a_1 x + a_0$, where $f(0) = 274.3$, $f(5) = 278.2$, and $f(15) = 289$.

As in the previous problems, the information here gives rise to a system of three linear equations in the three unknowns a_2, a_1, and a_0. Using Gauss-Jordan elimination leads to the following unique solution

$$a_2 = .02, a_1 = .68, \text{ and } a_0 = 274.3,$$

so $f(x) = .02x^2 + .68x + 274.3$ (check that $f(5) = 278.2$, and $f(15) = 289$).

The following chart compares the predicted leading driving distances with the actual leading driving distances.

Year	Predicted Average Distances	Actual Average Distances	Driving Distance Leader
1990	$f(10) = 283.1$ yards	279.6 yards	Tom Purtzer
2000	$f(20) = 295.9$ yards	301.4 yards	John Daly

In general, interpolating polynomials of minimal degree are most useful for understanding behavior between given data points (such as in the case of 1990 in the previous example) but are not very valuable for predicting future data points. Although our studies do not address more effective prediction models, it is interesting to note that there are many other types of interpolation functions that do have such a focus.

EXERCISES 6.5

1. For each collection of data points, find the interpolating polynomial of minimal degree.
   a. $(1, 1)$ and $(3, 7)$
   b. $(-3, 2)$ and $(3, 2)$
   c. $(0, -1), (2, 0),$ and $(-5, -1)$
   d. $(-1, 3), (0, 3),$ and $(1, 3)$
   e. $(1, 2), (2, 3),$ and $(3, 4)$
   f. $(1, 0), (2, 3), (-2, 3), (0, 4)$
2. Describe all quadratic polynomials whose graphs pass through the points $(2, 1)$ and $(-1, 1)$.

6.6 MATRIX ARITHMETIC AND MATRIX ALGEBRA

We have seen that matrices are a versatile representation and manipulation tool and are especially useful in the study of linear systems of equations. In the next two sections, we investigate another approach to solving systems of linear equations through the employment of matrices. The main idea of this new method is to replace a given system of linear equations with a single linear "matrix equation," and then use developed matrix arithmetic properties to solve the matrix equation in a manner similar to the way linear equations over the real numbers are solved.

For example, consider the following system of three equations in three unknowns:

$$\begin{aligned} 2x_1 + x_2 - 2x_3 &= 0 \\ x_1 + x_2 - 2x_3 &= 1 \\ x_1 + 2x_2 + x_3 &= -1 \end{aligned} \qquad 6.6.1$$

Let $A = \begin{bmatrix} 2 & 1 & -2 \\ 1 & 1 & -2 \\ 1 & 2 & 1 \end{bmatrix}$ (coefficient matrix), $X = \begin{bmatrix} x_1 \\ x_2 \\ x_3 \end{bmatrix}$ (matrix of variables), and

$B = \begin{bmatrix} 0 \\ 1 \\ -1 \end{bmatrix}$ (matrix of constants).

In an effort to represent the system in terms of one linear matrix equation of the form $AX = B$ (in analogy to a linear equation $ax = b$ over the real numbers), we define a matrix multiplication that gives this matrix equation the appropriate meaning.

First, to achieve this matrix equality, the product AX must be a 3×1 matrix if it is to equal the 3×1 matrix of constants B. Thus, if we define the 3×1 matrix AX by:

$$AX = \begin{bmatrix} 2x_1 + x_2 - 2x_3 \\ x_1 + x_2 - 2x_3 \\ x_1 + 2x_2 + x_3 \end{bmatrix},$$

then the equation $AX = B$ gives the three linear equations from the original system.

Each element of the matrix product is a sum of products from the given matrices. For example, the entry $2x_1 + x_2 - 2x_3$ in the first row and first column of AX is determined from a sum of products using the first row of B, $[\,2\ 1\ -2\,]$, and the first column of X, $\begin{bmatrix} x_1 \\ x_2 \\ x_3 \end{bmatrix}$ (in this case, the only column). Likewise, the entry $x_1 + x_2 - 2x_3$ in the second row and first column of AX is obtained from a sum of products using the second row of B, $[\,1\ 1\ -2\,]$, and the first column of X, $\begin{bmatrix} x_1 \\ x_2 \\ x_3 \end{bmatrix}$. Finally, the entry $x_1 + 2x_2 + x_3$ in the third row and first column of AX is found from a sum of products using the third row of B, $[\,1\ 2\ 1\,]$, and the first column of X, $\begin{bmatrix} x_1 \\ x_2 \\ x_3 \end{bmatrix}$.

Checking a Solution. Using Gauss-Jordan elimination, it follows that the previous system (6.6.1) has the unique solution $X = \begin{bmatrix} -1 \\ 2/5 \\ -4/5 \end{bmatrix}$. This can be verified by showing that the product $AX = \begin{bmatrix} 2 & 1 & -2 \\ 1 & 1 & -2 \\ 1 & 2 & 1 \end{bmatrix} \begin{bmatrix} -1 \\ 2/5 \\ -4/5 \end{bmatrix}$ equals the matrix $B = \begin{bmatrix} 0 \\ 1 \\ -1 \end{bmatrix}$. (Check.)

Classroom Problems. Determine the systems of equations represented by the matrix equation $AX = B$: **a.** $A = \begin{bmatrix} 2 & -1 \\ 5 & 7 \end{bmatrix}$, $X = \begin{bmatrix} x_1 \\ x_2 \end{bmatrix}$, $B = \begin{bmatrix} 0 \\ 4 \end{bmatrix}$

b. $A = \begin{bmatrix} 2 & -3 & 4 \\ 5 & 0 & -1 \end{bmatrix}$, $X = \begin{bmatrix} x_1 \\ x_2 \\ x_3 \end{bmatrix}$, $B = \begin{bmatrix} 5 \\ 6 \end{bmatrix}$. ◆

More on Matrix Multiplication. Before we formally articulate a precise definition of general matrix multiplication, let's see how our informal definition of matrix multiplication can be employed in a "real world" context.

Example. A small manufacturer (located in Italy) of semicustom collectible knives produces straight-blade knives and folding knives. The production process for high-quality limited-production knives is labor intensive and has three primary components: (1) cutting (blade, springs, spacers, handle materials, etc.), shaping, and heat treating the steel (tempering); (2) assembling the parts; and (3) sharpening and finishing. The corresponding hours of work for each of these jobs is summarized in the following table:

	Cutting and Shaping	**Assembly**	**Sharpening and Finishing**
Straight knives	3 hours	2 hours	1 hour
Folding knives	5 hours	3 hours	2 hours

This data can be represented in a simplified form by using a matrix:

$$A = \begin{bmatrix} 3 & 2 & 1 \\ 5 & 3 & 2 \end{bmatrix}.$$

The knives are produced in two nearby locations of Italy (Maniago and Belluno), both of which have rich traditions in custom cutlery. The workers at the Maniago site are more experienced craftsmen and are correspondingly paid at a slightly higher hourly rate. The next table summarizes the hourly wages (in euros) paid for each of the three manufacturing components at the two locations:

	Maniago	**Belluno**
Cutting, shaping, and heat treating	8 euros	7 euros
Assembly	7 euros	6 euros
Sharpening and finishing	6 euros	5 euros

As before, the data from this table can be displayed in matrix form:

$$B = \begin{bmatrix} 8 & 7 \\ 7 & 6 \\ 6 & 5 \end{bmatrix}.$$

The cost of making the two different knife types in the two different locations can be determined from the information provided in the tables. For example, the labor cost of making a semicustom straight knife in the town of Maniago is $3 \cdot 8 + 2 \cdot 7 + 1 \cdot 6 = 44$ euros. By the way, such a knife would sell for approximately 150 euros, which is at the low end in semicustom knives.

Note that this cost computation is the entry in the first row, first column of the matrix product AB. In particular, by using the matrix multiplication previously described, we see that

$$AB = \begin{bmatrix} 3 & 2 & 1 \\ 5 & 3 & 2 \end{bmatrix} \begin{bmatrix} 8 & 7 \\ 7 & 6 \\ 6 & 5 \end{bmatrix} = \begin{bmatrix} 3 \cdot 8 + 2 \cdot 7 + 1 \cdot 6 & 3 \cdot 7 + 2 \cdot 6 + 1 \cdot 5 \\ 5 \cdot 8 + 3 \cdot 7 + 2 \cdot 6 & 5 \cdot 7 + 3 \cdot 6 + 2 \cdot 5 \end{bmatrix}$$

$$= \begin{bmatrix} 44 & 38 \\ 73 & 63 \end{bmatrix},$$

and thus the labor costs of making the two different knife types in two different locations are the entries of the matrix product.

Notice that the previous matrix multiplication calculation was dependent on the fact that the number of columns of A was the same as the number of rows of B. If this were not the case, then we would be unable to multiply the matrices by the given method. For example, we could not apply the same definition to compute the product CB, where $C = \begin{bmatrix} 1 & 0 \\ -2 & 3 \end{bmatrix}$ and $B = \begin{bmatrix} 8 & 7 \\ 7 & 6 \\ 6 & 5 \end{bmatrix}$. However, our definition would allow us to compute the products

$$BA = \begin{bmatrix} 8 & 7 \\ 7 & 6 \\ 6 & 5 \end{bmatrix} \begin{bmatrix} 3 & 2 & 1 \\ 5 & 3 & 2 \end{bmatrix}$$

$$= \begin{bmatrix} 8 \cdot 3 + 7 \cdot 5 & 8 \cdot 2 + 7 \cdot 3 & 8 \cdot 1 + 7 \cdot 2 \\ 7 \cdot 3 + 6 \cdot 5 & 7 \cdot 2 + 6 \cdot 3 & 7 \cdot 1 + 6 \cdot 2 \\ 6 \cdot 3 + 5 \cdot 5 & 6 \cdot 2 + 5 \cdot 3 & 6 \cdot 1 + 5 \cdot 2 \end{bmatrix} = \begin{bmatrix} 59 & 37 & 22 \\ 51 & 32 & 19 \\ 43 & 27 & 16 \end{bmatrix},$$

and $BC = \begin{bmatrix} 8 & 7 \\ 7 & 6 \\ 6 & 5 \end{bmatrix} \begin{bmatrix} 1 & 0 \\ -2 & 3 \end{bmatrix} = \begin{bmatrix} 8 \cdot 1 + 7 \cdot (-2) & 8 \cdot 0 + 7 \cdot 3 \\ 7 \cdot 1 + 6 \cdot (-2) & 7 \cdot 0 + 6 \cdot 3 \\ 6 \cdot 1 + 5 \cdot (-2) & 6 \cdot 0 + 5 \cdot 3 \end{bmatrix} = \begin{bmatrix} -6 & 21 \\ -5 & 18 \\ -4 & 15 \end{bmatrix}.$

Our definition of matrix multiplication emerged from a few concrete examples, and with those as templates, we are now prepared to give a general definition.

Definition of Matrix Multiplication. Let $A = [a_{ij}]$ be an $m \times n$ matrix and let $B = [b_{ij}]$ be an $n \times p$ matrix with real number entries. The product, denoted by $AB = [c_{ij}]$, is an $m \times p$ matrix where

$$c_{ij} = a_{i1}b_{1j} + a_{i2}b_{2j} + \cdots + a_{in}b_{nj} = \sum_{t=1}^{n} a_{it}b_{tj}.$$

If the number of columns of A were not equal to the number of rows of matrix B, then the product AB would be not defined. We have already seen examples where the product AB is defined but where BA is not defined. It is easy to see that if $A, B \in M_n(\mathbf{R})$, then the products AB and BA are both defined, although they might not be equal (see the following example). Conversely, if the products AB and BA are both defined, it is not generally true that $A, B \in M_n(\mathbf{R})$ (give a counterexample to the converse). It is interesting to note that if the product AB is a square matrix, then BA is defined and is also a square matrix (not necessarily of the same size). (Why?)

Matrix Multiplication Is Not a Commutative Operation. For example, if $A = \begin{bmatrix} 2 & 0 \\ 1 & -1 \end{bmatrix}$ and $B = \begin{bmatrix} 1 & 3 \\ 0 & -2 \end{bmatrix}$, then

$$AB = \begin{bmatrix} 2 & 0 \\ 1 & -1 \end{bmatrix}\begin{bmatrix} 1 & 3 \\ 0 & -2 \end{bmatrix} = \begin{bmatrix} 2 & 6 \\ 1 & 5 \end{bmatrix} \neq BA = \begin{bmatrix} 1 & 3 \\ 0 & -2 \end{bmatrix}\begin{bmatrix} 2 & 0 \\ 1 & -1 \end{bmatrix} = \begin{bmatrix} 5 & -3 \\ -2 & 2 \end{bmatrix}.$$

Exponent Notation. For $t \geq 1$ and $A \in M_n(\mathbf{R})$, write the product of A with itself t-times as $A^t = AA \cdots A$, and for $s \geq 1$ and $t \geq 1$, observe that $A^s A^t = A^{s+t}$.

Classroom Problems. For $a, b \in \mathbf{R}$, and an integer $t \geq 1$, we know that $(ab)^t = a^t b^t$. (Justify.) The analogous conclusion cannot be drawn for all matrices $A, B \in M_n(\mathbf{R})$. In particular, find matrices $A, B \in M_2(\mathbf{R})$, such that $(AB)^2 \neq A^2 B^2$. ◆

Although matrix multiplication is not a commutative operation, it is an associative operation.

Associative Property for Matrix Multiplication. For the matrices $A = \begin{bmatrix} 1 & -1 \\ 0 & 2 \\ -3 & 0 \end{bmatrix}$,

$B = \begin{bmatrix} 0 & 3 \\ 5 & -2 \end{bmatrix}$, and $C = \begin{bmatrix} 3 & 0 & 7 \\ -4 & 1 & 1 \end{bmatrix}$, note that the products $A(BC)$ and $(AB)C$ are defined. (Why?) Let's compute each product and compare the results.

$$A(BC) = \begin{bmatrix} 1 & -1 \\ 0 & 2 \\ -3 & 0 \end{bmatrix}\left(\begin{bmatrix} 0 & 3 \\ 5 & -2 \end{bmatrix}\begin{bmatrix} 3 & 0 & 7 \\ -4 & 1 & 1 \end{bmatrix}\right) = \begin{bmatrix} 1 & -1 \\ 0 & 2 \\ -3 & 0 \end{bmatrix}\begin{bmatrix} -12 & 3 & 3 \\ 23 & -2 & 33 \end{bmatrix}$$

$$= \begin{bmatrix} -35 & 5 & -30 \\ 46 & -4 & 66 \\ 36 & -9 & -9 \end{bmatrix};$$

$$(AB)C = \left(\begin{bmatrix} 1 & -1 \\ 0 & 2 \\ -3 & 0 \end{bmatrix}\begin{bmatrix} 0 & 3 \\ 5 & -2 \end{bmatrix}\right)\begin{bmatrix} 3 & 0 & 7 \\ -4 & 1 & 1 \end{bmatrix} = \begin{bmatrix} -5 & 5 \\ 10 & -4 \\ 0 & -9 \end{bmatrix}\begin{bmatrix} 3 & 0 & 7 \\ -4 & 1 & 1 \end{bmatrix}$$

$$= \begin{bmatrix} -35 & 5 & -30 \\ 46 & -4 & 66 \\ 36 & -9 & -9 \end{bmatrix}.$$

Our calculations demonstrate that $A(BC) = (AB)C$. This property is called the **associative property of matrix multiplication** and is true in general when the products are defined. For instance, the associative property of multiplication holds for (all elements of) the set $M_n(\mathbf{R})$. Although we will not establish the validity of the associative property in complete generality, it is worth noting that an elementary proof can be constructed once a matrix is realized as a special type of "linear" function.

Classroom Problems. Show that the associative property holds for all 2×2 matrices with real entries, i.e., verify that $A(BC) = (AB)C$ for all A, B, and C in $M_2(\mathbf{R})$. Direction: Let $A = \begin{bmatrix} a_{11} & a_{12} \\ a_{21} & a_{22} \end{bmatrix}$, $B = \begin{bmatrix} b_{11} & b_{12} \\ b_{21} & b_{22} \end{bmatrix}$, and $C = \begin{bmatrix} c_{11} & c_{12} \\ c_{21} & c_{22} \end{bmatrix}$, where all the entries are real numbers. ◆

Scalar Multiplication. Matrix multiplication is an operation defined for matrices possessing specific size requirements. Additionally, it is useful when studying matrix arithmetic and algebra to consider another type of multiplication called **scalar multiplication**. This product involves multiplying each entry of an $m \times n$ matrix by a real number.

More specifically, given a real number r, called a **scalar**, and a matrix $A = [a_{ij}]$ in $M_{m \times n}(\mathbf{R})$, the **scalar product of r and A** is defined to be the $m \times n$ matrix

$$rA = [ra_{ij}] \in M_{m \times n}(\mathbf{R}).$$

For some explicit examples, consider the following scalar products:

$$3 \begin{bmatrix} 1 & 0 \\ 3 & -2 \\ 0 & 4 \end{bmatrix} = \begin{bmatrix} 3 & 0 \\ 9 & -6 \\ 0 & 12 \end{bmatrix} \text{ and } -1 \begin{bmatrix} 2 & -5 \\ -3 & 0 \end{bmatrix} = \begin{bmatrix} -2 & 5 \\ 3 & 0 \end{bmatrix}.$$

In Exercises 6.6, we consider several properties of scalar multiplication.

6.6.2 Classroom Discussion: Additional Matrix Arithmetic Concepts. Exercise 4b of Section 6.4 asked you to show that the sum of two solutions of an m-equation, n-unknown homogeneous linear system of equations is also a solution. With a bit more preparation, we can verify this assertion using a **matrix algebra approach**.

In particular, suppose $X_1 = \begin{bmatrix} u_1 \\ u_2 \\ \vdots \\ u_n \end{bmatrix}$ and $X_2 = \begin{bmatrix} w_1 \\ w_2 \\ \vdots \\ w_n \end{bmatrix}$ are (not necessarily distinct) solutions of the system $AX = \begin{bmatrix} 0 \\ 0 \\ \vdots \\ 0 \end{bmatrix}$, i.e., $AX_1 = \mathbf{0}$ and $AX_2 = \mathbf{0}$, where $\mathbf{0}$

denotes the $n \times 1$ matrix

$$\begin{bmatrix} 0 \\ 0 \\ \vdots \\ 0 \end{bmatrix}.$$

It is natural to define the sum of the matrices X_1 and X_2 as follows:

$$X_1 + X_2 = \begin{bmatrix} u_1 \\ u_2 \\ \vdots \\ u_n \end{bmatrix} + \begin{bmatrix} w_1 \\ w_2 \\ \vdots \\ w_n \end{bmatrix} = \begin{bmatrix} u_1 + w_1 \\ u_2 + w_2 \\ \vdots \\ u_n + w_n \end{bmatrix},$$

and thus we wish to show that

$$A(X_1 + X_2) = \mathbf{0}. \qquad 6.6.3$$

To verify this, and various other matrix algebra statements, we need to develop additional matrix arithmetic properties. Analogous arithmetic properties were studied in the context of modular arithmetic, and in some cases we adapt arguments used for modular systems to matrix systems. ◆

Additive Structure. Extending the component-wise addition just considered leads to a general definition for matrix addition on the set $M_{m \times n}(\mathbf{R})$.

Definition of Matrix Addition. If $A = [a_{ij}]$ and $B = [b_{ij}]$ are both members of the set $M_{m \times n}(\mathbf{R})$, then their **sum** is defined to be the $m \times n$ matrix:

$$A + B = [a_{ij} + b_{ij}] \in M_{m \times n}(\mathbf{R}).$$

For example, if $A = \begin{bmatrix} 1 & 0 & -1 \\ 0 & 3 & 2 \end{bmatrix}$ and $B = \begin{bmatrix} 5 & 6 & 2 \\ -3 & -4 & 0 \end{bmatrix}$, then $A + B = \begin{bmatrix} 6 & 6 & 1 \\ -3 & -1 & 2 \end{bmatrix}.$

Note that matrix addition is only defined for matrices of the same size. Moreover, it is common to use the symbol + to represent the addition of real numbers and the addition of matrices (a similar convention was applied in modular addition).

Section 6.6 Matrix Arithmetic and Matrix Algebra

Commutative Property of Matrix Addition. If $A = [a_{ij}]$ and $B = [b_{ij}]$ are arbitrary members of $M_{m \times n}(\mathbf{R})$, then

$$A + B = [a_{ij}] + [b_{ij}] = [a_{ij} + b_{ij}] = [b_{ij} + a_{ij}]$$
$$= [b_{ij}] + [a_{ij}] = B + A. \text{ (Justify.)}$$

Associative Property of Matrix Addition. If $A = [a_{ij}]$, $B = [b_{ij}]$, and $C = [c_{ij}]$ are arbitrary members of $M_{m \times n}(\mathbf{R})$, then

$$A + (B + C) = [a_{ij}] + ([b_{ij}] + [c_{ij}]) = [a_{ij}] + [(b_{ij} + c_{ij})]$$
$$= [a_{ij} + (b_{ij} + c_{ij})] = [(a_{ij} + b_{ij}) + c_{ij}]$$
$$= [(a_{ij} + b_{ij})] + [c_{ij}] = ([a_{ij}] + [b_{ij}]) + [c_{ij}]$$
$$= (A + B) + C. \text{ (Justify.)}$$

Distributive Property: Matrix Multiplication Distributes over Matrix Addition. Let's first look at a concrete example. Compute $A(B + C)$ and compare it with $AB + AC$, where

$$A = \begin{bmatrix} 2 & 1 \\ 3 & 0 \\ -4 & 5 \end{bmatrix}, B = \begin{bmatrix} 3 & -8 \\ 0 & -2 \end{bmatrix}, C = \begin{bmatrix} 0 & 1 \\ -1 & 0 \end{bmatrix}.$$

$$A(B + C) = \begin{bmatrix} 2 & 1 \\ 3 & 0 \\ -4 & 5 \end{bmatrix} \left(\begin{bmatrix} 3 & -8 \\ 0 & -2 \end{bmatrix} + \begin{bmatrix} 0 & 1 \\ -1 & 0 \end{bmatrix} \right)$$

$$= \begin{bmatrix} 2 & 1 \\ 3 & 0 \\ -4 & 5 \end{bmatrix} \begin{bmatrix} 3 & -7 \\ -1 & -2 \end{bmatrix} = \begin{bmatrix} 5 & -16 \\ 9 & -21 \\ -17 & 18 \end{bmatrix}.$$

$$AB + AC = \begin{bmatrix} 2 & 1 \\ 3 & 0 \\ -4 & 5 \end{bmatrix} \begin{bmatrix} 3 & -8 \\ 0 & -2 \end{bmatrix} + \begin{bmatrix} 2 & 1 \\ 3 & 0 \\ -4 & 5 \end{bmatrix} \begin{bmatrix} 0 & 1 \\ -1 & 0 \end{bmatrix}$$

$$= \begin{bmatrix} 6 & -18 \\ 9 & -24 \\ -12 & 22 \end{bmatrix} + \begin{bmatrix} -1 & 2 \\ 0 & 3 \\ -5 & -4 \end{bmatrix} = \begin{bmatrix} 5 & -16 \\ 9 & -21 \\ -17 & 18 \end{bmatrix}.$$

Therefore, $A(B + C) = AB + AC$. This arithmetic property of matrices is called the **(left) distributive property** of matrix multiplication over addition and holds in general when the products are defined. Similarly, the matrix arithmetic property $(B + C)A = BA + CA$ is called the **(right) distributive property** of matrix multiplication over addition and also holds when the products are defined. For example, the left and right distributive properties are valid in the set $M_n(\mathbf{R})$. The verification of the left and right distributive properties for the set $M_2(\mathbf{R})$ is considered in Problem 6b of Exercises 6.6.

Classroom Problems. For $a, b \in \mathbf{R}$, it follows from the distributive property of the real numbers that $(a + b)^2 = a^2 + 2ab + b^2$. (Justify.) Give an example showing that the analogous result is not true in general for matrices $A, B \in M_2(\mathbf{R})$. ◆

Additive Identity. An $m \times n$ matrix consisting of all zeros is called a **zero matrix** and is denoted by the (bold) numeral **0**. The size of a zero matrix is determined by the context in which the matrix appears, and if there is an ambiguity, then the notation $\mathbf{0}_{m \times n}$ can be used.

Note that if $A \in M_{m \times n}(\mathbf{R})$, then $A + \mathbf{0} = A = \mathbf{0} + A$ (in this context, **0** is the $m \times n$ matrix whose entries are all zero). Thus, **0** is an **additive identity** of the set $M_{m \times n}(\mathbf{R})$, and an argument is needed to show that **0** is the **unique additive identity** of the set $M_{m \times n}(\mathbf{R})$. The approach used to show that [0] is the unique additive identity of $\mathbf{Z}_n$ can be applied to handle this situation.

Analysis. Suppose $L \in M_{m \times n}(\mathbf{R})$ such that $A + L = A = L + A$ for each matrix A in $M_{m \times n}(\mathbf{R})$. We know that $A + \mathbf{0} = A = \mathbf{0} + A$ for each matrix A in $M_{m \times n}(\mathbf{R})$, so

$$\mathbf{0} = L + \mathbf{0} \text{ (viewing } L \text{ as an additive identity)}$$
$$= L \text{ (viewing } \mathbf{0} \text{ as an additive identity).}$$

Therefore, **0** is the unique additive identity of the set $M_{m \times n}(\mathbf{R})$.

Additive Inverse. For a matrix $A \in M_{m \times n}(\mathbf{R})$, notice that $A + (-1)A = \mathbf{0} = (-1)A + A$. (Justify.) The matrix $(-1)A$ is called an **additive inverse** of the matrix A, and it is the unique $m \times n$ matrix where $A + (-1)A = \mathbf{0} = (-1)A + A$.

To verify this uniqueness assertion, we give an argument that is analogous to the one we gave for the uniqueness of additive inverses in $\mathbf{Z}_n$ (see Exercises 4.5, no. 2b).

Analysis. Suppose there is a matrix $B \in M_{m \times n}(\mathbf{R})$ such that $A + B = \mathbf{0} = B + A$. Then,

$$B = B + \mathbf{0} = B + [A + (-1)A] = (B + A) + (-1)A$$
$$= \mathbf{0} + (-1)A = (-1)A. \text{ (Justify.)}$$

The element $(-1)A$ is called **the additive inverse** of A in $M_{m \times n}(\mathbf{R})$ and is denoted by $-A$.

Return to Classroom Discussion 6.6.2. We are now prepared to complete the matrix algebra proof showing that the sum of two solutions of a homogeneous system is also a solution. Recall that it suffices to show that $A(X_1 + X_2) = \mathbf{0}$ (see equation 6.6.3), and this follows easily from the distributive property, since

$$A(X_1 + X_2) = AX_1 + AX_2 = \mathbf{0} + \mathbf{0} = \mathbf{0}. \quad ◆$$

Relating Homogeneous and Nonhomogeneous Systems. For a system $AX = B \neq \mathbf{0}$ ($A \in M_{m \times n}(\mathbf{R})$), the system $AX = \mathbf{0}$ is called the **associated homogeneous system**. The next three observations develop a solution relationship among these systems.

Observation 1. If X_1 is a solution of the system $AX = \mathbf{0}$ and $r \in \mathbf{R}$, then the scalar product rX_1 is also a solution of the system $AX = \mathbf{0}$. To verify this, use Problem 5a of Exercises 6.6 to show that

$$A(rX_1) = r(AX_1) = r\mathbf{0} = \mathbf{0}.$$

Observation 2. If $AX_1 = B$ (i.e., X_1 is a solution of the system $AX = B$) and $AX_0 = \mathbf{0}$ (i.e., X_0 is a solution of the system $AX = \mathbf{0}$), then

$$A(X_1 + X_0) = AX_1 + AX_0 = B + \mathbf{0} = B,$$

so $X_1 + X_0$ is a solution of the system $AX = B$.

Observation 3. If $AX = B \neq \mathbf{0}$ is a consistent system, and X_1 is a particular solution of this system, then any arbitrary solution X_a of $AX = B$ can be written as $X_a = X_1 + X_0$, where X_0 is some solution of the associated homogeneous system $AX = \mathbf{0}$.

Justification. If X_a is an arbitrary solution of the system $AX = B$, then the desired representation can be achieved by writing

$$X_a = X_1 + (X_a - X_1),$$

since $(X_a - X_1)$ is a solution of the associated homogeneous system (apply Observations 1 and 2).

Proving a Previous Result Using a Matrix Algebra Argument. In General Case 6.4.3, we remarked that a general m equation–n unknown system either has no solutions, a unique solution, or infinitely many solutions. This conclusion could be reached by analyzing the reduced row echelon forms of the augmented matrix that represents a given system (we did this in some special cases). Matrix algebra provides us with another means to justify this same conclusion.

6.6.4 Theorem. A linear system $AX = B$ of m equations and n unknowns either has no solution, exactly one solution, or infinitely many solutions.

Proof. If the system $AX = B$ has no solution or one solution, then the proof would be finished, so we may as well assume that the system $AX = B$ has at least two distinct solutions, say $X_1 \neq X_2$. Set $X_0 = (X_1 - X_2) \neq \mathbf{0}$ and observe that

$$A(X_1 - X_2) = AX_1 - AX_2 = B - B = \mathbf{0}.$$

Hence, X_0 is a nonzero solution of the associated homogeneous system $AX = \mathbf{0}$, and thus by Observation 1, rX_0 is also a solution of the system $AX = \mathbf{0}$ for each real number r. Finally, Observation 2 shows that $X_1 + rX_0$ is a solution of the system $AX = B$ for each $r \in \mathbf{R}$, and therefore the system $AX = B$ has infinitely many solutions.

We shall complete this section by introducing another useful concept in the study of matrices.

Transpose of a Matrix. Given the 2×3 matrix $A = \begin{bmatrix} 2 & 3 & -1 \\ -5 & 0 & 4 \end{bmatrix}$, the 3×2 matrix $A^t = \begin{bmatrix} 2 & -5 \\ 3 & 0 \\ -1 & 4 \end{bmatrix}$, whose columns are the rows of A, is called the **transpose of A** and is denoted by A^t. In general, for an $m \times n$ matrix $A = \begin{bmatrix} a_{11} & a_{12} & a_{13} & \cdots & a_{1n} \\ a_{21} & a_{22} & a_{23} & \cdots & a_{2n} \\ a_{31} & a_{32} & a_{33} & \cdots & a_{3n} \\ \vdots & \vdots & \vdots & & \vdots \\ a_{m1} & a_{m2} & a_{m3} & \cdots & a_{mn} \end{bmatrix}$, the transpose of A is the $n \times m$ matrix $A^t = \begin{bmatrix} a_{11} & a_{21} & a_{31} & \cdots & a_{m1} \\ a_{12} & a_{22} & a_{32} & \cdots & a_{m2} \\ a_{13} & a_{23} & a_{33} & \cdots & a_{m3} \\ \vdots & \vdots & \vdots & & \vdots \\ a_{1n} & a_{2n} & a_{3n} & \cdots & a_{mn} \end{bmatrix}$ whose columns are the rows of A.

6.6.5 Classroom Problems. Provide justifications for the following transpose properties:

1. If $A \in M_{m \times n}(\mathbf{R})$, then $(A^t)^t = A^t$.
2. If $A, B \in M_{m \times n}(\mathbf{R})$, then $(A + B)^t = A^t + B^t$.
3. If $r \in \mathbf{R}$ and $A \in M_{m \times n}(\mathbf{R})$, then $(rA)^t = rA^t$.
4. Find $A, B \in M_2(\mathbf{R})$ such that $(AB)^t \neq A^t B^t$.
5. If A, B are matrices such that the product AB is defined, then $(AB)^t = B^t A^t$ (provide justification for the 2×2 case). ◆

EXERCISES 6.6

1. Determine if $AB = BA$ for the following matrices:

   a. $A = \begin{bmatrix} 0 & 0 \\ 0 & 1 \end{bmatrix}$ and $B = \begin{bmatrix} 1 & 0 \\ 0 & 0 \end{bmatrix}$;

   b. $A = \begin{bmatrix} 1 & 0 & 2 \\ -1 & 3 & 0 \\ 0 & 5 & -4 \end{bmatrix}$ and $B = \begin{bmatrix} 0 & 1 & -1 \\ 2 & 0 & 0 \\ 3 & -2 & 0 \end{bmatrix}$;

   c. $A = \begin{bmatrix} 1 & 0 & 0 \\ 0 & 2 & 0 \\ 0 & 0 & 3 \end{bmatrix}$ and $B = \begin{bmatrix} -2 & 0 & 0 \\ 0 & 0 & 0 \\ 0 & 0 & 4 \end{bmatrix}$.

Terminology. The **main diagonal** of an $n \times n$ matrix $A = [a_{ij}]$ consists of the entries a_{ii} for $i = 1, \ldots, n$, and A is called a **diagonal matrix** if $a_{ij} = 0$ whenever $i \neq j$ (the entries off the diagonal are all zero). The matrices A and B in Problem 1c (above) are examples of 3×3 diagonal matrices.

2. **a.** If $A \in M_n(\mathbf{R})$ and A is a diagonal matrix, then $A^t = A$.
   b. Show that if $A, B \in M_n(\mathbf{R})$ are diagonal matrices, then $AB = BA$.

3. **Scalar Multiplication Properties.** Let $A, B \in M_{m \times n}(\mathbf{R})$, and $r, s \in \mathbf{R}$. Establish the following properties:
   a. $r(AB) = (rA)B = A(rB)$. (For this part, assume $A, B \in M_n(\mathbf{R})$.)
   b. $r(sA) = (rs)A$.
   c. $r(A + B) = rA + rB$.
   d. $(r + s)A = rA + sA$.
   e. If $rA = \mathbf{0}$, then $r = 0$ or $A = \mathbf{0}$. Direction: Indirect proof.

4. **a.** Find matrices $A, B, C \in M_2(\mathbf{R})$ such that $A(B + C) \neq (B + C)A$.
   b. Prove the left and right distributive properties for arbitrary matrices $A, B, C \in M_2(\mathbf{R})$. Show: $A(B + C) = AB + AC$ and $(B + C)A = BA + CA$.

5. **a.** Find *nonzero* matrices $A, B \in M_2(\mathbf{R})$ such that $AB = \mathbf{0}$.
   b. Find a *nonzero* matrix $A \in M_2(\mathbf{R})$ such that $A^2 = \mathbf{0}$.
   c. Find *nonzero* matrices $A, B, C \in M_2(\mathbf{R})$ such that $A \neq B$ and $AC = BC$.

6. **a.** Find an example of a *nondiagonal* matrix $A \in M_2(\mathbf{R})$ such that $A = A^t$.
   b. A matrix $A \in M_n(\mathbf{R})$ is called **symmetric** if $A = A^t$. Describe the symmetry of the entries of a symmetric matrix A.
   c. Let $A \in M_n(\mathbf{R})$. Show that $A + A^t$ is a symmetric matrix. Direction: Prove that $(A + A^t)^t = A + A^t$.
   d. Let $A, B \in M_n(\mathbf{R})$ be symmetric matrices. Prove that $A + B$ is a symmetric matrix. Direction: Show that $(A + B)^t = A + B$.
   e. Let $A \in M_{m \times n}(\mathbf{R})$. Show that AA^t and A^tA are symmetric matrices. Direction: Using the properties of the transpose of a matrix (see Classroom Problems 6.6.5), verify that $(AA^t)^t = AA^t$ and $(A^tA)^t = A^tA$.

7. **a.** Find an example of a *nonzero* matrix $A \in M_2(\mathbf{R})$ such that $A^t = -A$.
   b. A matrix $A \in M_n(\mathbf{R})$ is called **skew-symmetric** if $A^t = -A$. Show that the main diagonal entries of a skew-symmetric are all zero.
   c. Let $A \in M_n(\mathbf{R})$. Show that $A - A^t$ is a skew-symmetric matrix. Direction: Prove that $(A - A^t)^t = -(A - A^t)$.
   d. Let $A, B \in M_n(\mathbf{R})$ be skew-symmetric matrices. Prove that $A + B$ is a skew-symmetric matrix. Direction: Show that $(A + B)^t = -(A + B)$.

8. Show that any square matrix $A \in M_n(\mathbf{R})$ can be expressed as $A = S + T$, where S is a symmetric matrix and T is a skew-symmetric matrix. Direction: Use Exercises 6c and 7c.

6.7 MULTIPLICATIVE INVERSES: SOLVING THE MATRIX EQUATION $AX = B$

One of the first types of equations encountered by beginning algebra students is the linear equation $ax = b$, where $a \neq 0$ and b is a real number. Solving this equation, amounts to multiplying both sides of the equation by the multiplicative inverse of a, namely $a^{-1} = \frac{1}{a}$. In particular,

$$\frac{1}{a}(ax) = \frac{1}{a}b \Rightarrow \left(\frac{1}{a}a\right)x = \frac{b}{a} \text{ (real number multiplication is associative)}$$

$$\Rightarrow 1x = \frac{b}{a} \Rightarrow x = \frac{b}{a} \text{ (multiplicative identity).}$$

This section's main objective is to show that a corresponding method can be employed to solve certain matrix equations of the form $AX = B$. Our approach hinges upon developing matrix arithmetic concepts that are analogous to those used in the previous solution method.

Multiplicative Identity for Square Matrices. Given an arbitrary matrix $A \in M_2(\mathbf{R})$, where $A = \begin{bmatrix} a_{11} & a_{12} \\ a_{21} & a_{22} \end{bmatrix}$, observe that

$$\begin{bmatrix} a_{11} & a_{12} \\ a_{21} & a_{22} \end{bmatrix}\begin{bmatrix} 1 & 0 \\ 0 & 1 \end{bmatrix} = \begin{bmatrix} a_{11} & a_{12} \\ a_{21} & a_{22} \end{bmatrix} = \begin{bmatrix} 1 & 0 \\ 0 & 1 \end{bmatrix}\begin{bmatrix} a_{11} & a_{12} \\ a_{21} & a_{22} \end{bmatrix},$$

so the matrix $I_2 = \begin{bmatrix} 1 & 0 \\ 0 & 1 \end{bmatrix}$ is a multiplicative identity for the set $M_2(\mathbf{R})$. In fact, I_2 is the unique **multiplicative identity** for the set $M_2(\mathbf{R})$. To see this, we adapt the approach we previously used to show that $[0]$ is the unique additive identity of $\mathbf{Z}_n$ and that $[1]$ is the unique multiplicative identity of $\mathbf{Z}_n$.

Analysis. Suppose there is a 2×2 matrix J such that $AJ = A = JA$ for each matrix A in $M_2(\mathbf{R})$. We know that $AI_2 = A = I_2A$ for each matrix A in $M_2(\mathbf{R})$, so

$$I_2 = JI_2 \text{ (viewing } J \text{ as a multiplicative identity)}$$

$$= J \text{ (viewing } I_2 \text{ as a multiplicative identity).}$$

Using the same reasoning, it follows that the $n \times n$ matrix I_n that has 1s on the main diagonal and 0s everywhere else is the unique **multiplicative identity** for the set $M_n(\mathbf{R})$.

Multiplicative Inverses for Square Matrices. In our studies thus far, we have considered the notion of multiplicative inverses in some different mathematical systems. For example, we reached the following conclusions concerning multiplicative inverses for the systems $\mathbf{Z}$ = integers, $\mathbf{Q}$ = rationals, $\mathbf{R}$ = reals, $\mathbf{Z}_n$ = integers modulo n:

a. $z \in \mathbf{Z}$ has a multiplicative inverse in $\mathbf{Z} \Leftrightarrow z = \pm 1$.

Section 6.7 Multiplicative Inverses: Solving the Matrix Equation $AX = B$

b. $0 \neq q \in \mathbf{Q}$ has a multiplicative inverse in $\mathbf{Q}$, and $q^{-1} = \frac{1}{q} \in \mathbf{Q}$.

c. $0 \neq r \in \mathbf{R}$ has a multiplicative inverse in $\mathbf{R}$, and $r^{-1} = \frac{1}{r} \in \mathbf{R}$.

d. $[a] \in \mathbf{Z}_n$ has a multiplicative inverse in $\mathbf{Z}_n \Leftrightarrow \gcd(a, n) = 1$; in this case, the Euclidean algorithm is used to find integers s and t so that $1 = as + tn$. Consequently, $[1] = [as + tn] = [as] + [tn] = [a][s]$ in $\mathbf{Z}_n$, and thus $[a]^{-1} = [s]$.

It is important to note that in each of these systems, when a multiplicative inverse exists, it is unique.

In terms of matrices, our intent is to

a. Determine when a matrix $A \in M_n(\mathbf{R})$ has a **multiplicative inverse** $B \in M_n(\mathbf{R})$ that satisfies $AB = I_n = BA$; establish a method for finding that inverse.

b. Establish the uniqueness of matrix multiplicative inverses.

As usual, it is instructive to first look at some examples.

6.7.1 Examples. 1. Consider the matrices $A = \begin{bmatrix} 1 & 3 \\ 2 & 4 \end{bmatrix}$ and $B = \begin{bmatrix} -2 & 3/2 \\ 1 & -1/2 \end{bmatrix}$ and notice that

$$AB = \begin{bmatrix} 1 & 3 \\ 2 & 4 \end{bmatrix} \begin{bmatrix} -2 & 3/2 \\ 1 & -1/2 \end{bmatrix} = \begin{bmatrix} 1 & 0 \\ 0 & 1 \end{bmatrix} = I_2 = \begin{bmatrix} -2 & 3/2 \\ 1 & -1/2 \end{bmatrix} \begin{bmatrix} 1 & 3 \\ 2 & 4 \end{bmatrix} = BA.$$

The matrix B is a multiplicative inverse of the matrix A, likewise, the matrix A is a multiplicative inverse of the matrix B.

2. For a matrix $A \in M_n(\mathbf{R})$ to have a multiplicative inverse, it is clear that $A \neq \mathbf{0}$. (Why?) However, this is not sufficient to guarantee the existence of a multiplicative inverse for A, since the nonzero matrix $A = \begin{bmatrix} 1 & 2 \\ 2 & 4 \end{bmatrix}$ does not have a multiplicative inverse.

To substantiate this assertion, we mimic the indirect argument (from modular arithmetic) that was used to show $[3] \in \mathbf{Z}_9$ does not have a multiplicative inverse in $\mathbf{Z}_9$ (see Example 4.6.1). A key point of that proof was that $[3]$ is a divisor of zero in $\mathbf{Z}_9$, i.e., $[3][3] = [0]$ in $\mathbf{Z}_9$.

Justification. (Indirect): Assume there is a matrix $B \in M_2(\mathbf{R})$ such that $AB = I_2 = BA$. Consider the nonzero matrix $C = \begin{bmatrix} -2 & 1 \\ -2 & 1 \end{bmatrix}$, and notice that $CA = \mathbf{0}$. By multiplying both sides of the equality $AB = I_2$ on the left by C (recall that matrix multiplication is not commutative), we obtain the equality $C(AB) = CI_2 = C$. Thus, since matrix multiplication is associative, it follows that

$$C(AB) = (CA)B = \mathbf{0}B = \mathbf{0} = C.$$

This contradiction completes the proof. Therefore, A does not have a multiplicative inverse.

Before we determine conditions that a square matrix needs in order to have a multiplicative inverse and how to find that inverse, we borrow an argument from modular arithmetic (see Section 4.6.2) to show that matrix multiplicative inverses are unique (provided they exist).

6.7.2 Matrix Multiplicative Inverses Are Unique. If a nonzero matrix $A \in M_n(\mathbf{R})$ has a multiplicative inverse $B \in M_n(\mathbf{R})$ so that $AB = I_n = BA$, then it is unique.

Proof. Suppose there are matrices $B, C \in M_n(\mathbf{R})$ such that $AB = I_n = BA$ and $AC = I_n = CA$. We will show that $B = C$. Multiply each side of the equality $AB = AC$ on the left by B to obtain the equality $B(AB) = B(AC)$, and then use associativity to get $(BA)B = (BA)C$. By definition, $BA = I_n$, so $B = C$. Therefore, we have demonstrated that matrix multiplicative inverses are unique (provided they exist).

A critical point of this proof is that, by definition, matrix multiplicative inverses must commute. This subtlety in the definition of multiplicative inverses was not mentioned in our previous work on multiplicative inverses in $\mathbf{Z}_n$, since multiplication in that system is commutative.

Terminology and Notation. If a nonzero matrix $A \in M_n(\mathbf{R})$ has a (unique) multiplicative inverse $B \in M_n(\mathbf{R})$ with $AB = I_n = BA$, then A is **invertible** or **nonsingular**, and the unique matrix B is denoted by A^{-1}, i.e., $AA^{-1} = I_n = A^{-1}A$. Analogously, A^{-1} is invertible, and its unique inverse is the matrix $(A^{-1})^{-1} = A$. A square matrix that is not invertible is called **singular**.

Corollary. If $A, B \in M_n(\mathbf{R})$ are each invertible matrices (not necessarily inverses of each other), then AB is also an invertible matrix, and $(AB)^{-1} = B^{-1}A^{-1}$.

Proof. Suppose that $A, B \in M_n(\mathbf{R})$ are each invertible matrices, i.e., A^{-1} and B^{-1} exist. Note that the matrix $(B^{-1}A^{-1})$ is a multiplicative inverse of AB since

$$(AB)(B^{-1}A^{-1}) = A(BB^{-1})A = AI_n A^{-1} = AA^{-1} = I_n \text{ and}$$

$$(B^{-1}A^{-1})(AB) = B^{-1}(A^{-1}A)B = B^{-1}I_n B = B^{-1}B = I_n.$$

Therefore, as multiplicative inverses are unique, it follows that $(AB)^{-1} = B^{-1}A^{-1}$.

Classroom Problems. Let $A \in M_n(\mathbf{R})$. Prove the following statements:

a. If A is invertible, then A^n is invertible for each positive integer n, and $(A^n)^{-1} = (A^{-1})^n$. Direction: Show that $A^n(A^{-1})^n = I_n = (A^{-1})^n A^n$.

b. If A is invertible, then A^t is invertible and $(A^t)^{-1} = (A^{-1})^t$. Direction: Show that $A^t(A^{-1})^t = I_n = (A^{-1})^t A^t$ by using the transpose property $(AB)^t = B^t A^t$ (see no. 5 of Classroom Problems 6.6.5.)

Section 6.7   Multiplicative Inverses: Solving the Matrix Equation $AX = B$ 301

c. If A^t is invertible, then A is invertible. Direction: If A^t is invertible, then $(A^t)^t$ is invertible by part b. ◆

Finding Matrix Multiplicative Inverses. In Examples 6.7.1, no. 1, we observed that the matrices $A = \begin{bmatrix} 1 & 3 \\ 2 & 4 \end{bmatrix}$ and $B = \begin{bmatrix} -2 & 3/2 \\ 1 & -1/2 \end{bmatrix}$ are multiplicative inverses of each other, i.e., $B = A^{-1}$ and $AA^{-1} = I_2 = A^{-1}A$.

Questions.   1. Given the matrix A, how was the matrix A^{-1} determined?

2. More generally, given a matrix $A \in M_n(\mathbf{R})$, how do we know if the matrix A has an inverse, and if it does, how do we find it?

Our analysis of Question 1 leads to an answer for Question 2.

Given the matrix $A = \begin{bmatrix} 1 & 3 \\ 2 & 4 \end{bmatrix}$, we are interested in discovering a procedure for finding the inverse matrix A^{-1}. A simple approach is to let $A^{-1} = \begin{bmatrix} b_{11} & b_{12} \\ b_{21} & b_{22} \end{bmatrix}$ and attempt to solve the matrix equation

$$\begin{bmatrix} 1 & 3 \\ 2 & 4 \end{bmatrix} \begin{bmatrix} b_{11} & b_{12} \\ b_{21} & b_{22} \end{bmatrix} = \begin{bmatrix} 1 & 0 \\ 0 & 1 \end{bmatrix}.$$

Multiplying the matrices on the left-hand side of the equation gives

$$\begin{bmatrix} b_{11} + 3b_{21} & b_{12} + 3b_{22} \\ 2b_{11} + 4b_{21} & 2b_{12} + 4b_{22} \end{bmatrix} = \begin{bmatrix} 1 & 0 \\ 0 & 1 \end{bmatrix},$$

which gives rise (by the definition of matrix equality) to the following two systems of equations:

System 1	**System 2**
$b_{11} + 3b_{21} = 1$	$b_{12} + 3b_{22} = 0$
$2b_{11} + 4b_{21} = 0$	$2b_{12} + 4b_{22} = 1$

It is important to note that the matrix A is invertible (i.e., A^{-1} exists) if and only if Systems 1 and 2 are consistent. (Explain.) Actually, since inverses are unique, it follows that the matrix A is invertible if and only if Systems 1 and 2 each have exactly one solution.

To solve Systems 1 and 2, apply Gauss-Jordan elimination to the associated augmented matrices

$$\begin{bmatrix} 1 & 3 & 1 \\ 2 & 4 & 0 \end{bmatrix}, \begin{bmatrix} 1 & 3 & 0 \\ 2 & 4 & 1 \end{bmatrix};$$

and since each of these systems has the same coefficient matrix A, they can be solved concurrently by combining them to form the augmented matrix

$$\begin{bmatrix} 1 & 3 & 1 & 0 \\ 2 & 4 & 0 & 1 \end{bmatrix}.$$

Notation. The augmented matrix, where I_2 is appended to A, is denoted by $[A : I_2]$. Employing Gauss-Jordan elimination to the augmented matrix $[A : I_2]$ gives

$$\begin{bmatrix} 1 & 3 & 1 & 0 \\ 2 & 4 & 0 & 1 \end{bmatrix} \xrightarrow{(-2)R1+R2} \begin{bmatrix} 1 & 3 & 1 & 0 \\ 0 & -2 & -2 & 1 \end{bmatrix} \xrightarrow{(-1/2)R2} \begin{bmatrix} 1 & 3 & 1 & 0 \\ 0 & 1 & 1 & -1/2 \end{bmatrix}$$

$$\xrightarrow{(-3)R2+R1} \begin{bmatrix} 1 & 0 & -2 & 3/2 \\ 0 & 1 & 1 & -1/2 \end{bmatrix},$$

and thus the solution to System 1 is $b_{11} = -2$, $b_{21} = 1$; the solution to System 2 is $b_{12} = 3/2$, $b_{22} = -1/2$. Hence, A is invertible and

$$A^{-1} = \begin{bmatrix} b_{11} & b_{12} \\ b_{21} & b_{22} \end{bmatrix} = \begin{bmatrix} -2 & 3/2 \\ 1 & -1/2 \end{bmatrix}.$$

Conclusion. The matrix A is invertible if and only if $[A : I_2]$ is row equivalent to $[I_2 : B]$, and in this case, $A^{-1} = B$. Furthermore, $[A : I_2]$ is row equivalent to $[I_2 : B]$ if and only if A is row equivalent to I_2, so the matrix A is invertible if and only if A can be transformed into I_2 by a finite number of elementary row operations.

Using the same reasoning as in the 2×2 case, it follows more generally that a matrix $A \in M_n(\mathbf{R})$ is invertible if and only if A is row equivalent to I_n.

We have previously demonstrated (through an indirect argument) that the matrix $A = \begin{bmatrix} 1 & 2 \\ 2 & 4 \end{bmatrix}$ of Examples 6.7.1 no. 2 does not have a multiplicative inverse. A direct approach to confirm that A is not invertible (singular) can be accomplished through elementary row operations. In particular,

$$\begin{bmatrix} 1 & 2 \\ 2 & 4 \end{bmatrix} \xrightarrow{(-2)R1+R2} \begin{bmatrix} 1 & 2 \\ 0 & 0 \end{bmatrix},$$

and thus A is not row equivalent to I_2.

Alternate Method for Finding the Inverse of a 2×2 Matrix. We can use the fact that a 2×2 matrix $A = \begin{bmatrix} a_{11} & a_{12} \\ a_{21} & a_{22} \end{bmatrix}$ is invertible if and only if A is row equivalent to I_2 to obtain an elementary test for invertibility and a shorthand method for finding A^{-1}.

Analysis. Let's start by assuming A is invertible and see what conditions are forced upon the entries of A. Since $A \sim I_2$, it follows that $a_{11} \neq 0$ or $a_{21} \neq 0$. (Why?)

Section 6.7 Multiplicative Inverses: Solving the Matrix Equation $AX = B$

Without loss of generality, we may assume that $a_{11} \neq 0$ (if not, interchange rows 1 and 2 and relabel). Note that

$$\begin{bmatrix} a_{11} & a_{12} & 1 & 0 \\ a_{21} & a_{22} & 0 & 1 \end{bmatrix} \xrightarrow{1/a_{11}} \begin{bmatrix} 1 & a_{12}/a_{11} & 1/a_{11} & 0 \\ a_{21} & a_{22} & 0 & 1 \end{bmatrix} \xrightarrow{(-a_{21})R1+R2}$$

$$\begin{bmatrix} 1 & a_{12}/a_{11} & 1/a_{11} & 0 \\ 0 & (-a_{21})(a_{12}/a_{11}) + a_{22} & -a_{21}/a_{11} & 1 \end{bmatrix}$$

$$= \begin{bmatrix} 1 & a_{12}/a_{11} & 1/a_{11} & 0 \\ 0 & (a_{11}a_{22} - a_{12}a_{21})/a_{11} & -a_{21}/a_{11} & 1 \end{bmatrix},$$

and $(a_{11}a_{22} - a_{12}a_{21})/a_{11} \neq 0$, as $A \sim I_2$. (Why?) Furthermore, $(a_{11}a_{22} - a_{12}a_{21})/a_{11} \neq 0$ if and only if $a_{11}a_{22} - a_{12}a_{21} \neq 0$. (Why?)

Continuing with Gauss-Jordan elimination gives

$$\begin{bmatrix} 1 & a_{12}/a_{11} & 1/a_{11} & 0 \\ 0 & (a_{11}a_{22} - a_{12}a_{21})/a_{11} & -a_{21}/a_{11} & 1 \end{bmatrix} \xrightarrow{a_{11}/(a_{11}a_{22}-a_{12}a_{21})R2}$$

$$\begin{bmatrix} 1 & a_{12}/a_{11} & 1/a_{11} & 0 \\ 0 & 1 & -a_{21}/(a_{11}a_{22} - a_{12}a_{21}) & a_{11}/(a_{11}a_{22} - a_{12}a_{21}) \end{bmatrix} \xrightarrow{(-a_{12}/a_{11})R2+R1}$$

$$\begin{bmatrix} 1 & 0 & (1/a_{11}) + (-a_{12}/a_{11})[-a_{21}/(a_{11}a_{22} - a_{12}a_{21})] & (-a_{12}/a_{11})[a_{11}/(a_{11}a_{22} - a_{12}a_{21})] \\ 0 & 1 & -a_{21}/(a_{11}a_{22} - a_{12}a_{21}) & a_{11}/(a_{11}a_{22} - a_{12}a_{21}) \end{bmatrix}$$

$$= \begin{bmatrix} 1 & 0 & a_{22}/(a_{11}a_{22} - a_{12}a_{21}) & -a_{12}/(a_{11}a_{22} - a_{12}a_{21}) \\ 0 & 1 & -a_{21}/(a_{11}a_{22} - a_{12}a_{21}) & a_{11}/(a_{11}a_{22} - a_{12}a_{21}) \end{bmatrix}.$$

Therefore, if A is invertible, then $(a_{11}a_{22} - a_{12}a_{21}) \neq 0$ and

$$A^{-1} = \frac{1}{(a_{11}a_{22} - a_{12}a_{21})} \begin{bmatrix} a_{22} & -a_{12} \\ -a_{21} & a_{11} \end{bmatrix}.$$

Conversely, if $(a_{11}a_{22} - a_{12}a_{21}) \neq 0$, then

$$\begin{bmatrix} a_{11} & a_{12} \\ a_{21} & a_{22} \end{bmatrix} \frac{1}{(a_{11}a_{22} - a_{12}a_{21})} \begin{bmatrix} a_{22} & -a_{12} \\ -a_{21} & a_{11} \end{bmatrix} = \begin{bmatrix} 1 & 0 \\ 0 & 1 \end{bmatrix} \text{(verify)},$$

so A is invertible.

Summary. Given a 2×2 matrix $A = \begin{bmatrix} a_{11} & a_{12} \\ a_{21} & a_{22} \end{bmatrix}$, the number $(a_{11}a_{22} - a_{12}a_{21})$ is called the **determinant of A** and is denoted by **det(A)**. We have just proved:

6.7.3 A is invertible if and only if $\det(A) \neq 0$, and when $\det(A) \neq 0$, then

$$A^{-1} = \frac{1}{\det(A)} \begin{bmatrix} a_{22} & -a_{12} \\ -a_{21} & a_{11} \end{bmatrix}.$$

Revisiting Some Examples. Previously we showed that the matrix $A = \begin{bmatrix} 1 & 3 \\ 2 & 4 \end{bmatrix}$ was invertible and used Gauss-Jordan elimination to find its inverse. Notice that

$$\det(A) = 1 \cdot 4 - 3 \cdot 2 = (-2) \neq 0,$$

so by 6.7.3 we see that

$$A^{-1} = \frac{1}{-2} \begin{bmatrix} 4 & -3 \\ -2 & 1 \end{bmatrix} = \begin{bmatrix} -2 & 3/2 \\ 1 & -1/2 \end{bmatrix}$$

(agreeing with our earlier calculation).

Also, we considered the matrix $A = \begin{bmatrix} 1 & 2 \\ 2 & 4 \end{bmatrix}$ and demonstrated that it is not invertible. This follows directly from 6.7.3, since $\det(A) = 1 \cdot 4 - 2 \cdot 2 = 0$.

Determinant of an $n \times n$ Matrix. Although we will not pursue the study of determinants for general $n \times n$ matrices, we do note that determinants are an important tool in the study of matrices, especially for theoretical purposes (it is difficult to compute $\det(A)$ for large n). Moreover, with the general definition of determinant, A is invertible if and only if $\det(A) \neq 0$.

Classroom Problems. Determine whether the following matrices are invertible or not invertible, and if invertible, find the inverse. For the 2×2 matrices given, use characterization 6.7.3.

a. $\begin{bmatrix} 0 & 1 \\ 1 & 0 \end{bmatrix}$
b. $\begin{bmatrix} 2 & -2 \\ 3 & 3 \end{bmatrix}$
c. $\begin{bmatrix} 2 & -2 \\ -3 & 3 \end{bmatrix}$
d. $\begin{bmatrix} 1 & 1 & 0 \\ 2 & 1 & 1 \\ 3 & 2 & 2 \end{bmatrix}$

e. $\begin{bmatrix} 1 & 1 & 2 \\ 2 & 1 & 3 \\ 1 & 1 & 2 \end{bmatrix}$
f. $\begin{bmatrix} 2 & 0 & 0 \\ 0 & 1 & 0 \\ 0 & 0 & 3 \end{bmatrix}$. ◆

Square Systems $AX = B$ Having Unique Solutions. We have laid the necessary groundwork so that we may characterize when a square system $AX = B$ (A is a square matrix) has a unique solution and then determine that unique solution.

Section 6.7   Multiplicative Inverses: Solving the Matrix Equation $AX = B$

6.7.4 Theorem. Let $AX = B$ be a system of linear equations, where $A \in M_n(\mathbf{R})$,

$$X = \begin{bmatrix} x_1 \\ x_2 \\ \vdots \\ x_n \end{bmatrix}, B = \begin{bmatrix} b_1 \\ b_2 \\ \vdots \\ b_n \end{bmatrix}, \text{ and } b_i \in \mathbf{R} \text{ for each } 1 \leq i \leq n.$$ The system $AX = B$ has a unique solution if and only if A is invertible. In this case, the unique solution is given by $X = A^{-1}B$.

Proof. ($\Rightarrow$) Assume the system $AX = B$ has a unique solution. In terms of Gauss-Jordan elimination, this means that the following augmented matrices are row equivalent:

$$[A : B] = \begin{bmatrix} a_{11} & a_{12} & \cdots & a_{1n} & b_1 \\ a_{21} & a_{22} & \cdots & a_{2n} & b_2 \\ & & \vdots & & \\ a_{n1} & a_{n2} & \cdots & a_{nn} & b_n \end{bmatrix} \sim [I_n : C] = \begin{bmatrix} 1 & 0 & \cdots & 0 & c_1 \\ 0 & 1 & \cdots & 0 & c_2 \\ & & \vdots & & \\ 0 & 0 & \cdots & 1 & c_n \end{bmatrix},$$

where C is some $n \times 1$ matrix. Hence, A is row equivalent to I_n, and thus from our previous considerations, A is invertible.

($\Leftarrow$) For this direction, assume A is invertible. Hence, A is row equivalent to I_n, and so the augmented matrix $[A : B]$ is row equivalent to the augmented matrix $[I_n : C]$ for some $n \times 1$ matrix C. Therefore, the system $AX = B$ has exactly one solution.

The next corollary follows immediately from the theorem if the matrix B of constants is taken to be the $n \times 1$ zero matrix.

Classroom Problems. Use result 6.7.3 and Theorem 6.7.4 to find the unique solution of the system

$$\begin{aligned} 2x_1 + 2x_2 &= 3 \\ 4x_1 - x_2 &= -1 \end{aligned};$$

in matrix notation, $\begin{bmatrix} 2 & 2 \\ 4 & -1 \end{bmatrix} \begin{bmatrix} x_1 \\ x_2 \end{bmatrix} = \begin{bmatrix} 3 \\ -1 \end{bmatrix}.$ ◆

6.7.5 Corollary. (Same notation as in Theorem 6.7.4.) A homogeneous system $AX = \mathbf{0}$ has the unique solution $X = \mathbf{0}$ if and only if A is invertible. (Justify.)

The next result consolidates our work concerning the invertibility of a square matrix.

6.7.6 Invertible Matrices. The following statements are equivalent for a matrix $A \in M_n(\mathbf{R})$:

1. A is invertible.
2. A is row equivalent to I_n.
3. $AX = B$ has a unique solution for any $n \times 1$ matrix B.
4. $AX = \mathbf{0}$ has a unique solution.
5. $\det(A) \neq 0$ (we only established this equivalence for $n = 2$).
6. A^t is invertible.

EXERCISES 6.7

1. Which of the following matrices are invertible or singular? Find the inverse of those that are invertible.

   a. $\begin{bmatrix} -2 & 3/2 \\ 1 & -1/2 \end{bmatrix}$

   b. $\begin{bmatrix} 1 & -2 \\ 3 & 6 \end{bmatrix}$

   c. $\begin{bmatrix} 1 & 2 & 2 \\ 1 & 1 & 2 \\ 1 & 2 & 2 \end{bmatrix}$

   d. $\begin{bmatrix} 1 & 0 & 1 \\ 2 & 1 & 1 \\ 3 & 2 & 0 \end{bmatrix}$

   e. $\begin{bmatrix} -2 & 0 & 0 \\ 0 & -3 & 0 \\ 0 & 0 & 4 \end{bmatrix}$

2. Characterize when a diagonal matrix $A = \begin{bmatrix} a_{11} & 0 & 0 & \cdots & 0 \\ 0 & a_{22} & 0 & \cdots & 0 \\ 0 & 0 & a_{33} & \cdots & 0 \\ \vdots & & & & \vdots \\ 0 & 0 & 0 & \cdots & a_{nn} \end{bmatrix}$ is invertible, and find its inverse.

3. Find an example of a *nondiagonal* matrix $A \in M_2(\mathbf{R})$ such that $A^2 = I_2$. Conclude that $A = A^{-1}$.
4. Find an example of a *nondiagonal* invertible matrix $A \in M_2(\mathbf{R})$ such that the entries of both A and A^{-1} are integers, i.e., $A, A^{-1} \in M_2(\mathbf{Z})$. Direction: Use result 6.7.3.
5. a. Find *nonzero* matrices $A, B, C \in M_2(\mathbf{R})$ such that $AB = AC$ and $B \neq C$.
   b. Let $A, B, C \in M_n(\mathbf{R})$. Show that if A is invertible and $AB = AC$, then $B = C$.
6. a. Find *nonzero* matrices $A, B \in M_2(\mathbf{R})$ such that $AB = \mathbf{0}$.
   b. Let $A, B \in M_n(\mathbf{R})$. Show that if A is invertible and $AB = \mathbf{0}$, then $B = \mathbf{0}$.
7. a. Find a *nonzero* matrix $A \in M_2(\mathbf{R})$ such that $A \neq I_2$ and $A^2 = A$.
   b. Let $A \in M_2(\mathbf{R})$. Show that if $A \neq I_2$ and $A^2 = A$, then A is singular. Direction: Use an indirect proof.

6.8 CODING WITH MATRICES

In Chapter 4 (Section 4.7, Application 3), we investigated elementary coding theory (cryptology) with a focus on linear codes. Our work on matrix arithmetic has provided us with powerful new tools for encoding and decoding messages, and we shall now describe this coding process in detail.

Section 6.8 Coding with Matrices

As in linear codes, the following table assigns an integer to each letter of the alphabet. In addition, it is useful to represent a blank space _ by the integer 26.

A	B	C	D	E	F	G	H	I	J	K	L	M
0	1	2	3	4	5	6	7	8	9	10	11	12
N	O	P	Q	R	S	T	U	V	W	X	Y	Z
13	14	15	16	17	18	19	20	21	22	23	24	25

To illustrate the matrix coding process, let's encode and then decode the famous quotation of the mathematician/philosopher René Descartes (1596–1650), **"I think, therefore I am"** (in Latin, "Cogito, ergo sum").

For this particular example, we encode the message with the 3×3 invertible matrix

$$A = \begin{bmatrix} -1 & -1 & 1 \\ 2 & 2 & -1 \\ 3 & 4 & -2 \end{bmatrix},$$

and then decode it with the 3×3 inverse matrix,

$$A^{-1} = \begin{bmatrix} 0 & 2 & -1 \\ 1 & -1 & 1 \\ 2 & 1 & 0 \end{bmatrix} \text{ (check that } AA^{-1} = I_3 = A^{-1}A\text{).}$$

You will soon discover that any 3×3 invertible matrix could be used in the encoding and decoding scheme we are about to describe.

First, let's break the sentence into segments of length three (including spaces between words) and associate each three-component segment with the corresponding integer assignments specified in the table. If necessary, we add spaces at the end to assure that each segment has three components.

$$\begin{bmatrix} I & \_ & T \\ 8 & 26 & 19 \end{bmatrix} \begin{bmatrix} H & I & N \\ 7 & 8 & 13 \end{bmatrix} \begin{bmatrix} K & \_ & T \\ 10 & 26 & 19 \end{bmatrix} \begin{bmatrix} H & E & R \\ 7 & 4 & 17 \end{bmatrix}$$

$$\begin{bmatrix} E & F & O \\ 4 & 5 & 14 \end{bmatrix} \begin{bmatrix} R & E & \_ \\ 17 & 4 & 26 \end{bmatrix} \begin{bmatrix} I & \_ & A \\ 8 & 26 & 0 \end{bmatrix} \begin{bmatrix} M & \_ & \_ \\ 12 & 26 & 26 \end{bmatrix}.$$

Next we convert the ordered three-component blocks of integers into an ordered list of 3×1 column matrices:

$$\begin{bmatrix} 8 \\ 26 \\ 19 \end{bmatrix} \begin{bmatrix} 7 \\ 8 \\ 13 \end{bmatrix} \begin{bmatrix} 10 \\ 26 \\ 19 \end{bmatrix} \begin{bmatrix} 7 \\ 4 \\ 17 \end{bmatrix} \begin{bmatrix} 4 \\ 5 \\ 14 \end{bmatrix} \begin{bmatrix} 17 \\ 4 \\ 26 \end{bmatrix} \begin{bmatrix} 8 \\ 26 \\ 0 \end{bmatrix} \begin{bmatrix} 12 \\ 26 \\ 26 \end{bmatrix}.$$

Encoding. To encode the message, we multiply the matrix A times each of the 3×1 column matrices to obtain a new set of ordered 3×1 column matrices:

$$A \begin{bmatrix} 8 \\ 26 \\ 19 \end{bmatrix} = \begin{bmatrix} -15 \\ 49 \\ 90 \end{bmatrix}, A \begin{bmatrix} 7 \\ 8 \\ 13 \end{bmatrix} = \begin{bmatrix} -2 \\ 17 \\ 27 \end{bmatrix}, A \begin{bmatrix} 10 \\ 26 \\ 19 \end{bmatrix} = \begin{bmatrix} -17 \\ 53 \\ 96 \end{bmatrix}, A \begin{bmatrix} 7 \\ 4 \\ 17 \end{bmatrix} = \begin{bmatrix} 6 \\ 5 \\ 3 \end{bmatrix}$$

$$A \begin{bmatrix} 4 \\ 5 \\ 14 \end{bmatrix} = \begin{bmatrix} 5 \\ 4 \\ 4 \end{bmatrix}, A \begin{bmatrix} 17 \\ 4 \\ 26 \end{bmatrix} = \begin{bmatrix} 5 \\ 16 \\ 15 \end{bmatrix}, A \begin{bmatrix} 8 \\ 26 \\ 0 \end{bmatrix} = \begin{bmatrix} -34 \\ 68 \\ 128 \end{bmatrix}, A \begin{bmatrix} 12 \\ 26 \\ 26 \end{bmatrix} = \begin{bmatrix} -12 \\ 50 \\ 88 \end{bmatrix}.$$

Now that the message is encoded, the ordered list of columns can be rewritten as an ordered list of three-component blocks of integers and then finally transmitted in the following list form:

$$-15, 49, 90, -2, 17, 27, -17, 53, 96, 6, 5, 3,$$
$$5, 4, 4, 5, 16, 15, -34, 68, 128, -12, 50, 88.$$

Decoding. The encoding process can easily be reversed (decoding) if the receiver first converts the ordered sequence of integers into an ordered list of 3×1 column matrices and then multiplies each of these 3×1 matrices by A^{-1}. To see why this action reverses the process, observe the following implications:

$$A \begin{bmatrix} 8 \\ 26 \\ 19 \end{bmatrix} = \begin{bmatrix} -15 \\ 49 \\ 90 \end{bmatrix} \Rightarrow A^{-1} \left(A \begin{bmatrix} 8 \\ 26 \\ 19 \end{bmatrix} \right) = A^{-1} \begin{bmatrix} -15 \\ 49 \\ 90 \end{bmatrix}$$

$$\Rightarrow \begin{bmatrix} 8 \\ 26 \\ 19 \end{bmatrix} = A^{-1} \begin{bmatrix} -15 \\ 49 \\ 90 \end{bmatrix}.$$

After multiplying each of the 3×1 column matrices obtained from the transmitted sequence by A^{-1}, the receiver then rewrites the 3×1 column matrices as three-component segments (1×3 matrices) and translates back into letters:

$$\begin{bmatrix} 8 & 26 & 19 \\ I & \_ & T \end{bmatrix} \begin{bmatrix} 7 & 8 & 13 \\ H & I & N \end{bmatrix} \begin{bmatrix} 10 & 26 & 19 \\ K & \_ & T \end{bmatrix} \begin{bmatrix} 7 & 4 & 17 \\ H & E & R \end{bmatrix}$$

$$\begin{bmatrix} 4 & 5 & 14 \\ E & F & O \end{bmatrix} \begin{bmatrix} 17 & 4 & 26 \\ R & E & \_ \end{bmatrix} \begin{bmatrix} 8 & 26 & 0 \\ I & \_ & A \end{bmatrix} \begin{bmatrix} 12 & 26 & 26 \\ M & \_ & \_ \end{bmatrix}.$$

This encoding/decoding process can be applied with invertible matrices of other sizes. For instance, if the encoding matrix A was a 2×2 invertible matrix, then we would simply break the message to be encoded into segments of length two and proceed as in the 3×3 case.

Classroom Problems. Use the 3×3 invertible matrix $A = \begin{bmatrix} 2 & 1 & -1 \\ 0 & 3 & -2 \\ 0 & 0 & -1 \end{bmatrix}$ to encode the following quotation of Blaise Pascal, 1623–1662: **"It is not certain that everything is uncertain."** ◆

EXERCISES 6.8

1. Encode the following statements with the given invertible matrices:
   a. **"Beauty is truth, truth beauty, that is all"** (from the poem, "Ode on a Grecian Urn" by John Keats, 1795–1821). Encoding matrix: $A = \begin{bmatrix} 2 & 1 \\ 1 & 1 \end{bmatrix}$.
   b. **"Factory windows are always broken"** (from the poem "Factory Windows Are Always Broken" by Vachel Lindsay, 1879–1931). Encoding matrix: $A = \begin{bmatrix} 1 & 0 & 1 \\ 0 & 2 & 1 \\ 1 & 1 & 2 \end{bmatrix}$.

2. The following number sequences represent encoded statements. Decode each one by finding the inverse of the given encoding matrix and applying the method described in this section.
   a. 20, 34, 16, 21, 44, 62, 19, 23, 22, 33, 7, 10, 27, 28, 2, 4, 32, 54, 17, 34, 21, 39, 34, 42, 44, 70, 16, 27, 20, 26. Encoding matrix: $A = \begin{bmatrix} 1 & 1 \\ 1 & 2 \end{bmatrix}$.
   b. 26, 48, 30, 32, 30, 38, 45, 29, 36, 27, 35, 36, 48, 30, 26, 48, 34, 30, 22, 37, 37, 23, 24, 27, 48, 48, 52. Encoding matrix: $A = \begin{bmatrix} 1 & 1 & 0 \\ 1 & 0 & 1 \\ 0 & 1 & 1 \end{bmatrix}$. (This message is engraved on the tombstone of the mathematician David Hilbert, 1862–1943.)

Glossary

Abundant integer: A positive integer is called abundant if $\sum_{d|n} d > 2n$; see Section 3.5.

***a* congruent to *b* modulo *n*:** If integers a and b have the same residue (mod n), then a is said to be congruent to b modulo n. This is denoted by $a \equiv b \pmod{n}$; see Section 4.4.

***a* ≡ *b* (mod *n*):** For a fixed integer $n \geq 2$ and integers a and b, $a \equiv b \pmod{n} \Leftrightarrow n \mid (a - b)$; see Section 4.4.

Addition (mod *n*): For elements $[a]$ and $[b]$ in $\mathbf{Z}_n$, addition (mod n) is defined as $[a] + [b] = [a + b]$; see Section 4.4.

Additive identity in $\mathbf{M}_{m \times n}(\mathbf{R})$: The zero matrix $\mathbf{0}$, is the unique element of $\mathbf{M}_{m \times n}(\mathbf{R})$ with the property $A + \mathbf{0} = A = \mathbf{0} + A$, for each $A \in \mathbf{M}_{m \times n}(\mathbf{R})$; see Section 6.6.

Additive identity in $\mathbf{Z}_n$: The element $[0] \in \mathbf{Z}_n$ is the unique element of $\mathbf{Z}_n$ with the property $[0] + [a] = [a] = [a] + [0]$, for each integer a; see Section 4.5.

Additive inverse in $\mathbf{M}_{m \times n}(\mathbf{R})$: For a matrix $A \in \mathbf{M}_{m \times n}(\mathbf{R})$, the matrix $(-1)A$ is called the additive inverse of the matrix A and has the property $A + (-1)A = \mathbf{0} = (-1)A + A$. The additive inverse of A in $\mathbf{M}_{m \times n}(\mathbf{R})$ is denoted by $-A$; see Section 6.6.

Additive inverses in $\mathbf{Z}_n$: For an element $[a]$ in $\mathbf{Z}_n$, the element $[-a]$ is called the additive inverse of a and has the property: $[a] + [-a] = [0] = [-a] + [a]$. The additive inverse of $[a]$ in $\mathbf{Z}_n$ is denoted by $-[a]$; see Section 4.5.

Arithmetic sequence/common difference: A sequence $a_1, a_2, a_3, \ldots$, where each term a_n is obtained by adding a fixed real number d (called the common difference) to the previous term a_{n-1} (where $n > 1$). Also called an arithmetic progression; see Section 1.3.

Associative property of addition in $\mathbf{Z}_n$: The property $[a] + ([b] + [c]) = ([a] + [b]) + [c]$ for each integer a, b, and c; see Section 4.5.

Associative property of matrix addition: The property $A + (B + C) = (A + B) + C$ for each A, B, and C in $\mathbf{M}_{m \times n}(\mathbf{R})$; see Section 6.6.

312   Glossary

Associative property of matrix multiplication: The property that $A(BC) = (AB)C$, when the matrix products are defined; see Section 6.6.

Associative property of multiplication irrn Z_n: The property $[a]([b][c]) = ([a][b])[c]$ for each integer a, b, and c; see Section 4.5.

Augmented matrix: The augmented matrix of a general system of m linear equations in n unknowns is the $m \times (n + 1)$ matrix consisting of the coefficients and the constants of the system; see Section 6.4.

Base b expanded form of an integer L/ base b place value: Let b be an integer with $b \geq 2$. If L is a positive integer, then there exists some integer $n \geq 0$ such that L can be expressed uniquely in the base b expanded form $L = r_n b^n + r_{n-1} b^{n-1} + \ldots + r_2 b^2 + r_1 b^1 + r_0 b^0$, where $r_i \in \{0, 1, 2, \ldots, b - 1\}$ for each i and $r_n \neq 0$. The base b place value notation for L is given by $L = (r_n r_{n-1} \ldots r_2 r_1 r_0)_b$; see Section 3.4.

Binomial theorem/binomial coefficients: $(x + y)^n = \sum_{r=0}^{n} \binom{n}{r} x^{n-r} y^r$ for each positive integer n. The coefficients $\binom{n}{r}$ are called binomial coefficients; see Section 1.10.

Canonical form: Ordering the distinct prime factors of any integer $n > 1$ in increasing magnitude $p_1 < p_2 < \cdots < p_s$, the prime factorization of n can be written uniquely in the canonical form $n = p_1^{m_1} p_2^{m_2} \cdots p_s^{m_s}$; see Section 2.6 (for modified canonical form, see Section 2.7).

Coefficient matrix: Given a system of m linear equations in n unknowns, the $m \times n$ matrix that consists of the coefficients; see Section 6.4.

Combination/number of combinations: A combination of r objects from a set A of n objects is a subset of A consisting of r objects. The number of such combinations is $\binom{n}{r} = \frac{n!}{r!(n-r)!}$; see Section 1.9.

Common divisor: A nonzero integer d is a common divisor of integers a and b if d is both a divisor of a and a divisor of b (i.e., if $a = dx$ and $b = dy$ for some integers x and y); see Section 2.5.

Common multiple: An integer m is a common multiple of integers a and b if m is both a multiple of a and a multiple of b (i.e., if $m = ax$ and $m = by$ for some integers x and y); see Section 2.5.

Commutative property of addition in Z_n: $[a] + [b] = [b] + [a]$ for each integer a and b; see Section 4.5.

Commutative property of matrix addition: $A + B = B + A$ for each A and B in $M_{m \times n}(\mathbf{R})$; see Section 6.6.

Commutative property of multiplication in Z_n: $[a][b] = [b][a]$ for each integer a and b; see Section 4.5.

Composite number: If an integer $n > 1$ has more than two distinct positive factors, then it is a composite number (or just composite); see Section 2.6.

Consistent system: A system of linear equations having at least one solution; see Section 6.4.

Cryptology: The field of study pertaining to encryption (also called cryptography); see Section 4.7.

Deficient integer: A positive integer is deficient if $\sum_{d|n} d < 2n$; see Section 3.5.

Diagonal matrix: An $n \times n$ matrix $A = [a_{ij}]$ is a diagonal matrix if $a_{ij} = 0$ whenever $i \neq j$; see Section 6.6.

Diophantine equations: Equations having integer coefficients and integer solutions; see Section 3.2.

Distance formula: The distance between points $A = (x_1, y_1)$ and $B = (x_2, y_2)$ is given by $AB = \sqrt{(x_2 - x_1)^2 + (y_2 - y_1)^2}$; see Section 5.4.

Distributive property of matrices: The matrix arithmetic property $A(B + C) = AB + AC$ is called the left distributive property of matrix multiplication over addition and holds in general when the products are defined. The matrix arithmetic property $(B + C)A = BA + CA$ is called the right distributive property of matrix multiplication over addition and also holds when the products are defined; see Section 6.6.

Distributive property of multiplication over addition in Z_n: $[a]([b] + [c]) = [a][b] + [a][c]$ for each integer a, b, c; see Section 4.5.

Division/quotient/remainder: The determination of integers q and r so that $b = qa + r$ and $0 \leq r < a$ is called division of b by a with quotient q and remainder r and is denoted by $b \div a$ or $a\overline{)b}$; see Section 3.1.

Divisor (factor): A *nonzero* integer a is a divisor (factor) of an integer b if $b = ac$, for some integer c; in this situation, b is said to be divisible by a. If a is a divisor of b, we write $a \mid b$ and read this as a divides b; see Section 2.4.

Elementary row operations: Operations that do not affect the solution set of a system of linear equations. These operations are: interchange any two equations, multiply an equation by a nonzero constant, and add a multiple of an equation to another equation; see Section 6.4.

Equality of matrices: Two matrices $A = [a_{ij}]$ and $B = [b_{ij}]$ in the set $M_{m \times n}(\mathbf{R})$ are said to be equal if $a_{ij} = b_{ij}$ for each i and j; see Section 6.4.

Equivalent systems: Two systems of linear equations having the same solution set; see Section 6.4.

Even and odd integers: An integer b is even if it is divisible by 2, which means $b = 2q$ for some integer q. An integer b that is not even is odd; in this case, $b = 2q + 1$ for some integer q; see Section 3.1.

Factoring/factorization: Expressing a positive integer as a product of some of its factors. The resulting representation is called a factorization of the number; see Section 2.6.

Factor tree: A tree diagram displaying factorizations of a given positive integer n; see Section 2.6.

Fermat's Last Theorem: Fermat's claim that the equation $x^n + y^n = z^n$ has no positive integer solutions for all integers $n > 2$; see Section 5.10.

Fibonacci sequence: The sequence of positive integers defined by the recursive rule, $F_0 = 1, F_1 = 1$, and $F_n = F_{n-1} + F_{n-2}$ for $n \geq 2$; see Section 1.11.

Finite geometric series: The sum of finite geometric sequence $a, ra, r^2a, \ldots, r^{n-1}a$. This is denoted by $S_n = a + ar + ar^2 + \cdots + ar^{n-1}$; see Section 1.7.

Fundamental Counting Principle: If there are m_1 ways to complete task M_1, m_2 to complete task $M_2, \ldots$, and m_n ways to complete task M_n, then there are $m_1 m_2 \cdots m_n$ different ways for the tasks $M_1, M_2, M_3, \ldots$, and M_n to be collectively completed; see Section 1.9.

Gauss-Jordan elimination: An algorithmic procedure for finding the solution(s) of a system of linear equations by performing a finite number of elementary row operations; see Section 6.4.

Geometric sequence/common ratio/n^{th} term of a geometric sequence: A sequence where each successive term is obtained by multiplying the previous term by a fixed number is called a geometric sequence, and the fixed number multiplier is called the common ratio. The nth term of a geometric sequence $a, ra, r^2a, \ldots, r^{n-1}a, \ldots$ with common ratio r is given by $a_n = r^{n-1}a$; see Section 1.7.

Golden ratio (golden mean, golden section): The number $\phi = \frac{1+\sqrt{5}}{2} \approx$ 1.61803398875 (also called golden mean or golden section); see Section 1.11.

Golden rectangle: A rectangle whose length-to-width ratio is approximately equal to the golden ratio; see Section 1.11.

Greatest common divisor: Let a and b be integers, not both zero. Then, $\gcd(a,b) = d$ provided:

1. d is a positive integer;
2. $d \mid a$ and $d \mid b$;
3. if $g \mid a$ and $g \mid b$, where g is a positive integer, then $g \leq d$; see Section 2.5.

Homogeneous system/associated homogeneous system: For a system $AX = B \neq \mathbf{0}$ ($A \in M_{m \times n}(\mathbf{R})$), the system $AX = \mathbf{0}$ is called the associated homogeneous system; see Section 6.6.

Hyperbola/center of hyperbola/standard form of hyperbola: Let F_1 and F_2 be two fixed points in the plane. A hyperbola H with foci at F_1 and F_2 is defined to be the set of points in the plane $H = \{P : |PF_1 - PF_2| = k\}$, where k is a fixed real number. The midpoint of the segment $\overline{F_1F_2}$ is called the center of the hyperbola. The standard form of the equation of the hyperbola with center at (0, 0), foci $F_1 = (-c, 0)$, $F_2 = (c, 0)$, where $c > 0$ and $k = 2a$ is: $\frac{x^2}{a^2} - \frac{y^2}{b^2} = 1$, where $b = \sqrt{c^2 - a^2}$; see Section 5.8.

Idempotent: An element $[a]$ in $\mathbf{Z}_n$ with the property that $[a]^2 = [a]$; see Section 4.6.

Inconsistent system: A system of linear equations that has no solution; see Section 6.4.

Integers: $\mathbf{Z} = \{\ldots, -3, -2, -1, 0, 1, 2, 3, \ldots\}$; see Section 2.4.

Integers (mod n): The set $\mathbf{Z}_n = \{[0]_n, [1]_n, [2]_n, \ldots, [n-1]_n\}$ consisting of n-distinct residue classes (mod n); see Section 4.4.

Interpolating polynomial of minimal degree: A unique polynomial, whose graph passes through a finite set of specified points; see Section 6.5.

Irrational numbers: Real numbers having a nonrepeating decimal expansion; see Section 6.3.

Kaprekar's constant: In 1949, the Indian mathematician D.R. Kaprekar discovered the properties of the number 6,174, which is called Kaprekar's constant; see Section 4.7.

Lattice points: Points in the plane having integer coordinates; see Section 3.3.

Least common multiple: Let a and b be nonzero integers. Then, $\text{lcm}(a, b) = m$ provided:

1. m is a positive integer,
2. $a \mid m$ and $b \mid m$,
3. if $a \mid c$ and $b \mid c$, where c is a positive integer, then $m \leq c$; see Section 2.5.

Matrix: A rectangular array of real numbers; see Section 6.4.

Matrix addition: If $A = [a_{ij}]$ and $B = [b_{ij}]$ are both members of the set $M_{m \times n}(\mathbf{R})$, then their sum is defined to be the $m \times n$ matrix: $A + B = [a_{ij} + b_{ij}] \in M_{m \times n}(\mathbf{R})$; see Section 6.6.

Matrix multiplication: Let $A = [a_{ij}]$ be an $m \times n$ matrix and $B = [b_{ij}]$ be an $n \times p$ matrix with real number entries. The product, denoted by $AB = [c_{ij}]$, is an $m \times p$ matrix, where $c_{ij} = a_{i1}b_{1j} + a_{i2}b_{2j} + \cdots + a_{in}b_{nj} = \sum_{t=1}^{n} a_{it}b_{tj}$; see Section 6.6.

$M_{m \times n}(\mathbf{R})$: The set of all $m \times n$ matrices with real number entries is denoted by $M_{m \times n}(\mathbf{R})$. If $m = n$ (square matrices), then the notation $M_n(\mathbf{R})$ is used; see Section 6.4.

Mersenne primes: Primes of the form $2^m - 1$; see Section 3.5.

m^{th} power: A positive integer $n > 1$ is an mth power if there is a positive integer a such that $n = a^m$; see Section 2.6.

Multiple: An integer b is a multiple of an integer a if $b = ca$, for some integer c; see Section 2.4.

Multiplication (mod n): For elements $[a]$ and $[b]$ in $\mathbf{Z}_n$ multiplication (mod n) is defined as $[a] \cdot [b] = [a \cdot b]$; see Section 4.4.

Multiplicative identity for the set $M_n(\mathbf{R})$: The $n \times n$ matrix I_n, having 1s on the main diagonal and 0s everywhere else. This matrix has the property that $A I_n = I_n A = A$ for each $A \in M_n(\mathbf{R})$; see Section 6.7.

Multiplicative identity in $\mathbf{Z}_n$: $[1][a] = [a][1] = [a]$ for each integer a; see Section 4.5.

Multiplicative inverses in $\mathbf{Z}_n$: A nonzero element $[a]$ in $\mathbf{Z}_n$ has a multiplicative inverse if there exists a nonzero $[b]$ in $\mathbf{Z}_n$ so that $[a][b] = [1]$ (and likewise $[b]$ has a multiplicative inverse in $\mathbf{Z}_n$); see Section 4.6.

n factorial: The definition of n factorial is given by $n! = n(n-1)(n-2)(n-3) \cdot 2 \cdot 1$; see Section 1.8.

Nilpotent: An element $[a]$ of $\mathbf{Z}_n$ is nilpotent if there is some positive integer t so that $[a]^t = [0]$; see Section 4.6.

Nonnegative integers (whole numbers): $\mathbf{Z}^+ \cup \{0\} = \{0, 1, 2, 3, \ldots\}$; see Section 2.4.

Nonzero-divisor in $\mathbf{Z}_n$: An element $[a]$ in $\mathbf{Z}_n$ is a nonzero-divisor if there does not exist a nonzero $[b]$ in $\mathbf{Z}_n$ so that $[a][b] = [0]$ in $\mathbf{Z}_n$ (this terminology is restricted to nonzero elements of $\mathbf{Z}_n$); see Section 4.6.

Nonzero integers: $(\mathbf{Z} - \{0\}) = \{z \in \mathbf{Z} : z \neq 0\}$; see Section 2.4.

One-to-one function: A function f defined from a set X into a set Y is one-to-one or injective if $f(x_1) \neq f(x_2)$ whenever $x_1 \neq x_2$ (x_1, x_2 are elements of X); see Section 4.7.

Order or size of a matrix: A matrix with m rows and n columns is said to have order (or size) $m \times n$ (read, m by n). If $m = n$, then the matrix is a square matrix of size $n \times n$. The entry in the ith row and j^{th} column of the matrix A is called the $(ij)^{th}$ entry and is denoted by a_{ij}; see Section 6.4.

Pascal's triangle/Pascal's formula: The triangular array of numbers defined by Pascal's formula, $\binom{n+1}{r} = \binom{n}{r-1} + \binom{n}{r}$, where $1 \leq r \leq n$; see Section 1.9.

Perfect number: A perfect number is a positive integer n that is equal to the sum of all its positive divisors other than n itself; see Section 3.5.

Permutation/number of permutations: A permutation of r objects from a set of n objects is an ordered arrangement of the r objects. The number of such permutations equals $_nP_r = n(n-1) \cdots ((n-r) + 1)$; see Section 1.9.

Positive integers (natural numbers, whole numbers): $\mathbf{Z}^+ = \{z \in \mathbf{Z} : z > 0\}$; see Section 2.4.

Power set: For a set X, the set of all subsets of X. This is denoted by $\mathcal{P}(X)$; see Section 1.8.

Prime number: If an integer $n > 1$ has exactly two distinct positive factors, it is a prime number (or just prime); see Section 2.6.

Primitive Pythagorean triple (PPT)/Primitive Pythagorean triangle: A PT $\{x, y, z\}$ with $\gcd(x, y, z) = 1$ is a primitive Pythagorean triple (PPT), and a right triangle whose sides form a PPT is a primitive Pythagorean triangle; see Section 5.7.

Principle of mathematical induction/inductive hypothesis: For each positive integer n, let $S(n)$ be a statement that is either true or false. The statement $S(n)$ is true for each positive integer n, provided the following two conditions are satisfied:

a. **Basis step:** $S(1)$ is true.
b. **Inductive step:** If $S(k)$ is true for any positive integer k, then $S(k + 1)$ is true.

The hypothesis "If $S(k)$ true for any positive integer k" is called the inductive hypothesis; see Section 1.8.

Pythagorean triangle/Pythagorean triple (PT): A right triangle with integer length sides x, y, z is a Pythagorean triangle, and the set $\{x, y, z\}$ is a Pythagorean triple (PT); see Section 5.7.

Rational numbers: A number is rational if and only if it can be expressed as the quotient $\frac{a}{b}$ of integers a and b, where $b \neq 0$. Equivalently, a number is rational if and only if it has a repeating decimal representation (terminating decimal expansions are considered repeating, since they are zero from some finite point on). The set of all rational numbers is usually denoted by **Q**; see Section 6.3.

Recursive pattern/recursive rule or formula: A pattern where, after some explicit terms are specified, each subsequent term is defined in terms of a previous term or a combination of previous terms is a recursive pattern. A rule that describes the relationship between these consecutive terms is a recursive rule or formula; see Section 1.2.

Reduced row echelon form/leading one: An $m \times n$ matrix is in reduced row echelon form if (a) all rows consisting entirely of zeros (if there are any) are placed at the bottom of the matrix; (b) the first nonzero entry of each (nonzero) row is 1 (referred to as the leading one of that row); (c) given two nonzero consecutive rows j and $j + 1$, the leading one of row $j + 1$ is to the right of the leading one of row j; (d) if a column contains a leading one, then all other entries of that column are zero; see Section 6.4.

Relatively prime: Two positive integers a and b are said to be relatively prime if they have no common prime factor. Note that a and b are relatively prime if and only if $\gcd(a, b) = 1$; see Section 2.6.

Residue class of a (mod n): $[a]_n = \{x \in \mathbf{Z} : x \text{ and } a \text{ have the same residue modulo } n\}$; see Section 4.4.

Residue modulo n: For a fixed integer $n \geq 2$ and any integer a, the remainder resulting from division of a by n (abbreviated as $a(\bmod n)$); see Section 4.4.

Row equivalent matrices: Two matrices $A, B \in M_{m \times n}(\mathbf{R})$ are row equivalent (denoted by $A \sim B$) if one can be transformed into the other by a finite sequence of elementary row operations; see Section 6.4.

Scalar product: Given a real number r, called a scalar, and a matrix $A = [a_{ij}]$ in $M_{m \times n}(\mathbf{R})$, the scalar product of r and A is defined to be the $m \times n$ matrix $rA = [ra_{ij}] \in M_{m \times n}(\mathbf{R})$; see Section 6.6.

Self-inverses: Note that $[1]$ and $[n-1]$ always have multiplicative inverses in $\mathbf{Z}_n$, since $\gcd(1, n) = 1 = \gcd(n-1, n)$. In particular, they are their own inverses (self-inverses); see Section 4.8.

Sequence/finite sequence: A function f whose domain is the positive integers (into any other set) is called an infinite sequence (or simply a sequence); see Section 1.2. A function f whose domain is the set $N_n = \{1, 2, 3, \ldots, n\}$, where n is a positive integer is called a finite sequence; see Section 1.3.

Skew-symmetric matrix: A matrix $A \in M_n(\mathbf{R})$ is skew-symmetric if $A^t = -A$; see Section 6.6.

Square in $\mathbf{Z}_n$: An element $[c]$ in $\mathbf{Z}_n$ is a square in $\mathbf{Z}_n$ if the equation $X^2 - [c] = [0]$ has a solution $X = [a]$ in $\mathbf{Z}_n$; see Section 4.11.

Square matrix: A matrix of order $n \times n$; see Section 6.4.

Square number: A positive integer n is a square (or a perfect square) if there is a positive integer a so that $n = a^2$; see Section 2.6.

Strong induction: For each positive integer n, let $S(n)$ be a statement that is either true or false. The statement $S(n)$ is true for each positive integer n provided the following two conditions are satisfied:

  a. **Basis step:** $S(1)$ is true.
  b. **Inductive step:** If $S(t)$ is true for all positive integers $t \leq k$, then $S(k + 1)$ is true; see Section 1.11.

Symmetric matrix: A matrix $A \in M_n(\mathbf{R})$ is symmetric if $A = A^t$; see Section 6.6.

Transpose of A: The transpose of $m \times n$ matrix $A = [a_{ij}]$ is the $n \times m$ matrix $A^t = [a_{ji}]$, whose columns are the rows of A; see Section 6.6.

Triangle inequality: The length of a triangle's side is always less than the sum of the lengths of the other two sides; see Section 5.8.

Triangular number: The nth triangular number T_n is defined to be the number $T_n = 1 + 2 + 3 + \cdots + n$; see Section 1.5.

Twin primes: A pair of primes (p_t, p_{t+1}) separated by *one* composite, i.e., $p_{t+1} - p_t = 2$ are referred to as twin primes; see Section 3.5.

$\tau(n)$: Number of positive divisors of a positive integer n. The correspondence $n \to \tau(n)$ defines a function from $\mathbf{Z}^+$ into $\mathbf{Z}^+$. Note that $\tau(p^m) = m + 1$, where p is a prime number; see Section 2.6.

Well-Ordering Principle: Every nonempty set of positive integers has a smallest element; see Section 2.5.

Zero-divisor in $\mathbf{Z}_n$: A nonzero element $[a]$ in $\mathbf{Z}_n$ is a zero-divisor or a divisor of zero if there is a nonzero $[b]$ in $\mathbf{Z}_n$ so that $[a][b] = [0]$ (in this case, $[b]$ is also a zero-divisor); see Section 4.6.

Zero matrix: The $m \times n$ matrix consisting of all zeros. This is denoted by the (bolded) numeral **0**; see Section 6.6.

References

1. Adler, A., and Coury, J. *The Theory of Numbers—A Text and Source Book for Problems.* Jones and Bartlett Publishers, 1995.
2. Billstein, R, Williamson, J, et al. *MathThematics.* McDougal-Littell/Houghton Mifflin, 1999.
3. Billstein, R., Libeskind, S., and Lott, J. *A Problem Solving Approach to Mathematics for Elementary School Teachers*, 7th ed. Addison-Wesley, 2001.
4. Burton, D. *Elementary Number Theory*, 2nd ed. Wm. C. Brown Publishers, 1989.
5. Conference Board of Mathematical Sciences. *The Mathematical Education of Teachers.* Washington, DC: Mathematics Association of America, 2001.
6. Education Development Center, Newton, MA. *MathScape.* Creative Publications, 1998.
7. Edwards, H. *Fermat's Last Theorem.* Springer-Verlag, 1977.
8. Fraleigh, J. *A First Course in Abstract Algebra*, 7th ed. Addison-Wesley, 2003.
9. Honsberger, R. "Stories in Combinatorial Geometry." *The Two-Year College Mathematics Journal*, Vol. 10, No. 5 (1979): 344–347.
10. Lappan, G. et al. *Connected Mathematics Project.* Prentice Hall, 2004.
11. Larson, R., and Edwards, B. *Elementary Linear Algebra.* D. C. Heath and Company, 1991.
12. Leitzel, J. *A Call for Change: Recommendations for the Mathematical Preparation of Teachers of Mathematics.* Washington, DC: Mathematics Association of America, 2001.
13. McCoy, N. *Introduction to Modern Algebra.* Allyn and Bacon, Inc., 1975.
14. National Commission on Teaching and America's Future. *Before It's Too Late: A Report to the Nation from the National Commission on Mathematics and Science Teaching for the 21st Century.* Washington, DC: U.S. Department of Education, 2000.
15. National Council of Teachers of Mathematics. *Curriculum and Evaluation Standards for School Mathematics*, Reston, VA: National Council of Teachers of Mathematics, 1989.
16. National Council of Teachers of Mathematics. *Principles and Standards for School Mathematics.* Reston, VA: National Council of Teachers of Mathematics, 2000.
17. Niven, I., and Zuckerman, H. *An Introduction to the Theory of Numbers.* John Wiley & Sons, 1966.
18. Ore, O. *Invitation to Number Theory.* Random House, 1967.
19. Papick, I., Beem, J., Reys, B., and Reys, R. "Impact of the Missouri Middle Mathematics Project on the Preparation of Prospective Middle School Teachers." *Journal of Mathematics Teacher Education,* 2 (1999): 301–310.
20. Ribenboim, P. *The New Book of Prime Number Records.* Springer-Verlag, 1996.
21. Romberg, T, Meyer, M, et al. *Mathematics in Context.* Encyclopedia Britannica, 1998.
22. Rothbart, A. *The Theory of Remainders.* Janson Publications, 1995.
23. Smith, K. *The Nature of Mathematics*, 4th ed. Brooks/Cole Publishing, 1984.

24. Stein, S. *Mathematics: The Man-made Universe*, 2nd ed. W. H. Freeman and Company, 1969.
25. Weeks, A., and Adkins, J. *A Course in Geometry—Plane and Solid*. Ginn and Company, 1961.

Answers to (Most) Odd-Numbered Exercises

CHAPTER 1

1.2: Reflections on Classroom Connections: Representing Patterns

1. Recursive rule: $P_1 = 6$
   $P_n = P_{n-1} + 4$
   Explicit rule: $P_n = 4n + 2$

3. a.

n	$S(n)$
1	4
2	6
3	8
4	10
5	12
6	14

   b. $P_1 = 4$
   $P_n = P_{n-1} + 2$
   c. $P_n = 2n + 2$
   d. $17(2) + 2 = 36$
   e. $P_{24} = 50$ and $P_{25} = 52$, thus 25 tables are needed.

5. a. 9, 18, 27, 51, 81, 105
   b. $P_1 = 6$
   $P_n = P_{n-1} + 3$
   c. $P_n = 3n + 3$
   d. 78
   e. $P_{439} = 439 \cdot 3 + 3 = 1{,}320$

7. a. $F(n) = 7n - 1$
   b. $a_1 = 5$
   $a_n = a_{n-1} + 3$
   c. $a_1 = 5$
   $a_n = 3a_{n-1}$

1.3: Arithmetic Sequences

1. $\dfrac{f(8) - f(3)}{5} = \dfrac{24 - 9}{5} = 3$
$f(3) = a_1 + 3(3 - 1)$
$9 = a_1 + 6$
$a_1 = 3$
$a_n = a_{n-1} + 3$

1.5: Reflections on Classroom Connections: A Quadratic Sequence

1. **a.** $T_{3n+1} = 1 + 2 + \ldots + (3n + 1) = \dfrac{(3n + 2)(3n + 1)}{2}$
$= \dfrac{9n^2 + 9n + 2}{2}$
$= \dfrac{9n(n + 1) + 2}{2} = 9T_n + 1$

c. $T_n + T_{n+1} = \dfrac{n(n + 1)}{2} + \dfrac{(n + 1)(n + 2)}{2} = \dfrac{(n + 1)(2n + 2)}{2}$
$= (n + 1)^2$

1.6: Finite Arithmetic Sequences

1. **a.** $\dfrac{9(1 + 17)}{2} = 81$

1.7: Geometric Sequences

1. Problem 1.2
   a.

Square	# of rubas
1	1
2	2
3	4
4	8
5	16
6	32

   b. It doubles each time.
   c. 524,288; 536,870,912; 9.2×10^{18}
   d. 21st square
   e. $5.12; $5,242.88; $5,368,709.12

3. Solve the inequality $r^2 - r - 1 < 0$. The upper bound is $\dfrac{1+\sqrt{5}}{2}$ and the lower bound is $\dfrac{1-\sqrt{5}}{2}$.

1.8: Mathematical Induction

1. a. Inductive step:
$$1 + 3 + 5 + \ldots + (2n - 1) + (2n + 1) = n^2 + 2n + 1 = (n + 1)^2$$

b. Inductive step:
$$3 + 3^2 + \ldots + 3^n + 3^{n+1} = \frac{3^{n+1} - 3}{2} + 3^{n+1} = \frac{3 \cdot 3^{n+1} - 3}{2}$$
$$= \frac{3^{n+2} - 3}{2}$$

c. Basis step:
For $n = 5$
$2^5 = 32 > 25 = 5^2$
Inductive step:
Assume for $n = k$ that $2^k > k^2$
$$2^{k+1} = 2^k + 2^k > k^2 + k^2 > k^2 + 2k + 1 = (k + 1)^2$$

g. Inductive step:
Assume 4 is a factor of $5^k - 1$.
$5^{k+1} - 1 = (5^k - 1)5 - 4$, which can be shown to be divisible by 4.

1.9: Classroom Connection: Counting Tools

1. a. $8!$

b. $\dfrac{26!}{18!} = 26 \cdot 25 \cdot 24 \cdot 23 \cdot 22 \cdot 21 \cdot 20 \cdot 19 = 65{,}854{,}152{,}000$

3. $9 \cdot 10 \cdot 10 \cdot 10 \cdot 5 = 45{,}000$

5. $\dfrac{12!}{10! \cdot 2!} = \dfrac{12 \cdot 11}{2} = 66$

7. a. $2 \cdot 2 \cdot 2 = 8$

b. m^n

9. a. $\dbinom{n + 1}{n} = \dfrac{(n + 1)!}{n! \cdot 1!} = n + 1$

b. $\dbinom{n}{n - 2} = \dfrac{n!}{(n - 2)! \cdot 2!} = \dfrac{n(n - 1)}{2}$

11. a. Think of arranging 12 units by separating them into four categories using dividers. For example, $\underbrace{xxxx}_{x_1} | \underbrace{xxx}_{x_2} | \underbrace{}_{x_3} | \underbrace{xxxxx}_{x_4}$ would represent the solution $x_1 = 4, x_2 = 3, x_3 = 0, x_4 = 5$. Thus, you have a total of fifteen spots ($12 + 4 - 1$), twelve of which you must place units in. Using the notation of combinations, this would be $\dbinom{12 + 4 - 1}{12}$.

b. $\dbinom{37 + 5 - 1}{37}$

c. $\binom{r+n-1}{r}$

1.10: The Binomial Theorem

1. a.
$$(x+5)^8 = \binom{8}{0}x^8 + \binom{8}{1}x^7y + \binom{8}{2}x^6y^2$$
$$+ \ldots + \binom{8}{7}xy^7 + \binom{8}{8}y^8$$
$$= x^8 + 8x^7y + 28x^6y^2 + 56x^5y^3 + 70x^4y^4 + 56x^3y^5$$
$$+ 28x^2y^6 + 8xy^7 + y^8$$

c.
$$(-2x+5y)^5 = \binom{5}{0}(-2x)^5 + \binom{5}{1}(-2x)^4(5y) + \ldots + \binom{5}{5}(5y)^5$$
$$= -32x^5 + 400x^4y - 2{,}000x^3y^2 + 5{,}000x^2y^3$$
$$- 6{,}250xy^4 + 3{,}125y^5$$

3. $0 = (1-1)^n = \sum_{r=0}^{n}\binom{n}{r}1^{n-r}(-1)^r = \sum_{r=0}^{n}(-1)^r\binom{n}{r}$

1.11: The Fibonacci Sequence

1. Inductive step:
Assume $F_k < 2^k$ and $F_{k-1} < 2^{k-1}$.
Thus, $F_{k+1} = F_k + F_{k-1} < 2^k + 2^{k-1} < 2^k + 2^k = 2^{k+1}$.

CHAPTER 2

2.1: A Few Mathematical Questions Concerning the Periodical Cicadas

1. $1{,}998 - (13 \cdot 17) = 1{,}998 - 221 = 1{,}777$; $1{,}998 + (13 \cdot 17) = 1{,}998 + 221 = 2{,}219$. The last time they appeared was in 1777, and the next time they will appear is 2219.

2.4: Multiples and Divisors

1. Property 1: $a|b \Rightarrow b = ax$ for some $x \in \mathbf{Z}$, and $b|a \Rightarrow a = by$ for some $y \in \mathbf{Z}$. Thus, $a = by = (ax)y = axy \Rightarrow xy = 1 \Rightarrow$ either $x = y = -1$ or $x = y = 1$, and so $a = \pm b$.
Property 2: $a|b \Rightarrow b = ay$ for some $y \in \mathbf{Z} \Rightarrow bx = (ay)x = a(yx) \Rightarrow a|\,bx$.
Property 3: $a|b$ and $a|c \Rightarrow aw = b$ and $av = c$ for some $w, v \in \mathbf{Z}$. Thus, $bx + cy = (aw)x + (av)y = a(wx + vy)$;
Similarly, $bx - cy = (aw)x - (av)y = a(wx - vy)$, so $a|(bx \pm cy)$.
Property 5: $a|b$ and $b|c \Rightarrow ax = b$ and $by = c$ for some $x, y \in \mathbf{Z}$.

Thus, $c = by = (ax)y = a(xy) \Rightarrow a|c$.

Property 6: $a|b$ and $c|d \Rightarrow ax = b$ and $cy = d$ for some $x, y \in \mathbf{Z}$.
Thus, $bd = (ax)(cy) = ac(xy) \Rightarrow ac|bd$.

3.
$$(a-1)(a^{n-1} + a^{n-2} + \ldots + a^1 + 1) = a(a^{n-1} + a^{n-2} + \ldots + a^1 + 1)$$
$$- 1(a^{n-1} + a^{n-2} + \ldots + a^1 + 1) = (a^n + a^{n-1} + \ldots + a^2 + a^1)$$
$$- (a^{n-1} + a^{n-2} + \ldots + a^1 + 1) = a^n + (a^{n-1} - a^{n-1})$$
$$+ (a^{n-2} - a^{n-2}) + \ldots + (a^1 - a^1) - 1 = a^n - 1 \Rightarrow (a-1)|(a^n - 1).$$

2.6: The Fundamental Theorem of Arithmetic (FTA)

1. **a.** Composite: $1{,}274 = 2 \cdot 7^2 \cdot 13$
   b. Composite: $7{,}921 = 89^2$
   c. Composite: $6{,}561 = 3^8$
   d. Prime
   e. Composite: $11{,}111 = 41 \cdot 271$
   f. Prime

3. **a.** $\tau(13^{13}) = 13 + 1 = 14$.
   b. $\tau(2^5 \cdot 3^6 \cdot 17^4) = (5+1)(6+1)(4+1) = 210$.
   c. If m is greater than 0, then the equation's left side would have an 11 in its prime factorization while the right side would not. Similarly, if n is greater than 0, then the equation's right side would have a 23 in its prime factorization while the left side would not. This would contradict FTA, so we have $m = n = 0$.
   d. We have $21^{m_1} \cdot 29^{m_2} = 3^{m_1} \cdot 7^{m_1} \cdot 29^{m_2}$, so if m_1 is positive, then there is a 3 in the prime factorization of $21^{m_1} \cdot 29^{m_2}$, but the 3 is not in the prime factorization of $7^{n_1} \cdot 19^{n_2} \cdot 23^{n_3}$. Thus, by FTA, $21^{m_1} \cdot 29^{m_2} \neq 7^{n_1} \cdot 19^{n_2} \cdot 23^{n_3}$. The cases for m_2, n_1, n_2, and n_3 are similar.

5. $n > 1$ is an mth power $\Leftrightarrow n = a^m = (p_1^{m_1} p_2^{m_2} \cdots p_t^{m_t})^m = p_1^{m \cdot m_1} p_2^{m \cdot m_2} \cdots p_t^{m \cdot m_t}$, (where $p_1^{m_1} p_2^{m_2} \cdots p_t^{m_t}$ is the prime factorization of a) $\Leftrightarrow$ the prime factors of n occur in multiples of m.

2.8: Relations and Results Concerning lcm and gcd

1. **a.** $\operatorname{lcm}(a,b) = 2^4 \cdot 3^2 \cdot 5^3 \cdot 7^3 \cdot 13^2$ and $\gcd(a,b) = 2^1 \cdot 3^0 \cdot 5^2 \cdot 7^2 \cdot 13^0$.
   b. $a = 2^4 \cdot 7 \cdot 11, b = 2^3 \cdot 5 \cdot 11^2 \Rightarrow \operatorname{lcm}(a,b) = 2^4 \cdot 5 \cdot 7 \cdot 11^2$, and $\gcd(a,b) = 2^3 \cdot 5^0 \cdot 7^0 \cdot 11$.
   c. $a = 2^6 \cdot 31, b = 2{,}003 \Rightarrow \operatorname{lcm}(a,b) = 2^6 \cdot 31 \cdot 2{,}003$, and $\gcd(a,b) = 2^0 \cdot 31^0 \cdot 2{,}003^0$.
   d. $a = 7 \cdot 11 \cdot 13, b = 11 \cdot 9{,}091 \Rightarrow \operatorname{lcm}(a,b) = 7 \cdot 11 \cdot 13 \cdot 9{,}091$, and $\gcd(a,b) = 7^0 \cdot 11 \cdot 13^0 \cdot 9{,}091^0$.
   e. $a = 2^4 \cdot 23 \cdot 47, b = 2^4 \cdot 1{,}151 \Rightarrow \operatorname{lcm}(a,b) = 2^4 \cdot 23 \cdot 47 \cdot 1{,}151$, and $\gcd(a,b) = 2^4 \cdot 23^0 \cdot 47^0 \cdot 1{,}151^0$.

3. **a.** No. $\gcd(a,b)$ must divide $\operatorname{lcm}(a,b)$, but here 4 does not divide 6.
   b. Yes. Let $a = 4$ and $b = 8$.

c. $d|m \Rightarrow \gcd(d,m) = d$ and $\text{lcm}(d,m) = m$. Conversely, let a and b be positive integers such that $\gcd(a,b) = d$ and $\text{lcm}(a,b) = m$. Since $\gcd(a,b)$ always divides $\text{lcm}(a,b)$, we have that $d|m$.

5. **a.** Let $d = \gcd(a,b)$, so $a = dw$ and $b = dv$ for some integers w, v. Thus, $1 = ax + by = (dw)x + (dv)y = d(wx + vy) \Rightarrow d|1 \Rightarrow d = 1$.
   b. No. $2 = 1 \cdot 1 + 1 \cdot 1$, but $\gcd(1,1) = 1 \neq 2$.

7. Let $p_1^{m_1} p_2^{m_2} \cdots p_t^{m_t}$ and $p_1^{n_1} p_2^{n_2} \cdots p_t^{n_t}$ be the modified canonical prime factorization of a and b, respectively. It follows that $a^2 = (p_1^{m_1} p_2^{m_2} \cdots p_t^{m_t})^2 = p_1^{2m_1} p_2^{2m_2} \cdots p_t^{2m_t}$ and $b^2 = (p_1^{n_1} p_2^{n_2} \cdots p_t^{n_t})^2 = p_1^{2n_1} p_2^{2n_2} \cdots p_t^{2n_t}$, so $\gcd(a^2, b^2) = p_1^{\min\{2m_1, 2n_1\}} p_2^{\min\{2m_2, 2n_2\}} \cdots p_t^{\min\{2m_t, 2n_t\}} = p_1^{2 \cdot \min\{m_1, n_1\}} p_2^{2 \cdot \min\{m_2, n_2\}} \cdots p_t^{2 \cdot \min\{m_t, n_t\}} = (p_1^{\min\{m_1, n_1\}} p_2^{\min\{m_2, n_2\}} \cdots p_t^{\min\{m_t, n_t\}})^2 = [\gcd(a,b)]^2 = d^2$.

9. Let $d_1 = \gcd(c, a)$ and $d_2 = \gcd(c, b)$, so $d_1 x = a$, $d_1 y = c$, $d_2 w = c$, and $d_2 v = b$ for some integers $x, y, w,$ and v. $c|(a+b) \Rightarrow cm = a + b = d_1 x + b$ for some integer $m \Rightarrow d_1 y m = d_1 x + b \Rightarrow d_1(ym - x) = b \Rightarrow d_1 | b$.
   Similarly, $c|(a+b) \Rightarrow cn = a + b = a + d_2 v$ for some integer $n \Rightarrow d_2 y n = a + d_2 v \Rightarrow d_2(yn - v) = a \Rightarrow d_2 | a$. Thus, d_1 and d_2 are common divisors of a and b, so d_1 and d_2 both divide $\gcd(a,b) = 1$. Therefore, $d_1 = d_2 = 1$.

CHAPTER 3

3.2: The Euclidean Algorithm

1. **a.** $35{,}784 = 29 \cdot 1{,}233 + 27 \Rightarrow q = 1{,}233$ and $r = 27$.
   b. $525 = 757 \cdot 0 + 525 \Rightarrow q = 0$ and $r = 525$.
   c. $-47{,}893 = 38 \cdot 1{,}261 + 25 \Rightarrow q = 1{,}261$ and $r = 25$.
   d. $-52 = 91 \cdot (-1) + 39 \Rightarrow q = -1$ and $r = 39$.

3. **a.** Case 1: $a = 2q$ for some integer $q \Rightarrow a^2 = 4q^2 \Rightarrow a^2$ is divisible by 4.
   Case 2: $a = 2q + 1$ for some integer $q \Rightarrow a^2 = 4q^2 + 4q + 1 \Rightarrow a^2$ has a remainder of 1 when divided by 4.
   b. Case 1: $a = 3q$ for some integer $q \Rightarrow a^2 = 9q^2 = 3(3q^2) \Rightarrow a^2$ is divisible by 3.
   Case 2: $a = 3q + 1$ for some integer $q \Rightarrow a^2 = 9q^2 + 6q + 1 = 3(3q^2 + 2q) + 1 \Rightarrow a^2$ has a remainder of 1 when divided by 3.
   Case 3: $a = 3q + 2$ for some integer $q \Rightarrow a^2 = 9q^2 + 12q + 4 = 3(3q^2 + 4q + 1) + 1 \Rightarrow a^2$ has a remainder of 1 when divided by 3.

5. ($\Rightarrow$) Suppose $\tau(n)$ is odd. Let $p_1^{m_1} p_2^{m_2} \cdots p_s^{m_s}$ be the canonical prime factorization of n. Since $\tau(n) = (m_1 + 1)(m_2 + 1) \cdots (m_s + 1)$ is odd, $m_i + 1$ must be odd for each $i = 1, 2, \ldots, s$, so m_i must be even for each $i = 1, 2, \ldots, s$. Thus, n must be a square.
   ($\Leftarrow$) Suppose n is a square. It follows that m_i must be even for each $i = 1, 2, \ldots, s$, so $m_i + 1$ must be odd for each $i = 1, 2, \ldots, s$.
   Thus, $\tau(n) = (m_1 + 1)(m_2 + 1) \cdots (m_s + 1)$ is odd.

7. **a.** $111 = 87 \cdot 1 + 24$
   $87 = 24 \cdot 3 + 15$
   $24 = 15 \cdot 1 + 9$
   $15 = 9 \cdot 1 + 6$
   $9 = 6 \cdot 1 + 3$
   $6 = 3 \cdot 2 + 0$
   $\gcd(87, 111) = 3 = 111 \cdot 11 + 87 \cdot (-14)$

Answers to (Most) Odd-Numbered Exercises   329

  b. $2{,}480 = 585 \cdot 4 + 140$
  $585 = 140 \cdot 4 + 25$
  $140 = 25 \cdot 5 + 15$
  $25 = 15 \cdot 1 + 10$
  $15 = 10 \cdot 1 + 5$
  $10 = 5 \cdot 2 + 0$
  $\gcd(585, 2{,}480) = 5 = 585 \cdot (-195) + 2{,}480 \cdot 46$
  c. $8{,}128 = 496 \cdot 16 + 192$
  $496 = 192 \cdot 2 + 112$
  $192 = 112 \cdot 1 + 80$
  $112 = 80 \cdot 1 + 32$
  $80 = 32 \cdot 2 + 16$
  $32 = 16 \cdot 2 + 0$
  $\gcd(496, 8{,}128) = 16 = 496 \cdot (-213) + 8{,}128 \cdot 13$
  d. $89{,}523 = 10{,}285 \cdot 8 + 7{,}243$
  $10{,}285 = 7{,}243 \cdot 1 + 3{,}042$
  $7{,}243 = 3{,}042 \cdot 2 + 1{,}159$
  $3{,}042 = 1{,}159 \cdot 2 + 724$
  $1{,}159 = 724 \cdot 1 + 435$
  $724 = 435 \cdot 1 + 289$
  $435 = 289 \cdot 1 + 146$
  $289 = 146 \cdot 1 + 143$
  $146 = 143 \cdot 1 + 3$
  $143 = 3 \cdot 47 + 2$
  $3 = 2 \cdot 1 + 1$
  $2 = 1 \cdot 2 + 0$
  $\gcd(10{,}285, 89{,}523) = 1 = 10{,}285 \cdot (-30{,}047) + 89{,}523 \cdot 3{,}452$

9. Any length can be measured with these straightedges (see Section 3.4 for justification). The conclusion does not follow for straightedges of length 3^n since 2 cannot be measured this way. Adding the condition that up to two of each straightedge can be used would let every length be measured.

11. **a.** Multiple solutions exist. Example: $4 = 3 \cdot 3 + 5(-1)$ represents (1) filling the 3-gallon jug, (2) emptying it into the 5-gallon jug, (3) filling the 3-gallon jug again, (4) emptying 2 of the 3 gallons from the 3-gallon jug into the 5-gallon jug and emptying the remaining gallon into the device, (5) filling the 3-gallon jug again, and (6) emptying the 3-gallon jug into the device for a total of 4 gallons in the device.
    b. No. Since $\gcd(6, 9) = 3$ and 3 does not divide 4, there cannot exists integers x and y such that $4 = 6x + 9y$.

3.3: Applications of the Representation gcd(a, b) = ax + by

1. If $s = 1$, then $p | a_1$, and we are done. Let $k \geq 1$, and suppose that if $a_1, a_2, \ldots, a_k$ are k positive integers and $p | a_1 a_2 \cdots a_k$, then $p | a_i$ for some $1 \leq i \leq k$. Now suppose $b_1, b_2, \ldots, b_k, b_{k+1}$ are $k + 1$ positive integers and $p | (b_1 b_2 \cdots b_k) b_{k+1}$. By Corollary 3.3.2 we know that $p | b_1 b_2 \cdots b_k$ or $p | b_{k+1}$. If $p | b_1 b_2 \cdots b_k$, we

330 Answers to (Most) Odd-Numbered Exercises

have by the induction hypothesis that $p|b_i$ for some $1 \leq i \leq k$. If not, we have $p|b_{k+1}$, so either way we have $p|b_i$ for some $1 \leq i \leq k+1$. Thus, by induction, we have that if $a_1, a_2, \ldots, a_s$ are positive integers and $p|a_1 a_2 \cdots a_s$, then $p|a_i$ for some $1 \leq i \leq s$.

3. **a.** Since $\gcd(a, b) = 1$, there exist integers x and y such that $1 = ax + by$. Thus, we have

$$1 = (a+b)x - bx + (a+b)y - by = (a+b)(x+y) - ay - bx$$
$$= (a+b)(x+y) - ay(ax+by) - bx(ax+by)$$
$$= (a+b)(x+y) - ab(x^2+y^2) + xy(b^2-a^2)$$
$$= (a+b)(x+y) - ab(x^2+y^2) + xy(b-a)(b+a)$$
$$= (a+b)[x+y+xy(b-a)] + ab(-x^2-y^2)$$
$$= (a+b)u + (ab)v,$$

so there exist integers u and v such that $1 = (a+b)u + (ab)v$, so $\gcd(a+b, ab) = 1$.

b. We have integers x, y, z, and w such that $1 = ax + by$ and $1 = az + cw$. Thus, $1 = (ax+by)(az+cw) = a^2xz + acwx + abyz + bcwy = a(axz + cwx + byz) + (bc)(wy) = au + (bc)v$, so there exist integers u and v such that $1 = au + (bc)v$, and $\gcd(a, bc) = 1$.

3.4: Place Value

1. **a.** $2{,}004 = (2202020)_3$
   b. $30{,}235 = (351551)_6$
   c. $(1100110011)_2 = 819 = (11234)_5$
   d. $276{,}664 = (98d0c)_{13}$
   e. $(564a017)_{11} = 9{,}896{,}003 = (20552678)_9$
   f. Varies

3.

+	0	1	2	3	4	5	6	7	8	9	a
0	0	1	2	3	4	5	6	7	8	9	a
1	1	2	3	4	5	6	7	8	9	a	$(10)_{11}$
2	2	3	4	5	6	7	8	9	a	$(10)_{11}$	$(11)_{11}$
3	3	4	5	6	7	8	9	a	$(10)_{11}$	$(11)_{11}$	$(12)_{11}$
4	4	5	6	7	8	9	a	$(10)_{11}$	$(11)_{11}$	$(12)_{11}$	$(13)_{11}$
5	5	6	7	8	9	A	$(10)_{11}$	$(11)_{11}$	$(12)_{11}$	$(13)_{11}$	$(14)_{11}$
6	6	7	8	9	a	$(10)_{11}$	$(11)_{11}$	$(12)_{11}$	$(13)_{11}$	$(14)_{11}$	$(15)_{11}$
7	7	8	9	a	$(10)_{11}$	$(11)_{11}$	$(12)_{11}$	$(13)_{11}$	$(14)_{11}$	$(15)_{11}$	$(16)_{11}$
8	8	9	a	$(10)_{11}$	$(11)_{11}$	$(12)_{11}$	$(13)_{11}$	$(14)_{11}$	$(15)_{11}$	$(16)_{11}$	$(17)_{11}$
9	9	a	$(10)_{11}$	$(11)_{11}$	$(12)_{11}$	$(13)_{11}$	$(14)_{11}$	$(15)_{11}$	$(16)_{11}$	$(17)_{11}$	$(18)_{11}$
a	A	$(10)_{11}$	$(11)_{11}$	$(12)_{11}$	$(13)_{11}$	$(14)_{11}$	$(15)_{11}$	$(16)_{11}$	$(17)_{11}$	$(18)_{11}$	$(19)_{11}$

•	0	1	2	3	4	5	6	7	8	9	a
0	0	0	0	0	0	0	0	0	0	0	0
1	0	1	2	3	4	5	6	7	8	9	a
2	0	2	4	6	8	A	$(11)_{11}$	$(13)_{11}$	$(15)_{11}$	$(17)_{11}$	$(19)_{11}$
3	0	3	6	9	$(11)_{11}$	$(14)_{11}$	$(17)_{11}$	$(1a)_{11}$	$(22)_{11}$	$(25)_{11}$	$(28)_{11}$
4	0	4	8	$(11)_{11}$	$(15)_{11}$	$(19)_{11}$	$(22)_{11}$	$(26)_{11}$	$(2a)_{11}$	$(33)_{11}$	$(37)_{11}$
5	0	5	a	$(14)_{11}$	$(19)_{11}$	$(23)_{11}$	$(28)_{11}$	$(32)_{11}$	$(37)_{11}$	$(41)_{11}$	$(46)_{11}$
6	0	6	$(11)_{11}$	$(17)_{11}$	$(22)_{11}$	$(28)_{11}$	$(33)_{11}$	$(39)_{11}$	$(44)_{11}$	$(4a)_{11}$	$(55)_{11}$
7	0	7	$(13)_{11}$	$(1a)_{11}$	$(26)_{11}$	$(32)_{11}$	$(39)_{11}$	$(45)_{11}$	$(51)_{11}$	$(58)_{11}$	$(64)_{11}$
8	0	8	$(15)_{11}$	$(22)_{11}$	$(2a)_{11}$	$(37)_{11}$	$(44)_{11}$	$(51)_{11}$	$(59)_{11}$	$(66)_{11}$	$(73)_{11}$
9	0	9	$(17)_{11}$	$(25)_{11}$	$(33)_{11}$	$(41)_{11}$	$(4a)_{11}$	$(58)_{11}$	$(66)_{11}$	$(74)_{11}$	$(82)_{11}$
a	0	a	$(19)_{11}$	$(28)_{11}$	$(37)_{11}$	$(46)_{11}$	$(55)_{11}$	$(64)_{11}$	$(73)_{11}$	$(82)_{11}$	$(91)_{11}$

a. $\dfrac{\begin{array}{r}(6a892a)_{11}\\+(47a65)_{11}\end{array}}{(745894)_{11}}$

b. $\dfrac{\begin{array}{r}(a18a9)_{11}\\\times(23a)_{11}\end{array}}{(21a40832)_{11}}$

3.5: Prime Thoughts

1. a. $f(x) = x^3 - 1 = (x - 1)(x^2 + x + 1)$ is prime $\Leftrightarrow x = 2$.
   b. $f(x) = x^4 - 1 = (x - 1)(x^3 + x^2 + x + 1)$ is prime $\Leftrightarrow x = 2$.
   c. $f(x) = x^2 - 9 = (x - 3)(x + 3)$ is prime $\Leftrightarrow x = 4$.
   d. $f(x) = x^2 - 6x + 9 = (x - 3)^2$ has no prime values (it is always a square).
   e. $f(x) = x^2 + x + 6$ has no prime values since if x is either odd or even, f is even.
   f. $f(x) = x^2 + 2x - 15 = (x - 3)(x + 5)$ is prime $\Leftrightarrow x = 4$.
   g. $f(x) = x^2 + 4x - 21 = (x - 3)(x + 7)$ is prime $\Leftrightarrow x = 4$.

3. $f(x) = x^2 - 4 = (x - 2)(x + 2) = 13p \Rightarrow x - 2 = 13$ or $x + 2 = 13 \Rightarrow x = 15$ or $x = 11 \Rightarrow p = x + 2 = 15 + 2 = 17$ or $p = x - 2 = 11 - 2 = 9$. Thus, the only possible value for x so that $f(x) = 13p$ and p is prime is $x = 15$, so $p = 17$.

5. a. We have $\sigma(n) = 1 + p + p^2 + \ldots + p^{2m}$. Note that p^i for $i \geq 1$ is either always even (in the case that $p = 2$) or always odd. The number of terms in the sum $p + p^2 + \ldots + p^{2m}$ is even, so in either case ($p = 2$ or not), the sum $p + p^2 + \ldots + p^{2m}$ is even. Thus, $\sigma(n) = 1 + p + p^2 + \ldots + p^{2m}$ is odd.
   b. Let $a = p_1^{m_1} p_2^{m_2} \cdots p_s^{m_s}$ be the canonical prime factorization of a, so $n = p_1^{2m_1} p_2^{2m_2} \cdots p_s^{2m_s}$.
   We have that $\sigma(n) = (1 + p_1 + \ldots + p_1^{2m_1}) \cdots (1 + p_s + \ldots + p_s^{2m_s})$, by part a, we saw that each term $1 + p_i + \ldots + p_i^{2m_i}$ was odd for $1 \leq i \leq s$. Thus, the product $\sigma(n) = (1 + p_1 + \ldots + p_1^{2m_1}) \cdots (1 + p_s + \ldots + p_s^{2m_s})$ is odd.
   c. No; $n = 2^3$, and $\sigma(n) = 1 + 2 + 4 + 8 = 15$.

d. If $n = a^2$, then $\sigma(n)$ is odd by part b. Since a perfect number k has the property that $\sigma(k) = 2k$ (i.e., $\sigma(k)$ is even), n cannot be perfect.

CHAPTER 4

4.4: Modular Arithmetic

1. **a.** $29 - 501 = -472 = 8 \cdot (-59)$.
   b. $-53 - (-81) = 28 = 4 \cdot 7$.
   c. $309 - (-207) = 516 = 12 \cdot 43$.
   d. $-136 - 136 = -272 = 17 \cdot (-16)$.

3. **a.** In Z_9,
   $[45,321] \cdot [794] = [4+5+3+2+1] \cdot [7+9+4] = [15] \cdot [20] = [1+5] \cdot [2+0] = [6] \cdot [2] = [12] = [3]$, and $[35,974,874] = [3+5+9+7+4+8+7+4] = [47] = [4+7] = [11] = [2]$.
   b. In Z_9,
   $[3,086] + [8,829] + [1,045] = [3+0+8+6] + [8+8+2+9] + [1+0+4+5] = [17] + [27] + [10] = [1+7] + [2+7] + [1+0] = [8] + [9] + [1] = [18] = [0]$, and $[11,960] = [1+1+9+6+0] = [16] = [1+6] = [7]$.

5. **a.** $[a] = [b] \Leftrightarrow n|(b-a) \Leftrightarrow n|((b+c) - (a+c)) \Leftrightarrow [a+c] = [b+c]$.
   b. $[a] = [b] \Rightarrow n|(b-a) \Rightarrow n|(b-a)c \Rightarrow n|(bc-ac)$
   $\Rightarrow [ac] = [bc] \Rightarrow [a][c] = [b][c]$.

7. Let $f(x) = a_n x^n + a_{n-1} x^{n-1} + \ldots + a_1 x + a_0$. Now evaluate $f(a)$ and $f(b)$ and apply Problems 5a and b and 6a to $f(a)$ and $f(b)$.

9. **a.** Let $m = 2, n = 4, a = 2$, and $b = 6$.
   b. $[a] = [b]$ in Z_m and $[a] = [b]$ in $Z_n \Leftrightarrow m|(b-a)$ and $n|(b-a) \Leftrightarrow \text{lcm}(m,n)|(b-a)$.
   c. $\gcd(m,n) = 1 \Rightarrow \text{lcm}(m,n) = mn$. Thus, by part b, $mn|b-a \Rightarrow [a] = [b]$ in Z_{mn}.

4.5: Comparing Arithmetic Properties of Z and Z_n

1. **a.** In Z_{99}, $[84] + [16] + [93] + [5] = ([-15] + [16]) + ([-6] + [5]) = [1] + [-1] = [0]$.
   b. In Z_{47}, $[45]^5 = [-2]^5 = [(-2)^5] = [-32] = [13]$.
   c. In Z_{64}, $[60] \cdot ([59] + [4]) = [-4] \cdot [63] = [-4] \cdot [-1] = [4]$.

4.6: Multiplicative Inverses in Z_n

1. **a.** In Z_{25}, $[0], [5], [10], [15]$, and $[20]$ are the only elements with no inverses.
   b. In Z_{19}, $[0]$ is the only element with no inverse.
   c. In $Z_{25}, [1], [7], [11], [13], [17], [19]$, and $[23]$ are the only elements with inverses.

3. $[a], [b]$ have multiplicative inverses in $Z_n \Rightarrow [a][b]([b]^{-1}[a]^{-1}) = [a]([b][b]^{-1})$
   $[a^{-1}] = [a]([1])[a]^{-1} = [a][a]^{-1} = 1$, so $[a][b]$ has a multiplicative inverse in Z_n.

Answers to (Most) Odd-Numbered Exercises   333

5. **a.** Let $n = 8$ and $[a] = [2]$.
   b. $[a]$ and $[b]$ are nilpotent $\Rightarrow [a]^t = [b]^s = [0]$ for some positive integers t and s. It follows that $[a]^k = [b]^j = [0]$ for all integers $k \geq t$ and $j \geq s$. Thus, by the binomial theorem,

$$([a] + [b])^{t+s} = \sum_{i=0}^{t+s} \binom{t+s}{i} [a]^{t+s-i}[b]^i$$

$$= \sum_{i=0}^{s} \binom{t+s}{i} [a]^{t+s-i}[b]^i + \sum_{i=s+1}^{t+s} \binom{t+s}{i} [a]^{t+s-i}[b]^i.$$

In the first summand, $t + s - i \geq t$, so $[a]^{t+s-i} = [0]$ for each $0 \leq i \leq s$. In the second summand, $i > s$, so $[b]^i = [0]$ for each $i \geq s + 1$. Thus, $\sum_{i=0}^{s} \binom{t+s}{i}[a]^{t+s-i}[b]^i + \sum_{i=s+1}^{t+s} \binom{t+s}{i}[a]^{t+s-i}[b]^i = [0]$.
Therefore, $[a] + [b]$ is nilpotent.

4.7: Elementary Applications of Modular Arithmetic

1. **a.**
   $$\begin{array}{ccccc} 1,000 & 9,990 & 9,981 & 8,820 & 8,532 \\ -0001 & -0999 & -1,899 & -0288 & -2,358 \\ \hline 0999 & 8,991 & 8,082 & 8,532 & 6,174 \end{array} \rightarrow \rightarrow \rightarrow \rightarrow$$

   $$\begin{array}{ccccc} 9,832 & 7,443 & 9,963 & 6,642 & 7,641 \\ -2,389 & -3,447 & -3,699 & -2,466 & -1,467 \\ \hline 7,443 & 3,996 & 6,264 & 4,176 & 6,174 \end{array} \rightarrow \rightarrow \rightarrow \rightarrow$$

   b. 9,710
   c. 495 is a Kaprekar constant for three-digit numbers.

3. $G([s]) = [15]([s] - [8]) = [15][s] + [10]$.
   JKXR KWLV SOPK XEHM AJKJ XRKW LOKV IXKZ KXUX IXK
   $\rightarrow$
   9, 10, 23, 17, _, 10, 22, 11, 21, _, 18, 14, 15, 10, _, 23, 4, 7, 12, _, 0, 10, 9, 10, _, 23, 17, 10, 22, _, 11, 14, 10, 21, _, 8, 23, 10, 25, _, 10, 23, 20, 23, _, 8, 23, 10.
   Now apply G
   $\rightarrow$
   15, 4, 17, 5, _, 4, 2, 19, 13, _, 20, 12, 1, 4, _, 17, 18, 11, 8, _, 10, 4, 15, 4, _, 17, 5, 4, 2, _, 19, 12, 4, 13, _, 0, 17, 4, 21, _, 4, 17, 24, 17, _, 0, 17, 4
   $\rightarrow$
   PERFECT NUMBERS LIKE PERFECT MEN ARE VERY RARE

5. **a.** $-[10 \cdot 0 + 9 \cdot 1 + 8 \cdot 9 + 7 \cdot 5 + 6 \cdot 1 + 5 \cdot 0 + 4 \cdot 5 + 3 \cdot 1 + 2 \cdot 9] =$
   $-[163] = -[1 - 6 + 3] = -[-2] = [2]$, so the correct check digit is 2.
   b. $-[10 \cdot 0 + 9 \cdot 3 + 8 \cdot 4 + 7 \cdot 5 + 6 \cdot 3 + 5 \cdot 0 + 4 \cdot 9 + 3 \cdot 1 + 2 \cdot 8] =$
   $-[167] = -[1 - 6 + 7] = -[2] = [9]$, so the check digit is correct.
   c. $-[10 \cdot 0 + 9 \cdot 8 + 8 \cdot 2 + 7 \cdot 4 + 6 \cdot 7 + 5 \cdot 9 + 4 \cdot 8 + 3 \cdot 1 + 2 \cdot 6] =$
   $-[250] = -[2 - 5 + 0] = -[-3] = [3]$, so the correct check digit is 3.
   d. $-[10 \cdot 3 + 9 \cdot 5 + 8 \cdot 4 + 7 \cdot 0 + 6 \cdot 9 + 5 \cdot 0 + 4 \cdot 3 + 3 \cdot 3 + 2 \cdot 3] =$
   $-[188] = -[1 - 8 + 8] = -[1] = [10]$, so the check digit is correct.

7. Check 1: $[4, 361, 157, 018] = [-4 + 3 - 6 + 1 - 1 + 5 - 7 + 0 - 1 + 8] =$
   $[-2] = [9]$.

Check 2: $[4, 361, 156, 822] = [-4 + 3 - 6 + 1 - 1 + 5 - 6 + 8 - 2 + 2] = [0]$.

9. Designs may vary.

4.8: Fermat's Little Theorem and Wilson's Theorem

1. **a.** $[2]^{12} = [1] \Rightarrow [2]^{-1} = [2]^{11} = [7]$; the other inverses are calculated similarly.
   b. $[2]^{18} = [1] \Rightarrow [2]^{-1} = [2]^{17} = [10]$; the other inverses are calculated similarly.
   c. $[2]^{22} = [1] \Rightarrow [2]^{-1} = [2]^{21} = [12]$; the other inverses are calculated similarly.
3. We know for primes p that $[a]^p = [a]$ in $\mathbf{Z}_p$, so let $p = 5$, we see that for each integer a, $[a]^5 = [a]$ in $\mathbf{Z}_5$ (i.e., $5|(a^5 - a)$). If a is even, then a^5 is also even, so $a^5 - a$ is even and $2|(a^5 - a)$. If a is odd, then a^5 is also odd, so $a^5 - a$ is even and $2|(a^5 - a)$. Since both 2 and 5 divide $a^5 - a$, we have $10|(a^5 - a)$, so $[a]^5 = [a]$ in $\mathbf{Z}_{10}$.
5. In $\mathbf{Z}_{13}$, $[12!] = [-1] \cdot [-2] \cdot [-3] \cdot [-4] \cdot [-5] \cdot [-6] \cdot [6] \cdot [5] \cdot [4] \cdot [3] \cdot [2] \cdot [1] = [6!]^2 = [30]^2[24]^2 = [30]^2[12]^2[2]^2 = [4]^2[-1]^2[2]^2 = [16][4] = [3][4] = [12] = [-1]$. Thus, by the converse of Wilson's theorem, 13 is prime.
7. By Wilson's theorem, in $\mathbf{Z}_{13}$, $[-1] = [22!] = [22 \cdot 21][20!] = [-1][-2][20!] \Rightarrow [20!] = [-2]^{-1} = [21]^{-1} = [11]$.

4.9: Linear Equations Defined over $\mathbf{Z}_n$

1. **a.** $X = [4]^{-1}[9] = [3][9] = [27] = [5]$.
   b. No solutions exist, since $\gcd(8, 12) = 4$ does not divide 7.
   c. There are exactly $3 = \gcd(9, 15)$ solutions: $X = [4], [4 + 5] = [9]$, and $[4 + 10] = [14]$.
   d. No solutions exist since $\gcd(12, 30) = 6$ does not divide 14.
   e. There are exactly $7 = \gcd(21, 49)$ solutions, $X = [4], [4 + 7] = [11], [4 + 14] = [18], [4 + 21] = [25], [4 + 28] = [32], [4 + 42] = [46]$, and $[4 + 35] = [39]$.

4.10: Extended Studies: The Chinese Remainder Theorem

1. **a.** $X = [52]$.
   b. $X = [53]$.

4.11: Extended Studies: Quadratic Equations Defined over $\mathbf{Z}_n$

1. **a.** $X^2 - [9]X + [8] = X^2 - [9]X + [20] = (X - [4])(X - [5]) = [0] \Rightarrow X = [4], [5], [1]$.
   b. $X^2 - [2]X - [8] = X^2 - [2]X + [3] = (X - [1])(X - [2]) = [0] \Rightarrow X = [1], [2]$.
   c. $X^2 + [3]X = X(X + [3]) = [0] \Rightarrow X = [0], [6], [3]$.
   d. $X^2 + X + [1] = [0]$ has no solutions in $\mathbf{Z}_5$.
   e. $X^2 - X = X(X - [1]) = [0] \Rightarrow X = [0], [1], [5], [6]$.
   f. $X^2 = [0] \Rightarrow X = [0], [5], [10], [15], [20]$.
   g. $X^2 = [0] \Rightarrow X = [0], [9], [18]$.

h. $X^2 = [0] \Rightarrow X = [121t]$ for $t = 0, 1, 2, \ldots, 10$.
i. $X^2 = [0] \Rightarrow X = [0]$.
j. $X^2 - [1] = (X - [1])(X + [1]) = [0] \Rightarrow X = [1], [15], [7], [9]$.
k. $X^2 - [119] = [0] \Rightarrow X = [119]$.
l. $X^2 + [1] = X^2 - [-1] = X^2 - [16] = (X - [4])(X + [4]) = [0] \Rightarrow X = [4], [13]$.

3. Let $n = 50$; then the solutions to the equation $X^2 - [25]$ in $\mathbf{Z}_{50}$ are $[5], [15], [25], [35],$ and $[45]$.

CHAPTER 5

5.1: The Significance of Daryl's Measurements and Related Geometry

Note: In the solutions to these exercises we are using the following definition of a parallelogram: A parallelogram is a quadrilateral in which opposite sides are parallel. Two lines are parallel if they are always the same distance apart. Let $ABCD$ be a quadrilateral.

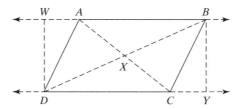

1. Let $ABCD$ be a quadrilateral, and let X be the point of intersection of the diagonals of $ABCD$. We have $AX = XC$, $BX = XD$, $m(\angle AXD) = m(\angle CXB)$, and $m(\angle AXB) = m(\angle CXD)$. Now use the SAS postulate to show that $AB = CD$ and $AD = CB$. Now $AB = CD$ implies that AD and BC are always the same distance apart and so are parallel. We may also conclude by SSS $m(\angle ABC) = m(\angle ADC)$ and $m(\angle DAB) = m(\angle BCD)$. Then $m(\angle BCY) = m(\angle WAD)$ and so $m(\angle WDA) = m(\angle CBY)$. Then the ASA postulate implies that $WD = BY$. Hence AB and DC are always the same distance apart and so are parallel. Thus, $ABCD$ is a parallelogram.

The converse is also true. Let $ABCD$ be a parallelogram, so that $WD = BY$ and $AB = DC$. Since AB is also parallel to DC, $m(\angle ABD) = m(\angle CDB)$, and $m(\angle BAC) = m(\angle DCA)$. It follows that $m(\angle ABC) = m(\angle CDA)$, and $m(\angle DAB) = m(\angle BCD)$. Then $m(\angle BCY) = m(\angle WAD)$ and so ASA implies that $AD = BC$. Let X be the point of intersection of the diagonals of $ABCD$, and use the ASA postulate to conclude $AX = XC$ and $BX = XD$.

3. Let $ABCD$ be a parallelogram and suppose (without loss of generality) that $\angle DAB$ is a right angle. Since $AB \| DC$, we have that $\angle ADC$ is a right angle. Use the same reasoning to show the other two angles are right angles.

5. Let $ABCD$ be a parallelogram with $AC = BD$. Since the diagonals bisect each other (call the point of intersection X), we have $AX = BX = CX = DX$, so the triangles $\triangle AXB$ and $\triangle AXD$ are isosceles. Thus, $m(\angle ABD) = m(\angle BAC)$, and $m(\angle ADB) = m(\angle CAD)$. We have $m(\angle AXB) + m(\angle AXD) = 180°$, $2 \cdot m(\angle BAC) + m(\angle AXB) = 180°$, and $2 \cdot m(\angle CAD) + m(\angle AXD) = 180°$. Use this to show that $m(\angle BAD) = 90°$.

5.3: Reflections on Classroom Connections: The Pythagorean Theorem and Its Converse

1. **a.** Use the ideas from parts a and b of Theorem 5.3.3 and separate into two cases: Case1: The angle opposite c is obtuse (use the idea from part a); and Case 2: The angle opposite c is acute (use the idea from part b).
   b. Position $\triangle BAC$ so that C is at the origin and B is directly above A (i.e., B and A have the same x-coordinate; call it t.) Justify that B must be above the line $y = 0$ and A must be below the same line. The distance from B to the line $y = 0$ is $\sqrt{a^2 - t^2}$, and the distance from A to the line $y = 0$ is $\sqrt{b^2 - t^2}$. Notice that $\sqrt{a^2 - t^2} + \sqrt{b^2 - t^2} = c \Rightarrow (\sqrt{a^2 - t^2} + \sqrt{b^2 - t^2})^2 = a^2 + b^2 + 2(\sqrt{a^2 - t^2} \cdot \sqrt{b^2 - t^2} - t^2) = c^2$. Use the fact that $a^2 + b^2 = c^2$ to conclude that $\sqrt{a^2 - t^2} \cdot \sqrt{b^2 - t^2} = t^2$. Now use the fact that the slope of the line passing through C and B is $\frac{\sqrt{a^2 - t^2}}{t}$, and the slope of the line passing through C and A is $\frac{-\sqrt{b^2 - t^2}}{t}$.

3. Let the altitude from C to $\overline{AB}$ intersect $\overline{AB}$ at the point X, and let $t = CX$ and $x = AX$. Use the fact that $x^2 + t^2 = 7^2$ and $(13 - x)^2 + t^2 = 9^2$ to get $t = \sqrt{\frac{14{,}355}{676}}$.

5. Proof 1: Note that the new triangle is similar to the old triangle with a similarity factor of r.
   Proof 2: Show that ra, rb, and rc satisfy $(ra)^2 + (rb)^2 = (rc)^2$.

7. **a.** No such triangle exists. Look at the two cases where 2 is either the length of a leg or the length of the hypotenuse.
   b. $a = 7$, $b = 24$, $c = 25$.
   c. No such triangle exists. Check possible values of a and b in $a^2 + b^2 = 7^2$.
   d. $a = 20$, $b = 21$, $c = 29$.
   e. $a = 5$, $b = 12$, $c = 13$.
   f. $a = 7$, $b = 24$, $c = 25$.
   g. $a = 5$, $b = 12$, $c = 13$.
   h. No such triangle exists. Look at the equation $a^2 + b^2 = c^2$ and see what happens if a and b are even or if b and c are even.
   i. No such triangle exists. Look at the equation $a^2 + b^2 = c^2$ and see what happens if $a, b,$ and c are odd.

9. In the diagram we see that if $a \neq b$, then the area of the large square $(a + b)^2$ is strictly greater than the sum of the areas of the four rectangles with sides of length a and b ($4ab$). Thus, we see that $\frac{a+b}{2} > \sqrt{ab}$. The two quantities will be equal only when the area of the small square in the center of the figure is zero (i.e., when $a = b$).

5.5: An Extension of the Pythagorean Theorem: The Law of Cosines

1. $b^2 = a^2 + c^2 - 2ac \cos \beta$, and $c^2 = a^2 + b^2 - 2ab \cos \gamma$.
3. $b \approx 24.7$, $\alpha \approx 34.9°$, and $\gamma \approx 25.1°$.
5. Let the parallelogram be called $ABCD$, and suppose we know $AD(= CB)$, DB, and $m(\angle ADB)$. Use the law of cosines to find the length of $AB(= CD)$.

Now use the law of cosines to find $m(\angle BAD)$. Now we know $m(\angle ADC) = 180° - m(\angle BAD)$. Now use the law of cosines to find AC.

5.6: Integer Distances in the Plane

1. Solve the equation $1 = \sqrt{(x_2 - x_1)^2 + [(3x_2 + 2) - (3x_1 + 2)]^2}$ for $|x_2 - x_1|$ to get $|x_2 - x_1| = \frac{1}{\sqrt{10}}$, then choose an x_1 and let $x_n = x_1 + n\frac{1}{\sqrt{10}}$ for $n > 1$.
3. If there were four such points, the distance between two of the points on the same side of the square would be an integer n, but the distance between two points on opposite ends of the diagonal of the square would be $n\sqrt{2}$, a contradiction.
5. Example: $(0,0), (0,60), (144,0), (-144,0), (25,0), (-25,0), (80,0),$ and $(-80,0)$.

5.7: Pythagorean Triples: Positive Integer Solutions to $x^2 + y^2 = z^2$

1. a. $m = 17, n = 2 \Rightarrow x = 285, y = 68,$ and $z = 293$.
       $m = 34, n = 1 \Rightarrow x = 1{,}155, y = 68,$ and $z = 1{,}157$.
    b. $m = 40, n = 3 \Rightarrow x = 1{,}591, y = 240,$ and $z = 1{,}609$.
       $m = 24, n = 5 \Rightarrow x = 551, y = 240,$ and $z = 601$.
       $m = 15, n = 8 \Rightarrow x = 161, y = 240,$ and $z = 289$.
       $m = 120, n = 1 \Rightarrow x = 14{,}399, y = 240,$ and $z = 14{,}401$.
    c. $m = 37, n = 2 \Rightarrow x = 1{,}365, y = 148,$ and $z = 1{,}373$.
       $m = 74, n = 1 \Rightarrow x = 5{,}475, y = 148,$ and $z = 5{,}477$.
    d. $m = 128, n = 1 \Rightarrow x = 16{,}383, y = 256,$ and $z = 16{,}385$.

5.8: Extended Studies: Further Investigations into Integer Distance Point Sets—a Theorem of Erdös

1. By the triangle inequality, $AB + BC > AC$. This gives $\frac{(AB+BC-AC)}{2} > 0$, so $\frac{(AB+BC+AC)}{2} > AC$. Repeat the argument for AB and BC.
3. False. Let $a = 3, b = 4,$ and $c = 5$.

5.9: Extended Studies: Additional Questions Concerning Pythagorean Triples

1. a. $r = 25, s = 3 \Rightarrow m = 14, n = 11 \Rightarrow x = 75, y = 308,$ and $z = 317$.
       $r = 75, s = 1 \Rightarrow m = 38, n = 37 \Rightarrow x = 75, y = 2{,}812,$ and $z = 2{,}813$.
    b. $r = 289, s = 1 \Rightarrow m = 145, n = 144 \Rightarrow x = 289, y = 41{,}760,$ and $z = 41{,}761$.
    c. $r = 313, s = 1 \Rightarrow m = 157, n = 156 \Rightarrow x = 313, y = 48{,}984,$ and $z = 48{,}985$.
    d. $r = 1{,}155, s = 1 \Rightarrow m = 578, n = 577 \Rightarrow x = 1{,}155, y = 667{,}012,$ and $z = 667{,}013$.
       $r = 385, s = 3 \Rightarrow m = 194, n = 191 \Rightarrow x = 1{,}155, y = 74{,}108,$ and $z = 74{,}117$.
       $r = 231, s = 5 \Rightarrow m = 118, n = 113 \Rightarrow x = 1{,}155, y = 26{,}668,$ and $z = 26{,}693$.
       $r = 165, s = 7 \Rightarrow m = 86, n = 79 \Rightarrow x = 1{,}155, y = 13{,}588,$ and $z = 13{,}637$.
       $r = 105, s = 11 \Rightarrow m = 58, n = 47 \Rightarrow x = 1{,}155, y = 5{,}452,$ and $z = 5{,}573$.
       $r = 77, s = 15 \Rightarrow m = 46, n = 31 \Rightarrow x = 1{,}155, y = 2{,}852,$ and $z = 3{,}077$.
       $r = 55, s = 21 \Rightarrow m = 38, n = 17 \Rightarrow x = 1{,}155, y = 1{,}292,$ and $z = 1{,}733$.
       $r = 35, s = 33 \Rightarrow m = 34, n = 1 \Rightarrow x = 1{,}155, y = 68,$ and $z = 1{,}157$.

338 Answers to (Most) Odd-Numbered Exercises

3. **a.** $m = 6, n = 5 \Rightarrow x = 9, y = 60,$ and $z = 61$.
   b. There are none since $43 \neq 4t + 1$.
   c. There are none since $3 \neq 4t + 1$ and 3 is a prime divisor of 145.
   d. There are none since $19 \neq 4t + 1$ and 19 is a prime divisor of 4,199.

5. We have $P = 2m(m + n) \Rightarrow P/2 = m(m + n)$, which cannot be prime since $m \geq 2$.

7. $A = 2,730$, PPT's $\{28, 195, 197\}$, and $\{60, 91, 109\}$.

9. If the triangle is a 5, 12, 13 triangle, then it is easily checked that the perimeter equals the area. If $P = A$, then $P = 2m(m + n) = rs\frac{(r+s)}{2}\frac{(r-s)}{2} = A$, where m, n, r, and s are as in the earlier parts of this section (i.e., we have $r = m + n$ and $s = m - n$). Thus, we get $2m(m + n) = (m + n)(m - n)\frac{(2m)}{2}\frac{(2n)}{2}$, and so $2 = n(m - n)$. Now consider the possibilities for n and m and find the resulting Pythagorean triple.

CHAPTER 6

6.3: Rational and Irrational Numbers

1. **a.** $\begin{array}{l} 1,000,000x = 516,277.\overline{277} \\ 1,000x = 516.\overline{277} \end{array} \Rightarrow 999,000x = 515,761 \Rightarrow x = \frac{515,761}{999,000}$

   b. $\begin{array}{l} 10x = 9.\overline{9} \\ x = .\overline{9} \end{array} \Rightarrow 9x = 9 \Rightarrow x = 1.$

   c. $\begin{array}{l} 10^8 x = 22,222,273.\overline{73} \\ 10^6 x = 222,222.\overline{73} \end{array} \Rightarrow 99,000,000x = 22,000,051 \Rightarrow x = \frac{22,000,051}{99,000,000}.$

3. Suppose $\sqrt[3]{2} = \frac{a}{b}$ with $a, b \in \mathbf{N}$; then $2b^3 = a^3$. The number of 2s in the factorization of b^3 is $3m$ for some positive integer m, and the number of 2s in the factorization of a^3 is $3n$ for some positive integer n, but $2b^3$ has a total of $3m + 1$ 2s in its factorization. Since $3m + 1 \neq 3n$, there are different numbers of 2s in the factorizations of $2b^3$ and a^3, which contradicts the FTA. Thus, $\sqrt[3]{2}$ is irrational.

5. Suppose $\sqrt{\frac{2}{3}} = \frac{a}{b}$ with $a, b \in \mathbf{N}$; then $2b^2 = 3a^2$. The number of 2s in each of the factorizations of a^2 and b^2 is even, so the number of 2s in the factorization of $2b^2$ is odd while the number of 2s in the factorization of $3a^2$ is even. This contradicts the FTA, so $\sqrt{\frac{2}{3}}$ is irrational.

6.4: Systems of Linear Equations

1. $x = $ amount in stock fund; $y = $ amount in fixed annuity.

   $\begin{array}{l} x + y = 2,500 \\ .06x + .03y = 1,200 \end{array} \Rightarrow \begin{array}{l} .06x + .06y = 1,500 \\ .06x + .03y = 1200 \end{array} \Rightarrow .03y = 300 \Rightarrow y = 10,000$
   and $x = 15,000$.

Answers to (Most) Odd-Numbered Exercises

3. a.
$$\begin{bmatrix} 2 & -7 & 8 \\ 3 & 4 & -6 \end{bmatrix} \xrightarrow{-\frac{3}{2}R_1+R_2} \cdots \xrightarrow{\frac{14}{29}R_2+R_1} \cdots \xrightarrow{\frac{1}{2}R_1} \cdots$$

$$\xrightarrow{\frac{2}{29}R_2} \begin{bmatrix} 1 & 0 & -\frac{10}{29} \\ 0 & 1 & -\frac{36}{29} \end{bmatrix} \Rightarrow x = -\frac{10}{29},\ y = -\frac{36}{29}.$$

b.
$$\begin{bmatrix} -\frac{1}{5} & \frac{2}{3} & \frac{1}{30} \\ -3 & 10 & \frac{1}{2} \end{bmatrix} \xrightarrow{30R_1} \cdots \xrightarrow{-\frac{1}{2}R_1+R_2} \cdots \xrightarrow{-\frac{1}{6}R_1} \begin{bmatrix} 1 & -\frac{10}{3} & -\frac{1}{6} \\ 0 & 0 & 0 \end{bmatrix}$$

$$\Rightarrow x = \frac{10}{3}y - \frac{1}{6}.$$

c.
$$\begin{bmatrix} 1 & 1 & 4 \\ -2 & -3 & 2 \\ 5 & 1 & -1 \end{bmatrix} \xrightarrow{2R_1+R_2} \cdots \xrightarrow{-5R_1+R_3} \cdots \xrightarrow{-R_2} \cdots$$

$$\xrightarrow{4R_2+R_3} \begin{bmatrix} 1 & 1 & 4 \\ 0 & 1 & -10 \\ 0 & 0 & -61 \end{bmatrix} \Rightarrow \text{the system is inconsistent and therefore has no solutions.}$$

d.
$$\begin{bmatrix} 2 & 4 & -1 & 4 \\ -1 & -2 & 1 & -6 \\ 2 & -3 & 1 & -1 \end{bmatrix} \xrightarrow{\frac{1}{2}R_1+R_2} \cdots \xrightarrow{-R_1+R_3} \cdots \xrightarrow{\frac{1}{2}R_1} \cdots$$

$$\xrightarrow{R_2+R_1} \cdots \xrightarrow{-4R_2+R_3} \cdots \xrightarrow{2R_2} \cdots \xrightarrow{P(2,3)} \cdots$$

$$\xrightarrow{-\frac{1}{7}R_2} \cdots \xrightarrow{-2R_2+R_1} \begin{bmatrix} 1 & 0 & 0 & -\frac{12}{7} \\ 0 & 1 & 0 & -\frac{1}{7} \\ 0 & 0 & 1 & -8 \end{bmatrix} \Rightarrow x = -\frac{12}{7},\ y = -\frac{1}{7},\ z = -8.$$

e.
$$\begin{bmatrix} 3 & 0 & 2 & 0 \\ 1 & 8 & -4 & 0 \end{bmatrix} \xrightarrow{P(1,2)} \cdots \xrightarrow{-3R_1+R_2} \cdots \xrightarrow{\frac{1}{3}R_2+R_1} \cdots$$

$$\xrightarrow{-\frac{1}{24}R_2} \begin{bmatrix} 1 & 0 & \frac{2}{3} & 0 \\ 0 & 1 & -\frac{7}{12} & 0 \end{bmatrix} \Rightarrow x_1 = -\frac{2}{3}x_3,\ x_2 = \frac{7}{12}x_3.$$

f.
$$\begin{bmatrix} 0 & -1 & 2 & 0 \\ 3 & 9 & 0 & 0 \\ 1 & -1 & -1 & 0 \end{bmatrix} \xrightarrow{P(1,3)} \cdots \xrightarrow{-3R_1+R_2} \cdots \xrightarrow{12R_3+R_2} \cdots \xrightarrow{P(2,3)}$$

5. a.

$$\begin{bmatrix} 1 & t & 3 \\ -1 & 2 & 1 \end{bmatrix} \xrightarrow{R_1+R_2} \cdots \xrightarrow{-\frac{t}{2+t}R_2+R_1} \cdots \xrightarrow{\frac{1}{2+t}R_2} \begin{bmatrix} 1 & 0 & \frac{6-t}{2+t} \\ 0 & 1 & \frac{4}{2+t} \end{bmatrix}.$$ If $t \neq 2$, then there exists a unique solution.

b.

$$\begin{bmatrix} -1 & t & 2 \\ t & -9 & 5 \end{bmatrix} \xrightarrow{tR_1+R_2} \cdots \xrightarrow{-\frac{t}{t^2-9}R_2+R_1} \cdots \xrightarrow{-R_1} \cdots$$

$$\xrightarrow{\frac{1}{t^2-9}R_2} \begin{bmatrix} 1 & 0 & \frac{5t+18}{t^2-9} \\ 0 & 1 & \frac{2t+5}{t^2-9} \end{bmatrix}.$$ If $t \neq \pm 3$, then there exists a unique solution.

6.5: Polynomial Curve Fitting: An Application of Systems of Linear Equations

1. a. $f(x) = a_1 x + a_0$

$$\begin{bmatrix} 1 & 1 & 1 \\ 3 & 1 & 7 \end{bmatrix} \xrightarrow{-3R_1+R_2} \cdots \xrightarrow{-\frac{1}{2}R_1} \cdots \xrightarrow{-R_2+R_1} \begin{bmatrix} 1 & 0 & 3 \\ 0 & 1 & -2 \end{bmatrix}$$

$$\Rightarrow a_1 = 3, a_0 = -2.$$

b. $f(x) = a_1 x + a_0$

$$\begin{bmatrix} -3 & 1 & 2 \\ 3 & 1 & 2 \end{bmatrix} \xrightarrow{R_1+R_2} \cdots \xrightarrow{\frac{1}{2}R_2} \cdots \xrightarrow{-R_2+R_1} \cdots$$

$$\xrightarrow{-\frac{1}{3}R_1} \begin{bmatrix} 1 & 0 & 0 \\ 0 & 1 & 2 \end{bmatrix} \Rightarrow a_1 = 0, a_0 = 2.$$

c. $f(x) = a_2 x^2 + a_1 x + a_0$

$$\begin{bmatrix} 25 & -5 & 1 & -1 \\ 4 & 2 & 1 & 0 \\ 0 & 0 & 1 & -1 \end{bmatrix} \xrightarrow{-R_3+R_2} \cdots \xrightarrow{-R_3+R_1} \cdots \xrightarrow{\frac{4}{25}R_1+R_2} \cdots$$

$$\xrightarrow{\frac{5}{14}R_2} \cdots \xrightarrow{5R_2+R_1} \cdots \xrightarrow{\frac{1}{25}R_1} \begin{bmatrix} 1 & 0 & 0 & \frac{1}{14} \\ 0 & 1 & 0 & \frac{5}{14} \\ 0 & 0 & 1 & -1 \end{bmatrix}$$

$$\Rightarrow a_2 = \frac{1}{14}, a_1 = \frac{5}{14}, a_0 = -1.$$

Answers to (Most) Odd-Numbered Exercises 341

d. $f(x) = a_2x^2 + a_1x + a_0$

$$\begin{bmatrix} 1 & 1 & 1 & 3 \\ 1 & -1 & 1 & 3 \\ 0 & 0 & 1 & 3 \end{bmatrix} \xrightarrow{-R_1+R_2} \cdots \xrightarrow{-R_3+R_1} \cdots$$

$$\xrightarrow{-\frac{1}{2}R_2} \cdots \xrightarrow{-R_2+R_1} \begin{bmatrix} 1 & 0 & 0 & 0 \\ 0 & 1 & 0 & 0 \\ 0 & 0 & 1 & 3 \end{bmatrix} \Rightarrow a_2 = 0, a_1 = 0, a_0 = 3.$$

e. $f(x) = a_2x^2 + a_1x + a_0$

$$\begin{bmatrix} 1 & 1 & 1 & 2 \\ 4 & 2 & 1 & 3 \\ 9 & 3 & 1 & 4 \end{bmatrix} \xrightarrow{-4R_1+R_2} \cdots \xrightarrow{-9R_1+R_3} \cdots \xrightarrow{-3R_2+R_3} \cdots$$

$$\xrightarrow{3R_3+R_2} \cdots \xrightarrow{-\frac{1}{2}R_2} \cdots \xrightarrow{-R_2+R_1} \cdots \xrightarrow{-R_3+R_1} \begin{bmatrix} 1 & 0 & 0 & 0 \\ 0 & 1 & 0 & 1 \\ 0 & 0 & 1 & 1 \end{bmatrix}$$

$\Rightarrow a_2 = 0, a_1 = 1, a_0 = 1.$

f. $f(x) = a_3x^3 + a_2x^2 + a_1x + a_0$

$$\begin{bmatrix} 1 & 1 & 1 & 1 & 0 \\ 8 & 4 & 2 & 1 & 3 \\ -8 & 4 & -2 & 1 & 3 \\ 0 & 0 & 0 & 1 & 4 \end{bmatrix} \xrightarrow{-8R_1+R_2} \cdots \xrightarrow{-R_4+R_3} \cdots \xrightarrow{-R_4+R_1} \cdots$$

$$\xrightarrow{8R_1+R_3} \cdots \xrightarrow{P(2,3)} \cdots \xrightarrow{7R_4+R_3} \cdots \xrightarrow{-\frac{1}{4}R_3} \cdots \xrightarrow{-6R_3+R_2}$$

$$\cdots \xrightarrow{-R_3+R_1} \cdots \xrightarrow{\frac{1}{12}R_2} \cdots \xrightarrow{-R_2+R_1} \begin{bmatrix} 1 & 0 & 0 & 0 & \frac{5}{4} \\ 0 & 1 & 0 & 0 & -\frac{1}{4} \\ 0 & 0 & 1 & 0 & -5 \\ 0 & 0 & 0 & 1 & 4 \end{bmatrix}$$

$\Rightarrow a_3 = \frac{5}{4}, a_2 = -\frac{1}{4}, a_1 = -5, a_0 = 4.$

6.6: Matrix Arithmetic and Matrix Algebra

1. a. $AB = \begin{bmatrix} 0 & 0 \\ 0 & 1 \end{bmatrix} \begin{bmatrix} 1 & 0 \\ 0 & 0 \end{bmatrix} = \begin{bmatrix} 0 & 0 \\ 0 & 0 \end{bmatrix} = \begin{bmatrix} 1 & 0 \\ 0 & 0 \end{bmatrix} \begin{bmatrix} 0 & 0 \\ 0 & 1 \end{bmatrix} = BA.$

b.
$$AB = \begin{bmatrix} 1 & 0 & 2 \\ -1 & 3 & 0 \\ 0 & 5 & -4 \end{bmatrix} \begin{bmatrix} 0 & 1 & -1 \\ 2 & 0 & 0 \\ 3 & -2 & 0 \end{bmatrix} = \begin{bmatrix} 6 & \cdots & \cdots \\ \cdots & \cdots & \cdots \\ \cdots & \cdots & \cdots \end{bmatrix}$$

$$BA = \begin{bmatrix} 0 & 1 & -1 \\ 2 & 0 & 0 \\ 3 & -2 & 0 \end{bmatrix} \begin{bmatrix} 1 & 0 & 2 \\ -1 & 3 & 0 \\ 0 & 5 & -4 \end{bmatrix} = \begin{bmatrix} -1 & \cdots & \cdots \\ \cdots & \cdots & \cdots \\ \cdots & \cdots & \cdots \end{bmatrix}$$

$\Rightarrow AB \neq BA.$

c.

$$AB = \begin{bmatrix} 1 & 0 & 0 \\ 0 & 2 & 0 \\ 0 & 0 & 3 \end{bmatrix} \begin{bmatrix} -2 & 0 & 0 \\ 0 & 0 & 0 \\ 0 & 0 & 4 \end{bmatrix} = \begin{bmatrix} -2 & 0 & 0 \\ 0 & 0 & 0 \\ 0 & 0 & 12 \end{bmatrix}$$

$$= \begin{bmatrix} -2 & 0 & 0 \\ 0 & 0 & 0 \\ 0 & 0 & 4 \end{bmatrix} \begin{bmatrix} 1 & 0 & 0 \\ 0 & 2 & 0 \\ 0 & 0 & 3 \end{bmatrix} = BA.$$

3. a.

$$r(AB) = r\left(\left[\sum_{t=1}^{n} a_{it}b_{tj}\right]\right) = \left[r\left(\sum_{t=1}^{n} a_{it}b_{tj}\right)\right] = \left[\sum_{t=1}^{n}(ra_{it})b_{tj}\right] = (rA)B$$

$$= \left[\sum_{t=1}^{n} a_{it}(rb_{tj})\right] = A(rB).$$

b. $r(sA) = r([sa_{ij}]) = [r(sa_{ij})] = [(rs)a_{ij}] = (rs)A.$
c. $r(A + B) = r([a_{ij} + b_{ij}]) = [r(a_{ij} + b_{ij})] = [ra_{ij} + rb_{ij}] = rA + rB.$
d. $(r + s)A = [(r + s)a_{ij}] = [ra_{ij} + sa_{ij}] = rA + sA.$
e. Suppose $r \neq 0$ and $A \neq 0$; then $a_{ij} \neq 0$ for some entry a_{ij} of A. Thus, $ra_{ij} \neq 0$, so the entry ra_{ij} of rA is not, and $rA \neq 0$. Therefore, if $rA = 0$, then either $r = 0$ or $A = 0$.

5. a. $AB = \begin{bmatrix} 0 & 0 \\ 0 & 1 \end{bmatrix} \begin{bmatrix} 1 & 0 \\ 0 & 0 \end{bmatrix} = \begin{bmatrix} 0 & 0 \\ 0 & 0 \end{bmatrix}.$

b. $A^2 = \begin{bmatrix} 0 & 0 \\ 1 & 0 \end{bmatrix} \begin{bmatrix} 0 & 0 \\ 1 & 0 \end{bmatrix} = \begin{bmatrix} 0 & 0 \\ 0 & 0 \end{bmatrix}.$

c. $A = \begin{bmatrix} 1 & 2 \\ 3 & 4 \end{bmatrix}, B = \begin{bmatrix} 1 & 2 \\ 3 & 0 \end{bmatrix}, C = \begin{bmatrix} 1 & 0 \\ 0 & 0 \end{bmatrix} \Rightarrow AC = \begin{bmatrix} 1 & 2 \\ 3 & 4 \end{bmatrix} \begin{bmatrix} 1 & 0 \\ 0 & 0 \end{bmatrix}$

$= \begin{bmatrix} 1 & 0 \\ 3 & 0 \end{bmatrix} = \begin{bmatrix} 1 & 2 \\ 3 & 0 \end{bmatrix} \begin{bmatrix} 1 & 0 \\ 0 & 0 \end{bmatrix} = BC.$

7. a. $A = \begin{bmatrix} 0 & 1 \\ -1 & 0 \end{bmatrix} \Rightarrow A^t = \begin{bmatrix} 0 & -1 \\ 1 & 0 \end{bmatrix} = -A.$

b. $[a_{ij}]^t = [a_{ji}] = A^t = -A = [-a_{ij}] \Rightarrow a_{ii} = -a_{ii}$ for each $1 \leq i \leq n$; thus, $a_{ii} = 0$ for each $1 \leq i \leq n$.
c. $(A - A^t)^t = A^t - (A^t)^t = A^t - A = -(-A^t + A) = -(A - A^t).$
d. $(A + B)^t = A^t + B^t = -A + (-B) = -(A + B).$

6.7: Multiplicative Inverses: Solving the Matrix Equation $AX = B$

1. a. $\det\left(\begin{bmatrix} -2 & \frac{3}{2} \\ 1 & -\frac{1}{2} \end{bmatrix}\right) = -\frac{1}{2}$, so it is invertible with inverse $-2\begin{bmatrix} -\frac{1}{2} & -\frac{3}{2} \\ -1 & -2 \end{bmatrix}.$

b. $\det\left(\begin{bmatrix} 1 & -2 \\ 3 & 6 \end{bmatrix}\right) = 12$, so it is invertible with inverse $-\frac{1}{12}\begin{bmatrix} 6 & 2 \\ -3 & 1 \end{bmatrix}.$

c. $\begin{bmatrix} 1 & 2 & 2 \\ 1 & 1 & 2 \\ 1 & 2 & 2 \end{bmatrix}$ is not invertible since it is row equivalent to $\begin{bmatrix} 1 & 2 & 2 \\ 1 & 1 & 2 \\ 0 & 0 & 0 \end{bmatrix}$.

d. By Gaussian elimination, the augmented matrix $\begin{bmatrix} 1 & 0 & 1 & 1 & 0 & 0 \\ 2 & 1 & 1 & 0 & 1 & 0 \\ 3 & 2 & 0 & 0 & 0 & 1 \end{bmatrix}$ becomes the matrix $\begin{bmatrix} 1 & 0 & 0 & 2 & -2 & 1 \\ 0 & 1 & 0 & -3 & 3 & -1 \\ 0 & 0 & 1 & -1 & 2 & -1 \end{bmatrix}$, so the matrix is invertible with inverse $\begin{bmatrix} 2 & -2 & 1 \\ -3 & 3 & -1 \\ -1 & 2 & -1 \end{bmatrix}$.

e. By Gaussian elimination, the augmented matrix $\begin{bmatrix} -2 & 0 & 0 & 1 & 0 & 0 \\ 0 & -3 & 0 & 0 & 1 & 0 \\ 0 & 0 & 4 & 0 & 0 & 1 \end{bmatrix}$ becomes the matrix $\begin{bmatrix} 1 & 0 & 0 & -\frac{1}{2} & 0 & 0 \\ 0 & 1 & 0 & 0 & -\frac{1}{3} & 0 \\ 0 & 0 & 1 & 0 & 0 & \frac{1}{4} \end{bmatrix}$, so the matrix is invertible with inverse $\begin{bmatrix} -\frac{1}{2} & 0 & 0 \\ 0 & -\frac{1}{3} & 0 \\ 0 & 0 & \frac{1}{4} \end{bmatrix}$.

3. $A = \frac{1}{\sqrt{2}}\begin{bmatrix} 1 & -1 \\ 1 & 1 \end{bmatrix} \Rightarrow \left(\frac{1}{\sqrt{2}}\begin{bmatrix} 1 & -1 \\ 1 & 1 \end{bmatrix}\right)^2 = \frac{1}{2}\begin{bmatrix} 1 & -1 \\ 1 & 1 \end{bmatrix}\begin{bmatrix} 1 & -1 \\ 1 & 1 \end{bmatrix} = \frac{1}{2}\begin{bmatrix} 2 & 0 \\ 0 & 2 \end{bmatrix} = \begin{bmatrix} 1 & 0 \\ 0 & 1 \end{bmatrix}$, so $A^{-1} = A$.

5. a. $A = \begin{bmatrix} 1 & 0 \\ 0 & 0 \end{bmatrix}, B = \begin{bmatrix} 1 & 1 \\ 0 & 0 \end{bmatrix}, C = \begin{bmatrix} 1 & 1 \\ 1 & 1 \end{bmatrix} \Rightarrow AB = \begin{bmatrix} 1 & 0 \\ 0 & 0 \end{bmatrix}\begin{bmatrix} 1 & 1 \\ 0 & 0 \end{bmatrix} = \begin{bmatrix} 1 & 1 \\ 0 & 0 \end{bmatrix}$
$= \begin{bmatrix} 1 & 0 \\ 0 & 0 \end{bmatrix}\begin{bmatrix} 1 & 1 \\ 1 & 1 \end{bmatrix} = AC$.

b. Since A is invertible, we have $A^{-1}(AB) = A^{-1}(AC) \Rightarrow (A^{-1}A)B = (A^{-1}A)C \Rightarrow IB = IC \Rightarrow B = C$.

7. a. $A = \begin{bmatrix} 1 & 0 \\ 0 & 0 \end{bmatrix} \Rightarrow A^2 = \begin{bmatrix} 1 & 0 \\ 0 & 0 \end{bmatrix}\begin{bmatrix} 1 & 0 \\ 0 & 0 \end{bmatrix} = \begin{bmatrix} 1 & 0 \\ 0 & 0 \end{bmatrix} = A$.

b. Suppose A is not singular and that $A \neq I_n$; then $A^2 = A \Rightarrow A^{-1}(A^2) = A^{-1}A \Rightarrow A = I_n$, which is a contradiction.
Thus, A cannot have an inverse, so A is singular.

6.8: Coding with Matrices

1. a. $\begin{bmatrix} 2 & 1 \\ 1 & 1 \end{bmatrix} \left(\begin{bmatrix} 1 \\ 4 \end{bmatrix}, \begin{bmatrix} 0 \\ 20 \end{bmatrix}, \begin{bmatrix} 19 \\ 24 \end{bmatrix}, \begin{bmatrix} 26 \\ 8 \end{bmatrix}, \begin{bmatrix} 18 \\ 26 \end{bmatrix}, \begin{bmatrix} 19 \\ 17 \end{bmatrix}, \begin{bmatrix} 20 \\ 19 \end{bmatrix}, \begin{bmatrix} 7 \\ 26 \end{bmatrix}, \begin{bmatrix} 19 \\ 17 \end{bmatrix}, \begin{bmatrix} 20 \\ 19 \end{bmatrix}, \right.$
$\left. \begin{bmatrix} 7 \\ 26 \end{bmatrix}, \begin{bmatrix} 1 \\ 4 \end{bmatrix}, \begin{bmatrix} 0 \\ 20 \end{bmatrix}, \begin{bmatrix} 19 \\ 24 \end{bmatrix}, \begin{bmatrix} 26 \\ 19 \end{bmatrix}, \begin{bmatrix} 7 \\ 0 \end{bmatrix}, \begin{bmatrix} 19 \\ 26 \end{bmatrix}, \begin{bmatrix} 8 \\ 18 \end{bmatrix}, \begin{bmatrix} 26 \\ 0 \end{bmatrix}, \begin{bmatrix} 11 \\ 11 \end{bmatrix} \right)$ gives the
message 6, 5, 20, 20, 62, 43, 60, 34, 62, 44, 55, 36, 29, 39, 40, 33, 55, 36, 59, 39, 40, 33, 6, 5, 20, 20, 62, 43, 71, 45, 14, 7, 64, 45, 34, 26, 52, 26, 33, 22.

b. $\begin{bmatrix} 1 & 0 & 1 \\ 0 & 2 & 1 \\ 1 & 1 & 2 \end{bmatrix} \left(\begin{bmatrix} 5 \\ 0 \\ 2 \end{bmatrix}, \begin{bmatrix} 19 \\ 14 \\ 17 \end{bmatrix}, \begin{bmatrix} 24 \\ 26 \\ 22 \end{bmatrix}, \begin{bmatrix} 8 \\ 13 \\ 3 \end{bmatrix}, \begin{bmatrix} 14 \\ 22 \\ 18 \end{bmatrix}, \begin{bmatrix} 26 \\ 0 \\ 17 \end{bmatrix}, \begin{bmatrix} 4 \\ 26 \\ 0 \end{bmatrix}, \begin{bmatrix} 11 \\ 22 \\ 0 \end{bmatrix}, \begin{bmatrix} 24 \\ 18 \\ 26 \end{bmatrix}, \right.$
$\left. \begin{bmatrix} 1 \\ 17 \\ 14 \end{bmatrix}, \begin{bmatrix} 10 \\ 4 \\ 13 \end{bmatrix} \right)$ gives the message 7, 2, 9, 36, 45, 67, 46, 74, 94, 11, 29, 27, 32, 62, 72, 43, 17, 60, 4, 52, 30, 11, 44, 33, 50, 62, 94, 15, 48, 46, 23, 21, 40.

Photo Credits

Page 2	From *Connected Mathematics: Say It with Symbol* by Glenda Lappan, James T. Fey, William M. Fitzgerald, Susan N. Friel, and Elizabeth Defanis Phillips. © 2004 by Michigan State University. Published by Pearson Education, Inc., publishing as Pearson Prentice Hall. Used by permission.
Page 3	From *Connected Mathematics: Say It with Symbol* by Glenda Lappan, James T. Fey, William M. Fitzgerald, Susan N. Friel, and Elizabeth Defanis Phillips. © 2004 by Michigan State University. Published by Pearson Education, Inc., publishing as Pearson Prentice Hall. Used by permission.
Page 10	From *MathScape: Seeing and Thinking Mathematically*, Course 1, Patterns in Numbers and Shapes. © 1991, Glencoe/McGraw-Hill. Reprinted by permission.
Page 11	From *MathScape: Seeing and Thinking Mathematically*, Course 1, Patterns in Numbers and Shapes. © 1991, Glencoe/McGraw-Hill. Reprinted by permission.
Page 12	Used by permission of John Lannin.
Page 13	Used by permission of John Lannin.
Page 13	From *MathScape: Seeing and Thinking Mathematically*, Course 1, Patterns in Numbers and Shapes. © 1991, Glencoe/McGraw-Hill. Reprinted by permission.
Page 14	From *MathScape: Seeing and Thinking Mathematically*, Course 1, Patterns in Numbers and Shapes. © 1991, Glencoe/McGraw-Hill. Reprinted by permission.
Page 16	From *Connected Mathematics: Frogs, Fleas and Painted Cubes* by Glenda Lappan, James T. Fey, William M. Fitzgerald, Susan N. Friel, and Elizabeth Defanis Phillips. © 2004 by Michigan State University. Published by Pearson Education, Inc., publishing as Pearson Prentice Hall. Used by permission.
Page 17	From *Connected Mathematics: Frogs, Fleas and Painted Cubes* by Glenda Lappan, James T. Fey, William M. Fitzgerald, Susan N. Friel, and Elizabeth Defanis Phillips. © 2004 by Michigan State University. Published by Pearson Education, Inc., publishing as Pearson Prentice Hall. Used by permission.
Page 26	From *Connected Mathematics: Growing, Growing, Growing* by Glenda Lappan, James T. Fey, William M. Fitzgerald, Susan N. Friel, and Elizabeth Defanis Phillips. © 2004 by Michigan State University. Published by Pearson Education, Inc., publishing as Pearson Prentice Hall. Used by permission.
Page 27	From *Connected Mathematics: Growing, Growing, Growing* by Glenda Lappan, James T. Fey, William M. Fitzgerald, Susan N. Friel, and Elizabeth Defanis Phillips. © 2004 by Michigan State University. Published by Pearson Education, Inc., publishing as Pearson Prentice Hall. Used by permission.
Page 28	From *Connected Mathematics: Growing, Growing, Growing* by Glenda Lappan, James T. Fey, William M. Fitzgerald, Susan N. Friel, and Elizabeth Defanis Phillips. © 2004 by Michigan State University. Published by Pearson Education, Inc., publishing as Pearson Prentice Hall. Used by permission.
Page 35–37	Reprinted with permission from *Mathematics in Context: Patterns and Symbols*. © 2003 by Encyclopaedia Britannica, Inc.
Page 38	Used by permission of the Office des Emissions de Timbres-Poste.
Page 52	*Math Thematics: Book 3*, by Rick Billstein and Jim Williamson. McDougal Littell, 1999. Reprinted by permission of McDougal Littell.
Page 53	*Math Thematics: Book 3*, by Rick Billstein and Jim Williamson. McDougal Littell, 1999. Reprinted by permission of McDougal Littell.
Page 63	Used by permission of the Office des Emissions de Timbres-Poste.
Page 65	Lee Jenkins Collection, University of Missouri. Used by permission.

Photo Credits

Page 66	From *Connected Mathematics: Prime Time* by Glenda Lappan, James T. Fey, William M. Fitzgerald, Susan N. Friel, and Elizabeth Defanis Phillips. © 2004 by Michigan State University. Published by Pearson Education, Inc., publishing as Pearson Prentice Hall. Used by permission.
Page 67	From *Connected Mathematics: Prime Time* by Glenda Lappan, James T. Fey, William M. Fitzgerald, Susan N. Friel, and Elizabeth Defanis Phillips. © 2004 by Michigan State University. Published by Pearson Education, Inc., publishing as Pearson Prentice Hall. Used by permission.
Page 68	From *Connected Mathematics: Prime Time* by Glenda Lappan, James T. Fey, William M. Fitzgerald, Susan N. Friel, and Elizabeth Defanis Phillips. © 2004 by Michigan State University. Published by Pearson Education, Inc., publishing as Pearson Prentice Hall. Used by permission.
Page 69	From *Connected Mathematics: Prime Time* by Glenda Lappan, James T. Fey, William M. Fitzgerald, Susan N. Friel, and Elizabeth Defanis Phillips. © 2004 by Michigan State University. Published by Pearson Education, Inc., publishing as Pearson Prentice Hall. Used by permission.
Page 88	From *Connected Mathematics: Prime Time* by Glenda Lappan, James T. Fey, William M. Fitzgerald, Susan N. Friel, and Elizabeth Defanis Phillips. © 2004 by Michigan State University. Published by Pearson Education, Inc., publishing as Pearson Prentice Hall. Used by permission.
Pages 116–117	Photos courtesy of the author.
Page 142	From *Connected Mathematics: Prime Time* by Glenda Lappan, James T. Fey, William M. Fitzgerald, Susan N. Friel, and Elizabeth Defanis Phillips. © 2004 by Michigan State University. Published by Pearson Education, Inc., publishing as Pearson Prentice Hall. Used by permission.
Page 147	*Math Thematics: Book 3*, by Rick Billstein and Jim Williamson. McDougal Littell, 1999. Reprinted by permission of McDougal Littell.
Pages 151–152	Photos courtesy of the author.
Page 155	Photo courtesy of the author.
Page 188	Travelers Express/MoneyGram. Used by permission.
Page 216	Photo courtesy of the author.
Page 218	*Math Thematics: Book 3*, by Rick Billstein and Jim Williamson. McDougal Littell, 1999. Reprinted by permission of McDougal Littell.
Page 219	*Math Thematics: Book 3*, by Rick Billstein and Jim Williamson. McDougal Littell, 1999. Reprinted by permission of McDougal Littell.
Page 220	*Math Thematics: Book 3*, by Rick Billstein and Jim Williamson. McDougal Littell, 1999. Reprinted by permission of McDougal Littell.
Page 221	*Math Thematics: Book 3*, by Rick Billstein and Jim Williamson. McDougal Littell, 1999. Reprinted by permission of McDougal Littell.
Page 238	*Math Thematics: Book 3*, by Rick Billstein and Jim Williamson. McDougal Littell, 1999. Reprinted by permission of McDougal Littell.
Page 255	Reprinted with permission from *Mathematics in Context: Comparing Quantities*. © 2003 by Encyclopaedia Britannica, Inc.
Page 256	Reprinted with permission from *Mathematics in Context: Comparing Quantities*. © 2003 by Encyclopaedia Britannica, Inc.
Page 257	Reprinted with permission from *Mathematics in Context: Comparing Quantities*. © 2003 by Encyclopaedia Britannica, Inc.
Page 261	Reprinted with permission from *Mathematics in Context: Comparing Quantities*, Teacher's Edition. © 1998 by Encyclopaedia Britannica, Inc.
Page 262	Reprinted with permission from *Mathematics in Context: Comparing Quantities*, Teacher's Edition. © 1998 by Encyclopaedia Britannica, Inc.
Page 275	Reprinted with permission from *Mathematics in Context: Comparing Quantities*, © 2003 by Encyclopaedia Britannica, Inc.

Index

The page number indicates the first time in book the definition of the term appears.

A
Abundant integer   141
Addition (mod n)   156
Additive identity
    in $M_{m \times n}(\mathbf{R})$ 294
    in $\mathbf{Z}_n$ 165
Additive inverse
    in $M_{m \times n}(\mathbf{R})$ 294
    in $\mathbf{Z}_n$ 165
Arithmetic sequence   14
    Finite arithmetic sequence   21
Associative property of addition
    in $M_{m \times n}(\mathbf{R})$ 293
    in $\mathbf{Z}_n$ 166
Associative property of matrix
    multiplication   290
Associative property of multiplication in
    $\mathbf{Z}_n$ 166
Augmented matrix   272

B
Base b expanded form of an integer L/
    base b place value   120
Binomial coefficients   49
Binomial Theorem   49

C
Canonical form   79
Cole, Frederick   139
Coefficient matrix   271
Combination/number of
    combinations   41
Common difference   14
Common divisor   74
Common multiple   73
Common ratio   24
Commutative property of addition
    in $M_{m \times n}(\mathbf{R})$ 293
    in $\mathbf{Z}_n$ 166
Commutative property of multiplication
    in $\mathbf{Z}_n$ 166

Composite number   76
Consistent system   278
Cryptology   178

D
Deficient integer   141
Diagonal matrix   297
Diophantine equation   105
Distance formula   226
Distributive property of matrices   293
Distributive property of multiplication
    over addition in $\mathbf{Z}_n$ 167
Division Algorithm   99
Divisor   71

E
Elementary row operations   268
Equality of matrices   272
Equivalent systems   268
Erdös, Paul   238
Euclid   102
Euler, Leonard   138
Even integers   101

F
Factoring/factorization   75
Factor   71
Factor tree   76
Fermat's Last Theorem   250
Fermat, Pierre de Fermat   251
Fibonacci sequence   55
Fundamental Counting Principle   40

G
Gauss, Carl Friedrich   63
Gauss-Jordan Elimination   276
Geometric sequence   24
    Finite geometric series   25
Golden ratio   61
Golden rectangle   61
Greatest common divisor   74

H
Homogeneous system   281
Hyperbola   239

I
Idempotent   174
Inconsistent system   278
Integers   71
Integers (mod n)   153
Interpolating polynomial of minimal degree   285
Irrational numbers   265

K
Kaprekar's constant   176

L
Lattice points   109
Least common multiple   74
Lucas, Edouard   139

M
Matrix   271
Matrix addition   292
Matrix multiplication   289
Mersenne, Marin   138
Mersenne primes   138
$M_n(\mathbf{R})$ 272
Multiple   71
Multiplication (mod n)   156
Multiplicative identity
    in $M_n(\mathbf{R})$ 298
    in $\mathbf{Z}_n$ 166
Multiplicative inverse
    in $M_n(\mathbf{R})$ 299
    in $\mathbf{Z}_n$ 169

N
n factorial   31
Non-zero-divisor in $\mathbf{Z}_n$ 171
Nilpotent   174

O
Odd integer   101
One-to-one function   179
Order or size of a matrix   271

P
Pascal, Blaise   38
Pascal's formula   44
Pascal's Triangle   38

Perfect number   135
Permutation/number of permutations   41
Power set   31
Prime number   76
Primitive Pythagorean Triangle   225
Primitive Pythagorean Triple (PPT)   231
Principle of Mathematical Induction   30
Pythagorean Triangle   225
Pythagorean Triple (PT)   230

Q
$\mathbf{Q}$ = rational numbers   263

R
$\mathbf{R}$ = real numbers   263
Rational numbers   263
Recursive rule or formula   5
Relatively prime   82
Reduced row echelon form   276
Residue class of a (mod n)   153
Residue modulo n 153
Row equivalent matrices   273

S
Scalar product   291
Sequence   6
    Finite sequence   21
Self-inverses   190
Skew-symmetric matrix   297
Square matrix   271
Square in $\mathbf{Z}_n$ 205
Square integer   81
Strong induction   58
Symmetric matrix   297

T
Transpose of A   296
Triangle inequality   238
Triangular number   20
Twin primes   129
$\tau(n)$ 80

W
Well-Ordering Principle   74
Wiles, Andrew   252

Z
$\mathbf{Z}$ = integers   71
Zero-divisor in $\mathbf{Z}_n$ 170
Zero matrix   294